$\prod_{p \mid n,\ p \text{ prime}}$	product over all distin[ct primes] p of n; 82
$\nu(n)$	number of positive divisors function; 86
$\sigma(n)$	sum of positive divisors function; 89
$\mu(n)$	Möbius mu-function; 96
$\left(\frac{a}{p}\right)$	Legendre symbol; 109
$\left(\frac{a}{n}\right)$	Jacobi symbol; 125
$\text{ord}_m a$	order of a modulo m; 128
$\binom{n}{k}$	number of combinations of n things taken k at a time; 264–265
$\text{ind}_r a$	index of a relative to r; 146
$\mathbf{R}$	the set of real numbers; 181
$\mathbf{Q}$	the set of rational numbers; 181
$\mathbf{R} - \mathbf{Q}$	the set of irrational numbers; 181
$0.a_1a_2 \cdots a_{N-1}\overline{a_Na_{N+1} \cdots a_{N+(\rho-1)}}$	eventually periodic decimal expansion with period length ρ; 185
$[a_0, a_1, a_2, \ldots, a_n]$	finite simple continued fraction expansion; 191
C_i	ith convergent; 195
$[a_0, a_1, a_2, \ldots]$	infinite simple continued fraction expansion; 202
$[a_0, a_1, a_2, \ldots, a_{N-1}, \overline{a_N, a_{N+1}, \ldots, a_{N+(\rho-1)}}]$	eventually periodic simple continued fraction expansion with period length ρ; 214
α'	conjugate of α; 218

Elementary Number Theory

Elementary Number Theory

James K. Strayer
Lock Haven University

Prospect Heights, Illinois

For information about this book, contact:
Waveland Press, Inc.
P.O. Box 400
Prospect Heights, Illinois 60070
(847) 634-0081
www.waveland.com

2002 reissued by Waveland Press, Inc.

ISBN 1-57766-224-5

Printed in the United States of America

7 6 5 4 3 2 1

Preface

Elementary Number Theory is intended to be used for a first course in number theory at the undergraduate level, preferably in the sophomore or junior year of the typical curriculum. The book arose from lecture notes prepared in 1986 while I was employed as a Graduate Assistant in the Department of Mathematics at The Pennsylvania State University. The notes were further refined at Lock Haven University in 1988; in 1989, the notes were converted to book form, and a preliminary draft of this book was used for an individualized instruction course at Lock Haven University. A second draft of the book was used at Lock Haven University in 1992. Having been tried and tested in the classroom at various stages of its development, *Elementary Number Theory* reflects many modifications either suggested directly by students or deemed appropriate because of students' responses in the classroom setting. My intention throughout the book is to address common errors and difficulties as clearly and effectively as possible.

Much effort has been given to the organization of the material in the text, in an attempt to achieve an orderly and natural progression of topics. The core material for a first course in number theory is presented in Chapters 1–4. Chapter 1 discusses the two basic concepts of elementary number theory—divisibility and factorization—and culminates in the Fundamental Theorem of Arithmetic. Most readers will be familiar with certain topics in the chapter, such as prime numbers, greatest common divisors, and least common multiples. Consequently, this chapter is intended to provide a straightforward introduction to elementary number theory through familiar concepts. Chapter 2 uses the relation of divisibility to develop the theory of congruences. Linear congruences in one variable are solved, and the Chinese Remainder Theorem, Wilson's Theorem, Fermat's Little Theorem, and Euler's Theorem are carefully studied. Chapter 3 covers the crucial functions of elementary number theory: $\phi(n)$ (the Euler phi-function), $\nu(n)$ (the number of divisors function), and $\sigma(n)$ (the sum of divisors function). In addition, perfect numbers and the Möbius Inversion Formula are discussed. Chapter 4 concludes the core material of the book with a look at quadratic residues and the unexpected result contained therein, the quadratic reciprocity law. The beauty and elegance of this law are unmatched in all of elementary number theory.

Chapters 5–8 of *Elementary Number Theory* present additional optional topics for a first course in number theory. Chapter 5 discusses primitive roots and culminates in the Primitive Root Theorem. The related concept of indices is also included. Chapter 6 introduces Diophantine equations and considers the solvability or unsolvability of several particular important equations. The conjecture commonly known as Fermat's Last Theorem is offered as a graphic illustration of how easily questions in number theory can be formulated and how tenaciously they sometimes resist proof. Chapter 7 develops continued fractions as alternate means for representing real numbers; the central result of this chapter is the proof that those real numbers that are representable as eventually periodic infinite simple continued fractions are precisely the quadratic irrational numbers. Finally, Chapter 8 investigates the application of elementary number theory in a simple pencil-and-paper game, the field of cryptography (in particular, the RSA Encryption System), primality testing algorithms, and the analysis of a special Diophantine equation known as Pell's Equation. The four applications of Chapter 8 may be interspersed throughout the course as the necessary material for each application is covered. The particular prerequisite for each of these four applications is as follows:

Section 8.1: Chapter 2
Section 8.2: Chapter 3
Section 8.3: Chapter 5
Section 8.4: Chapter 7

The interdependence of Chapters 1–7 is given by the chart below (with the core chapters appearing inside the dotted rectangle).

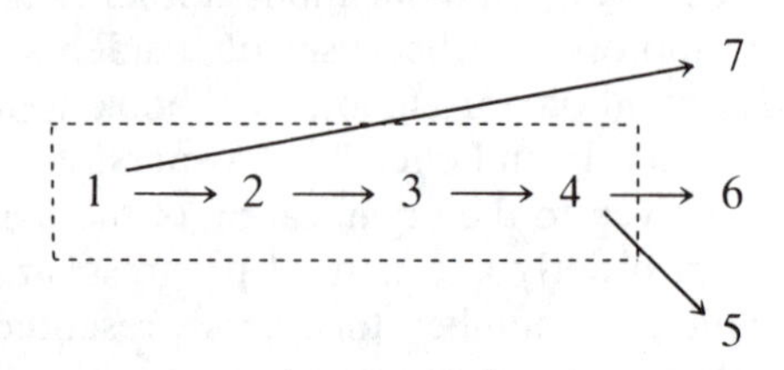

Inasmuch as *Elementary Number Theory* is to be used as a textbook for the learning of number theory, there are many examples and exercises. Virtually every definition is followed by examples; virtually every theorem is illustrated with a concrete example. Occasionally, even the proofs of important theorems are "walked through" by means of a concrete example, to help the student grasp the motivating ideas in these proofs. The exercises range from easy computations to difficult proofs and are carefully chosen to elucidate and/or complement the exposition. The organization of the exercises within each section is designed to enable the student to practice and perfect computational skills before attempting the more demanding theoretical exercises. The answers to approximately half of the computational exercises, as well as hints to several of the theoretical exercises, appear in the back of the book.

One other feature of this textbook deserves special mention. Each chapter of the book concludes with seven Student Projects, the first of which always involves the programming of a calculator or computer. These projects are intended to pique student interest in the material of that chapter through the solution of rather involved problems, many of which require supplementary reading. (Appropriate references are given.) In addition to reinforcing the concepts of elementary number theory and providing opportunities for ingenuity, these projects help familiarize the student with the mathematical literature available. Because of the increased difficulty of many of these problems, they are ideal as small-group assignments.

Every attempt has been made to make the book as versatile as possible in terms of prerequisites. Prior experience with the various proof techniques of mathematics (direct proof, proof by contraposition, and proof by contradiction) is assumed. Brief appendices have been included to cover mathematical induction, equivalence relations, abstract algebra, and the binomial theorem (as well as references for more complete discussions of these topics). More material than can reasonably be covered in a semester is included in the book, so as to accommodate differing instructor preferences.

Each chapter of the book, except the introduction, is divided into sections. The numbering of definitions, examples, proven results, and particular referenced equations proceeds sequentially throughout each chapter. The exercises are also numbered sequentially throughout each chapter. This scheme is intended to make it easier to find any particular referenced item.

Acknowledgments

I would like to express my sincere appreciation to Steve Quigley and Cathie Griffin, as well as David Hoyt, for their assistance and professionalism throughout the publication of this book. I would also like to thank the following reviewers of the book:

Paul M. Cook, II, Furman University
Richard L. Francis, Southeast Missouri State University
Elmo Moore, Humboldt State University
Steve Ligh, Southeastern Louisiana University
Don Redmond, Southern Illinois University
Stephen Richters, Saint Bonaventure University
Kimmo I. Rosenthal, Union College
Kenneth G. Valente, Colgate University
Burdette C. Wheaton, Mankato State University
Roy E. Worth, Georgia State University

Their informed suggestions greatly improved the final product. I must also thank the many students at Lock Haven University who shaped what the book was to become by offering their comments and criticisms. Finally, I wish to express my gratitude to Susan Handy for her unending support and encouragement; this book is dedicated to her.

James K. Strayer

Preface to the Student

Reading mathematics is an active endeavor. The complete comprehension of mathematical material frequently requires pencil-and-paper verifications or justifications on the side. Indeed, you will have many opportunities in this book to become an active participant in the unfolding of the concepts to be presented — through questions, challenges, and conjectures. This participation is crucial to understanding number theory and cannot be overemphasized. In general, mathematical literature is not meant to be read quickly; it is best to follow a slow pace, accompanied by careful thought.

I would like to relate a scenario that I often encounter while teaching number theory. In a typical number theory class, there is always at least one student who attempts only to understand the *statements* of the major results and disregards all proofs. Such a student thinks that the sole importance of a proof in mathematics is to establish a result. This is in fact the primary importance of a proof, but proofs also provide illustrations of *techniques* in mathematics, most of which will be useful in establishing future results. Hence, it is extremely important that all steps of a proof be understood, since such steps may be useful later. Your success in number theory will depend, in part, on the acquisition of certain theoretical skills (after all, it is number *theory*); these skills are more easily developed through an understanding of the statements of major results as well as the proofs of those results.

Throughout this book, you will find that the results are labeled with a variety of terms — proposition, theorem, lemma, corollary, and porism — which are intended to be descriptive. Meanings of these terms will now be provided. The fundamental mathematical result is called a *proposition.* The proofs of propositions usually follow from definitions in a relatively straightforward manner. A *theorem* is a result of crucial significance and importance. There will be many theorems in this book; inasmuch as the whole purpose of mathematics is to prove theorems, make sure that you completely understand each theorem and its proof. The proofs of theorems will usually be more involved than the proofs of propositions. Think of theorems as "greater" than propositions. A *lemma* (the plural is *lemmata*) is a result used to prove a proposition or a theorem. When you encounter a lemma, be prepared for a major result using the result of the lemma. A *corollary* is a result that follows easily or without much proof from a proposition or a theorem. Corollaries provide useful facts related to a major result. Finally, a *porism* is a result contained within the *proof* of a theorem or a proposition but not formulated explicitly in the statement of it. While proving a major result, another result is sometimes inadvertently proven; a porism stresses this associated result.

Part of the attraction of elementary number theory is the wealth of easily stated, unsolved problems, many of which have resisted proof for quite a long time. Many of these problems are given throughout the exposition; each such problem is labeled as a *conjecture.* Proving or disproving any of these conjectures will make you famous indeed. (The exceptions are the "prove/disprove" problems that are formulated as conjectures in the exercises.) In any event, the seeming simplicity of some of these conjectures will probably surprise you.

To gain and reinforce comprehension of the material in this book, you should attempt as many of the exercises as possible. Use of the Hints and Answers section in the back of the book is crucial in this regard. Usually, computational exercises should be attempted first, since the answers to about half of these exercises can be checked in the Hints and Answers section. The correctness of computational exercises involving a particular concept provides a much-needed boost into the successful completion of the theoretical exercises involving that concept. As you attempt more and more of the theoretical exercises, you will find certain techniques or "tricks" to be frequently useful. Many of these are highlighted in the exposition, used repeatedly in proofs, and/or pointed out in the Hints and Answers section. So by using the Hints and Answers section prudently, you will maximize your performance on the more demanding theoretical exercises and begin to be able to construct proofs on your own.

Each chapter of the book concludes with seven student projects. The first such project will always be a programming exercise for a calculator or computer. Many of the projects require supplementary reading of journal articles or sections of other number theory textbooks. The purpose of these projects is twofold: first, to reinforce the material presented, and second, to introduce you to the various literature available to the mathematician—particularly, the number theorist. Do not get discouraged if you have difficulty solving the problems in these projects; most of them are quite challenging. (On the positive side, you should feel very proud and elated when you solve one.)

Finally, it should not be forgotten that mathematics is a *human* endeavor; many of the results in this book are due to some of the most creative and ingenious minds of all time. Accordingly, brief biographical sketches of a few of these great mathematicians are interspersed throughout the text. I conclude by saying simply that I hope your journey through elementary number theory is a pleasurable one and that the material contained in this book motivates you to study number theory in more detail.

James K. Strayer

Contents

Introduction

Elementary number theory is essentially the study of the system of integers. The system of integers consists of the set of integers $\mathbf{Z} = \{\ldots, -3, -2, -1, 0, 1, 2, 3, \ldots\}$ and various properties of this set under the operations of addition and multiplication and under the usual ordering relation of "less than." These important properties of the integers are summarized below.

Closure property of addition: If $a, b \in \mathbf{Z}$, then $a + b \in \mathbf{Z}$.
Closure property of multiplication: If $a, b \in \mathbf{Z}$, then $ab \in \mathbf{Z}$.

Commutative property of addition: If $a, b \in \mathbf{Z}$, then $a + b = b + a$.
Commutative property of multiplication: If $a, b \in \mathbf{Z}$, then $ab = ba$.

Associative property of addition: If $a, b, c \in \mathbf{Z}$, then $(a + b) + c = a + (b + c)$.
Associative property of multiplication: If $a, b, c \in \mathbf{Z}$, then $(ab)c = a(bc)$.

Distributive property of multiplication over addition: If $a, b, c \in \mathbf{Z}$, then $a(b + c) = ab + ac$.

Additive identity property: If $a \in \mathbf{Z}$, then $a + 0 = 0 + a = a$.
Multiplicative identity property: If $a \in \mathbf{Z}$, then $a \cdot 1 = 1 \cdot a = a$.

Additive inverse property: If $a \in \mathbf{Z}$, then $a + (-a) = (-a) + a = 0$. If $a, b \in \mathbf{Z}$, then $a + (-b)$ is written $a - b$.

Zero property of multiplication: If $a \in \mathbf{Z}$, then $a \cdot 0 = 0 \cdot a = 0$.

Cancellation property of addition: If $a, b, c \in \mathbf{Z}$ and $a + b = a + c$, then $b = c$.
Cancellation property of multiplication: If $a, b, c \in \mathbf{Z}, a \neq 0$, and $ab = ac$, then $b = c$.

Trichotomy law: If $a \in \mathbf{Z}$, then exactly one of the following statements is true:
(i) $a < 0$
(ii) $a = 0$
(iii) $a > 0$

Properties of inequality:
(i) If $a, b, c \in \mathbf{Z}$ and $a < b$, then $a + c < b + c$.
(ii) If $a, b, c \in \mathbf{Z}$, $a < b$, and $c > 0$, then $ac < bc$.
(iii) If $a, b, c \in \mathbf{Z}$, $a < b$, and $c < 0$, then $ac > bc$.

Well-ordering property: Every nonempty set of positive integers contains a least element.

We hope that you are familiar, or at least comfortable, with these properties as a result of your past encounters with ordinary arithmetic. The properties above are taken as axioms of the system of integers in this book. These axioms are not independent; in other words, it is possible to prove some of the axioms from others. A rigorous development of the system of integers may be found in Fletcher and Patty (1992).

The well-ordering property is a fundamentally important axiom of the system of integers above. One powerful proof technique available to the number theorist (indeed, any mathematician) is that of *mathematical induction.* A review of the two essential forms of mathematical induction, called the first and second principles, may be found in Appendix A along with a few exercises and an appropriate reference. Both of these principles of mathematical induction are logically equivalent to the well-ordering principle. So, by accepting the well-ordering principle as an axiom of the system of integers, we obtain two forms of induction for our use. Mathematical induction will be used frequently in this book, especially in Chapter 7.

This short introductory chapter concludes by establishing the framework of the remainder of this book. Throughout, the results are labeled with a variety of terms: proposition, theorem, lemma, corollary, and porism. For the meanings of the terms and further explanation, refer to the Preface to the Student. More globally, mathematics breaks roughly into two disciplines: The *pure* discipline of mathematics is concerned primarily with theory, while the *applied* discipline is concerned primarily with applications. These disciplines within mathematics do not have clear-cut boundaries; indeed, there is much theory in applied mathematics and there are frequently many applications in pure mathematics. Elementary number theory falls into the pure discipline of mathematics. Elementary number theory does, however, have applications in other fields. One such field is computer science. The importance of computer science applications in today's technological world cannot be overemphasized. An excellent reference for applications of elementary number theory in computer science is Knuth (1981). (See also Student Project 7 in Chapter 8.) Another field based on the principles of elementary number theory is cryptography, the making and breaking of secret codes. A particular cryptographic system is discussed in Section 8.2. An excellent reference for cryptographic applications of elementary number theory is Konheim (1981).

1

Divisibility and Factorization

Elementary number theory is the study of the divisibility properties of the integers. Inasmuch as these divisibility properties form the basis for the study of more advanced topics in number theory, they can be thought of as forming the foundation for the entire area of number theory. It is both beautiful and appropriate that such a pure area of mathematics is derived from such a simple source. In addition, the main topics of this chapter are quite old, dating back to the Alexandrian Greek period of mathematics, which began approximately 300 B.C. In fact, most of the ideas discussed here appear in Euclid's *Elements.*

In this chapter, we develop the concept of divisibility and the related concept of factorization, which culminate in an extremely important property of the integers appropriately named the Fundamental Theorem of Arithmetic. In addition, we will encounter and highlight certain proof strategies that are used again and again in elementary number theory. These strategies form some of the important "tools of the trade" of the number theorist.

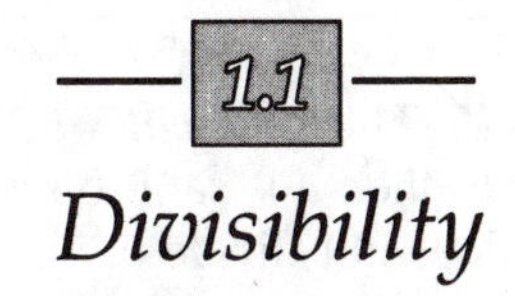

Divisibility

The fundamental relation connecting one integer to another is the notion of divisibility. In terms of long division, the divisibility relation means "divides evenly with zero remainder." Hence, the integer 3 divides the integer 6 since 3 divides evenly into 6 (two times) with zero remainder. Similarly, the integer 3 does not divide the integer 5, since 3 does not divide evenly into 5; 3 divides into 5 with a quotient of 1 and a remainder of 2. We now make the divisibility relation more mathematically precise.

Biography

Euclid of Alexandria (365?–275? B.C.)

Little is known of Euclid's life. The exact dates of his birth and death (as well as his birthplace and nationality) are unknown. Euclid is remembered mainly for his monumental work, entitled *Elements*, which was essentially a compilation of the mathematics of the classical Greek period that preceded him. This massive work comprised 13 books and contained 465 propositions. *Elements* emphasized the discipline of mathematics as a deductive science based on explicit axioms and, as such, has influenced the course of mathematics as has no other work. While much of *Elements* was devoted to geometry, Books VII–IX as well as portions of Book X were devoted to the theory of numbers. It is Euclid's proof of the infinitude of the prime numbers that is still used today. In addition, Euclid studied the division algorithm (culminating in the so-called Euclidean algorithm for the computation of greatest common divisors), perfect numbers, and Pythagorean triples. (All of the number-theoretic topics above will be discussed in this book.)

Definition 1: Let $a, b \in \mathbf{Z}$. Then *a divides b*, denoted $a \mid b$, if there exists $c \in \mathbf{Z}$ such that $b = ac$. If $a \mid b$, then a is said to be a *divisor* or *factor* of b. The notation $a \nmid b$ means that a does not divide b.

Example 1:

(a) $3 \mid 6$ since there exists $c \in \mathbf{Z}$ such that $6 = 3c$. Here, $c = 2$. Hence 3 is a divisor of 6.

(b) $3 \nmid 5$ since there does not exist $c \in \mathbf{Z}$ such that $5 = 3c$. Note that c would have to be $\frac{5}{3}$ here and $\frac{5}{3} \notin \mathbf{Z}$. Hence 3 is not a divisor of 5.

(c) $3 \mid -6$ since there exists $c \in \mathbf{Z}$ such that $-6 = 3c$. Here, $c = -2$. Hence 3 is a divisor of -6. In fact, it is easily seen that $\pm 3 \mid \pm 6$.

(d) If $a \in \mathbf{Z}$, then $a \mid 0$ since there exists $c \in \mathbf{Z}$ such that $0 = ac$. Here, $c = 0$. In other words, any integer divides 0.

(e) If $a \in \mathbf{Z}$ and $0 \mid a$, then there exists $c \in \mathbf{Z}$ such that $a = 0c$ if and only if $a = 0$. In other words, the only integer having zero as a divisor is zero.

We now make two remarks concerning the notation $a \mid b$. First of all, $a \mid b$ does not have the same meaning as either of the notations a/b and b/a. Note that $a \mid b$ is a statement about the relationship between two integers: It says that a divides into b evenly with no remainder. The notations a/b and b/a are interpreted, respectively, as $a \div b$ and $b \div a$ and, as such, are rational numbers. In other words, $a \mid b$ is a statement *about* numbers while a/b and b/a *are* numbers. Note, however, that $a \mid b$ does imply that a divides b exactly b/a times provided that $a \neq 0$.

The restriction that a not be zero in the implication above brings us to our second remark. You may feel uncomfortable about allowing zero as a divisor in (e) of Example 1 above. The phrase "division by zero is undefined" has probably been ingrained in your mind through constant repetition by many of your mathematics teachers. We say that $0 \mid 0$ here only because it is consistent with our definition of divisibility; it should not be interpreted as implying that 0 divides 0 exactly $0/0$ times. The form $0/0$ is said to be *indeterminate*: It has no meaning. The difference between the notation $0 \mid 0$ and $0/0$ is hence clear— the former notation has meaning (albeit minimal and only as a consequence of our definition), and the latter notation does not have meaning. We urge the reader to pause and carefully consider the remarks above before continuing.

The divisibility relation enjoys the following two properties, which are recorded as propositions.

Proposition 1.1: Let $a, b, c \in \mathbf{Z}$. If $a \mid b$ and $b \mid c$, then $a \mid c$.

Proof: Since $a \mid b$ and $b \mid c$, there exist $e, f \in \mathbf{Z}$ such that $b = ae$ and $c = bf$. Then

$$c = bf = (ae)f = a(ef)$$

and $a \mid c$. ■

Before continuing, note that we have just proven that the divisibility relation is transitive. If the word *transitive* is unfamiliar to you, see Appendix B [in particular, Example 3(b)] for a further discussion of relations and their properties.

Proposition 1.2: Let $a, b, c, m, n \in \mathbf{Z}$. If $c \mid a$ and $c \mid b$, then $c \mid ma + nb$.

Proof: Since $c \mid a$ and $c \mid b$, there exist $e, f \in \mathbf{Z}$ such that $a = ce$ and $b = cf$. Then

$$ma + nb = mce + ncf = c(me + nf)$$

and $c \mid ma + nb$. ■

A special case of Proposition 1.2 is important enough to be highlighted separately. We first give a special name to the expression $ma + nb$.

Definition 2: The expression $ma + nb$ in Proposition 1.2 is said to be an *integral linear combination* of a and b.

Proposition 1.2 says that an integer dividing each of two integers also divides any integral linear combination of those integers. This fact is extremely valuable in establishing theoretical results. Consider the special case where $m = n = 1$. We obtain the fact that, if an integer divides each of two integers, then it divides the sum of the integers. The case where $m = 1$ and $n = -1$ similarly yields the fact that, if an integer divides each of two integers, then it divides the difference of the two integers. We will use these facts repeatedly in this book, and you will find them particularly useful in several theoretical exercises.

We now introduce a function with the set of real numbers **R** as its domain and the set of integers **Z** as its range. This function will be used shortly to prove the major result on divisibility; it will also be useful in Chapters 4 and 7.

Definition 3: Let $x \in \mathbf{R}$. The *greatest integer function of* x, denoted $[x]$, is the greatest integer less than or equal to x.

The existence of a greatest integer less than or equal to a given real number follows from the well-ordering property of the integers discussed in the Introduction.

Example 2:

(a) If $a \in \mathbf{Z}$, then $[a] = a$ since the greatest integer less than or equal to any integer is the integer itself. It is easily seen that the converse of this statement is also true, namely that, if $[a] = a$ for some $a \in \mathbf{R}$, then $a \in \mathbf{Z}$.

(b) Since the greatest integer less than or equal to $\frac{5}{3}$ is 1, we have $[\frac{5}{3}] = 1$.

(c) Since the greatest integer less than or equal to π is 3, we have $[\pi] = 3$.

(d) Since the greatest integer less than or equal to $-\frac{5}{3}$ is -2, we have $[-\frac{5}{3}] = -2$.

(e) Since the greatest integer less than or equal to $-\pi$ is -4, we have $[-\pi] = -4$.

The following lemma is basically an immediate consequence of the definition of the greatest integer function, but a short proof is nonetheless provided for your inspection.

Lemma 1.3: Let $x \in \mathbf{R}$. Then $x - 1 < [x] \leq x$.

Proof: Since the greatest integer function of x is less than or equal to x, the second inequality is clear. For the first inequality, assume, by way of contradiction, that $x - 1 \geq [x]$. Then $[x] + 1 \leq x$ and

$$[x] < [x] + 1 \leq x$$

$$\begin{array}{rl} & 1 \quad \leftarrow\text{Quotient} \\ \text{Divisor}\rightarrow 3 & \overline{)5} \quad \leftarrow\text{Dividend} \\ & \underline{-3} \\ & 2 \quad \leftarrow\text{Remainder} \end{array}$$

Figure 1.1

Since $[x] + 1$ is an integer, this contradicts the fact that $[x]$ is the greatest integer less than or equal to x. Hence $x - 1 < [x]$. ■

When one integer (the divisor) is divided into another integer (the dividend) to obtain an integer quotient, an integer remainder is obtained. For example, the long division for 3 divided into 5 (shown in Figure 1.1) can be expressed more compactly via the equation

$$5 = 3 \cdot 1 + 2$$

or by the phrase "the dividend is equal to the divisor times the quotient plus the remainder." Note that the remainder is strictly less than the divisor. If the divisor in such a long division of integers is positive, the fact above is true in general as the following theorem illustrates.

Theorem 1.4: (The Division Algorithm) Let $a, b \in \mathbf{Z}$ with $b > 0$. Then there exist unique $q, r \in \mathbf{Z}$ such that

$$a = bq + r, \qquad 0 \le r < b$$

(Note that q stands for *quotient* and r stands for *remainder*.)

Proof: Let $q = [\frac{a}{b}]$ and $r = a - b[\frac{a}{b}]$. Then $a = bq + r$ is easily checked. It remains to show that $0 \le r < b$. By Lemma 1.3, we have that

$$\frac{a}{b} - 1 < \left[\frac{a}{b}\right] \le \frac{a}{b}$$

Multiplying all terms of this inequality by $-b$, we obtain

$$b - a > -b\left[\frac{a}{b}\right] \ge -a$$

Reversing the inequality and adding a to all terms gives

$$0 \le a - b\left[\frac{a}{b}\right] < b$$

which is precisely $0 \le r < b$ as desired; so, q and r as defined above have the desired properties. It remains to show the uniqueness of q and r. Assume that

$$a = bq_1 + r_1, \qquad 0 \le r_1 < b$$

and

$$a = bq_2 + r_2, \qquad 0 \le r_2 < b$$

We must show that $q_1 = q_2$ and $r_1 = r_2$. We have

$$0 = a - a = bq_1 + r_1 - (bq_2 + r_2) = b(q_1 - q_2) + (r_1 - r_2)$$

which implies that

$$r_2 - r_1 = b(q_1 - q_2) \tag{1}$$

Hence, $b \mid r_2 - r_1$. Now $0 \le r_1 < b$ and $0 \le r_2 < b$ imply $-b < r_2 - r_1 < b$; along with $b \mid r_2 - r_1$, we have $r_2 - r_1 = 0$ or $r_1 = r_2$. Now (1) becomes

$$0 = b(q_1 - q_2)$$

Since $b \neq 0$, we have that $q_1 - q_2 = 0$ or $q_1 = q_2$ as desired. ■

Note that $r = 0$ in the division algorithm if and only if $b \mid a$. Equivalently, a necessary and sufficient condition in Theorem 1.4 for the remainder in a division of integers to be zero is that the divisor evenly divide the dividend.

Given $a, b \in \mathbf{Z}$ with $b > 0$, the q and r of the division algorithm may be obtained by using the equations defining q and r in the first statement of the proof of the algorithm. (In fact, one would use precisely these equations to compute q and r on a standard calculator.) We illustrate with an example.

Example 3:

Find q and r as in the division algorithm if $a = -5$ and $b = 3$.
By the proof of Theorem 1.4, we have

$$q = \left[\frac{a}{b}\right] = \left[\frac{-5}{3}\right] = -2$$

and

$$r = a - b\left[\frac{a}{b}\right] = -5 - 3(-2) = 1$$

Please check for yourself that $a = bq + r$ and $0 \le r < b$.

A remark is in order here. When asked, "What is -5 divided by 3?" (as in Example 3 above), many students will respond "-1 with a remainder of -2" since $-\frac{5}{3} = -1 - \frac{2}{3}$. Although $q = -1$ and $r = -2$ satisfy $a = bq + r$, the condition that $0 \le r < b$ in Theorem 1.4 is no longer true. Hence, division in the context of the division algorithm must be performed carefully so that both desired conditions hold. More generalized versions of Theorem 1.4 that allow negative divisors are investigated in Exercise 15.

We conclude this section with a definition and an example of concepts with which you are probably already familiar.

Definition 4: Let $n \in \mathbf{Z}$. Then n is said to be *even* if $2 \mid n$ and n is said to be *odd* if $2 \nmid n$.

Example 4:

The set of even integers is given by $\{\ldots, -6, -4, -2, 0, 2, 4, 6, \ldots\}$; the set of odd integers is given by $\{\ldots, -7, -5, -3, -1, 1, 3, 5, 7, \ldots\}$.

Exercise 11 establishes important facts about even and odd integers that will be useful throughout this book; solving this exercise is greatly encouraged.

Exercise Set 1.1

1. Prove or disprove each statement below.
 (a) $6 \mid 42$
 (b) $4 \mid 50$
 (c) $16 \mid 0$
 (d) $0 \mid 15$
 (e) $14 \mid 997157$
 (f) $17 \mid 998189$

2. Find integers a, b, and c such that $a \mid bc$ but $a \nmid b$ and $a \nmid c$.

3. Find the unique integers q and r guaranteed by the division algorithm (Theorem 1.4) with each dividend and divisor below.
 (a) $a = 47$, $b = 6$
 (b) $a = 281$, $b = 13$
 (c) $a = 343$, $b = 49$
 (d) $a = -105$, $b = 10$
 (e) $a = -469$, $b = 31$
 (f) $a = -500$, $b = 28$

4. If $a, b \in \mathbf{Z}$, find a necessary and sufficient condition that $a \mid b$ and $b \mid a$.

5. Prove or disprove the following statements.
 (a) If a, b, c, and d are integers such that $a \mid b$ and $c \mid d$, then $a + c \mid b + d$.
 (b) If a, b, c, and d are integers such that $a \mid b$ and $c \mid d$, then $ac \mid bd$.
 (c) If a, b, and c are integers such that $a \nmid b$ and $b \nmid c$, then $a \nmid c$.

6. (a) Let $a, b, c \in \mathbf{Z}$ with $c \neq 0$. Prove that $a \mid b$ if and only if $ac \mid bc$.
 (b) Provide a counterexample to show why the statement of part (a) does not hold if $c = 0$.

7. Let $a, b \in \mathbf{Z}$ with $a \mid b$. Prove that $a^n \mid b^n$ for every positive integer n.

8. Let $n \in \mathbf{Z}$ with $n > 0$. Prove that $n \mid (n + 1)^n - 1$.

9. Let a, m and n be positive integers with $a > 1$. Prove that $a^m - 1 \mid a^n - 1$ if and only if $m \mid n$. [*Hint*: For the "if" direction, write $n = md$ with d a positive integer and use the factorization $a^{md} - 1 = (a^m - 1) \times (a^{m(d-1)} + a^{m(d-2)} + \cdots + a^m + 1)$.]

10. (a) Let $n \in \mathbf{Z}$. Prove that $3 \mid n^3 - n$.
 (b) Let $n \in \mathbf{Z}$. Prove that $5 \mid n^5 - n$.
 (c) Let $n \in \mathbf{Z}$. Is it true that $4 \mid n^4 - n$? Provide a proof or counterexample.

11. (a) Let $n \in \mathbf{Z}$. Prove that n is an even integer if and only if $n = 2m$ with $m \in \mathbf{Z}$.
 (b) Let $n \in \mathbf{Z}$. Prove that n is an odd integer if and only if $n = 2m + 1$ with $m \in \mathbf{Z}$.
 (c) Prove that the sum and product of two even integers are even.
 (d) Prove that the sum of two odd integers is even and that their product is odd.

(e) Prove that the sum of an even integer and an odd integer is odd and that their product is even.

12. Prove that the square of any odd integer is expressible in the form $8n + 1$ with $n \in \mathbf{Z}$.

13. Prove that the fourth power of any odd integer is expressible in the form $16n + 1$ with $n \in \mathbf{Z}$.

14. *(a)* Let x be a positive real number and let d be a positive integer. Prove that the number of positive integers less than or equal to x that are divisible by d is $\left[\frac{x}{d}\right]$.

(b) Find the number of positive integers not exceeding 500 that are divisible by 3.

(c) Find the number of positive integers between 200 and 500 that are divisible by 3.

15. [The following exercise presents two alternate versions of the division algorithm (Theorem 1.4). Both versions allow negative divisors; as such, they are more general than Theorem 1.4.]

(a) Let a and b be nonzero integers. Prove that there exist unique $q, r \in \mathbf{Z}$ such that

$$a = bq + r, \qquad 0 \le r < |b|$$

(b) Find the unique q and r guaranteed by the division algorithm of part (a) above with $a = 47$ and $b = -6$.

(c) Let a and b be nonzero integers. Prove that there exist unique $q, r \in \mathbf{Z}$ such that

$$a = bq + r, \qquad -\frac{|b|}{2} < r \le \frac{|b|}{2}$$

This algorithm is called the *absolute least remainder algorithm.*

(d) Find the unique q and r guaranteed by the division algorithm of part (c) above with $a = 47$ and $b = -6$.

Prime Numbers

Every integer greater than one has at least two positive divisors, namely, 1 and the integer itself. Those positive integers having no other positive divisors (and so exactly two positive divisors) are of crucial importance in number theory and are introduced now.

Definition 5: Let $p \in \mathbf{Z}$ with $p > 1$. Then p is said to be *prime* if the only positive divisors of p are 1 and p. If $n \in \mathbf{Z}$, $n > 1$, and n is not prime, then n is said to be *composite.*

Note that the positive integer 1 is neither prime nor composite by definition. The reason for disallowing 1 as a prime number is investigated in Exercise 66.

Example 5:

The prime numbers between 2 and 50 inclusive are 2, 3, 5, 7, 11, 13, 17, 19, 23, 29, 31, 37, 41, 43, and 47. Exercise 22 shows that all prime numbers except the integer 2 are odd and hence the integer 2 is the only even prime number.

Having introduced the concept of a prime number, a most fundamental question arises. Are there finitely many or infinitely many prime numbers? Before reading further, formulate your own conjecture; Theorem 1.6 will ultimately provide the answer to this question. We first prove a preliminary lemma.

Lemma 1.5: Every integer greater than 1 has a prime divisor.

Proof: Assume, by way of contradiction, that some integer greater than 1, say n, has no prime divisor. By the well-ordering property, we may assume that n is the least such integer. Now $n \mid n$; since n has no prime divisor, n is not prime. So n is composite and consequently there exist $a, b \in \mathbf{Z}$ such that $n = ab$, $1 < a < n$, and $1 < b < n$. Since $1 < a < n$, we have that a has a prime divisor, say p, so that $p \mid a$. But $a \mid n$ so we have that $p \mid n$ by Proposition 1.1 from which n has a prime divisor, a contradiction. So every integer greater than 1 has a prime divisor. ■

We may now answer our motivating question posed prior to Lemma 1.5. This result appears as Proposition 20 in Book IX of Euclid's *Elements*; study the proof carefully because it is a vivid illustration of great mathematical ingenuity.

Theorem 1.6: (Euclid) There are infinitely many prime numbers.

Proof: Assume, by way of contradiction, that there are only finitely many prime numbers, say $p_1, p_2, \ldots, p_n$. Consider the number $N = p_1 p_2 \cdots p_n + 1$. Now N has a prime divisor, say p, by Lemma 1.5. So $p = p_i$ for some i, $i = 1, 2, \ldots, n$. Then $p \mid N - p_1 p_2 \cdots p_n$ (by Proposition 1.2), which implies that $p \mid 1$, a contradiction. Hence, there are infinitely many prime numbers. ■

The point of ingenuity in the proof above is the construction of the number N; it is precisely consideration of this number that eventually results in the desired contradiction. You will find similar constructions useful throughout this book.

Given that there are infinitely many prime numbers, how can we go about finding these numbers? The following proposition is useful in this regard.

Proposition 1.7: Let n be a composite number. Then n has a prime divisor p with $p \leq \sqrt{n}$.

Proof: Since n is a composite number, there exist $a, b \in \mathbf{Z}$ such that $n = ab$, $1 < a < n$, $1 < b < n$, and, without loss of generality, $a \leq b$. Now $a \leq \sqrt{n}$. (If $a > \sqrt{n}$, we have

$$n = ab > \sqrt{n}\sqrt{n} = n$$

which is impossible.) By Lemma 1.5, we have that a has a prime divisor, say p, so that $p \mid a$. But $a \mid n$, so we have that $p \mid n$ by Proposition 1.1. Furthermore, $p \le a \le \sqrt{n}$; p is the desired prime divisor of n. ■

Proposition 1.7 yields a method for finding all prime numbers less than or equal to a specified integer $n > 1$ by producing a criterion satisfied by all composite numbers. If an integer greater than one fails to satisfy this criterion, then the integer cannot be composite and so is prime. This method is called the *sieve of Eratosthenes* and is one of several "sieve methods." We illustrate the sieve of Eratosthenes with an example.

Example 6:

Suppose that we wish to find all prime numbers less than or equal to 50. By Proposition 1.7, any composite number less than or equal to 50 must have a prime divisor less than or equal to $\sqrt{50} \approx 7.07$. The prime numbers less than or equal to 7.07 are 2, 3, 5, and 7. Hence, from a list of the integers from 2 to 50, we delete all multiples of 2, all multiples of 3, all multiples of 5, and all multiples of 7 (not including 2, 3, 5, and 7 themselves). Note that all such multiples are clearly composite.

	2	3	~~4~~	5	~~6~~	7	~~8~~	~~9~~	~~10~~
11	~~12~~	13	~~14~~	~~15~~	~~16~~	17	~~18~~	19	~~20~~
~~21~~	~~22~~	23	~~24~~	~~25~~	~~26~~	~~27~~	~~28~~	29	~~30~~
31	~~32~~	~~33~~	~~34~~	~~35~~	~~36~~	37	~~38~~	~~39~~	~~40~~
41	~~42~~	43	~~44~~	~~45~~	~~46~~	47	~~48~~	~~49~~	~~50~~

Any number remaining in the list is not divisible by 2, 3, 5, or 7 and, by Proposition 1.7, cannot be composite. So the numbers remaining in the list above are prime (compare these numbers with those in Example 5); the composite numbers have been "sieved out" of the list.

In addition to finding prime numbers, Proposition 1.7 can be used to determine an algorithm for testing whether a given positive integer $n > 1$ is prime or composite. One checks whether n is divisible by any prime number p with $p \le \sqrt{n}$. If it is divisible, then n is composite; if not, then n is prime. Such an algorithm is an example of a *primality test.* Note that this particular primality test is highly inefficient since it requires testing a given integer for divisibility by all prime numbers less than or equal to the square root of the integer. More useful primality testing algorithms exist — unfortunately, they require more number theory than we have developed at the present time. Two such primality tests are discussed in Section 8.3. The largest prime number known at this writing was discovered in 1999: $2^{6972593} - 1$. We will return to this prime number later. In addition, the primality or compositeness of any given integer between 101 and 9999 inclusive may be deduced using the grid inside the cover of this book; instructions for using this grid may be found in Table 1 of Appendix E.

Biography

Eratosthenes of Cyrene (276–194 B.C.)

Eratosthenes, a contemporary of Archimedes, is perhaps best remembered as the chief librarian at the University of Alexandria in ancient Egypt. Born in Cyrene on the southern coast of the Mediterranean Sea, he was gifted not only as a mathematician, but also as a philosopher, astronomer, poet, historian, and athlete. There is much speculation as to his nickname of "Beta." One theory offers that the nickname originated from the fact that he stood at least second in each of the branches of knowledge of his day. In 240 B.C., Eratosthenes made what was perhaps his most significant scientific contribution when he measured the circumference of the earth using a simple application of Euclidean geometry. His primary contribution to number theory is his famous sieve for finding prime numbers as discussed in the text.

How are prime numbers distributed among the positive integers? A solution of Exercise 17 suggests that the prime numbers are distributed more sparsely as you progress through larger and larger positive integers. The proposition below shows more precisely that there exist arbitrarily long sequences of consecutive positive integers containing no prime numbers. Equivalently, there exist arbitrarily large gaps between prime numbers.

Proposition 1.8: For any positive integer n, there are at least n consecutive composite positive integers.

Proof: Given the positive integer n, consider the n consecutive positive integers

$$(n+1)!+2,\ (n+1)!+3, \ldots, (n+1)!+n+1$$

Let i be a positive integer such that $2 \le i \le n+1$. Since $i \mid (n+1)!$ and $i \mid i$, we have

$$i \mid (n+1)!+i, \qquad 2 \le i \le n+1$$

by Proposition 1.2. So each of the n consecutive positive integers above is composite. ■

Example 7:

By the proof of Proposition 1.8 above, a sequence of eight consecutive composite positive integers is given by $9!+2,\ 9!+3, \ldots, 9!+9$ or, equivalently, 362882, 362883, ..., 362889. Note that there is no guarantee that the sequence of integers produced by the proof of Proposition 1.8 will be the least such occurrence of integers. For example, the least occurrence of eight consecutive composite positive integers is given by the sequence 114, 115, 116, 117, 118, 119, 120, and 121.

Since 2 is the only even prime number, the only consecutive prime numbers are 2 and 3. How many pairs of prime numbers differ by two as in 3 and 5, 5 and 7, 11 and 13, and so on? Such pairs of prime numbers are said to be *twin primes*. Unfortunately, the answer to this question is unknown. The operative conjecture asserts the existence of infinitely many twin primes, as stated below.

Conjecture 1: (Twin Prime Conjecture) There are infinitely many prime numbers p for which $p + 2$ is also a prime number.

The largest known pair of twin primes is $318032361 \cdot 2^{107001} \pm 1$, discovered in 2001 by D. Underbakke and P. Carmody.

The most famous result concerning the distribution of prime numbers is called the Prime Number Theorem. This theorem was conjectured in 1793 by Carl Friedrich Gauss but resisted proof until 1896 when two independent proofs were produced by J. Hadamard and C. J. de la Vallée Poussin. The Prime Number Theorem gives an estimate of the number of prime numbers less than or equal to a given positive real number x. The estimate improves as x gets large. We first define a function that counts prime numbers.

Definition 6: Let $x \in \mathbf{R}$ with $x > 0$. Then $\pi(x)$ is the function defined by

$$\pi(x) = |\{p : p \text{ prime}; 1 < p \leq x\}|$$

In this book, vertical bars enclosing a set (as in Definition 6 above) will denote the cardinality of the set. So, if x is a positive real number, then $\pi(x)$ is the number of prime numbers less than or equal to x.

Example 8:

From Example 5, we have that the number of prime numbers less than or equal to 50 is 15. So $\pi(50) = 15$.

We may now state the Prime Number Theorem.

Theorem 1.9: (Prime Number Theorem) $\lim_{x \to \infty} \frac{\pi(x) \ln x}{x} = 1.$

The Prime Number Theorem says that, for large x, the quantity $\frac{\pi(x) \ln x}{x}$ is close to 1. This is equivalent to saying that the quantity $\pi(x)$ may be approximated by $\frac{x}{\ln x}$. In other words, $\frac{x}{\ln x}$ is an estimate for $\pi(x)$ for large x. Table 1.1 gives a

Table 1.1

x	$\pi(x)$	$\frac{x}{\ln x}$	$\frac{\pi(x)\ln x}{x}$
10^3	168	144.8	1.161
10^4	1229	1085.7	1.132
10^5	9592	8685.9	1.104
10^6	78498	72382.4	1.084
10^7	664579	620420.7	1.071
10^8	5761455	5428681.0	1.061

comparison of $\pi(x)$ and $\frac{x}{\ln x}$ for increasingly larger values of x. Note how the ratio $\frac{\pi(x)\ln x}{x}$ gets closer and closer to 1 as $x \to \infty$.

The most direct proof of the Prime Number Theorem requires considerable complex analysis, which is beyond the scope of this book. A complete discussion of this analytic proof may be found in Apostol (1976). In 1949, Atle Selberg and Paul Erdös discovered a surprising new proof of the Prime Number Theorem. This new proof was termed "elementary" by the mathematical community; although lengthy and considerably more intricate than the original proof, the new one is accessible to anyone with a knowledge of calculus. The elementary proof may be found in Hardy and Wright (1979).

Many unsolved problems in number theory deal with integers that are expressible in certain forms. One of the most famous unsolved problems in all of number theory was conjectured by Christian Goldbach in a letter to the great Leonhard Euler.

Conjecture 2: (Goldbach, 1742) Every even integer greater than 2 can be expressed as the sum of two (not necessarily distinct) prime numbers.

As three quick illustrations of such expressions, consider $4 = 2 + 2$, $6 = 3 + 3$, and $8 = 3 + 5$. Goldbach's Conjecture (as Conjecture 2 has come to be known, appropriately enough) has been verified for all even integers less than $4 \cdot 10^{14}$. By experimentation with small even integers, the interested reader can discover that the representation of an even integer as the sum of two prime numbers may not be unique. (In fact, the number of such representations has been predicted in a formula by English mathematicians G. H. Hardy and J. E. Littlewood.) However, a proof that at least one such representation exists for every even integer greater than 2 remains elusive.

We conclude this section with a discussion of three unsolved problems concerning prime numbers expressible in certain forms and a related helpful remark for the exercises. The first form is named after a French monk, Father Marin Mersenne (1588–1648).

Definition 7: Any prime number expressible in the form $2^p - 1$ with p prime is said to be a *Mersenne prime.*

Example 9:

The first five Mersenne primes are 3 ($= 2^2 - 1$), 7 ($= 2^3 - 1$), 31 ($= 2^5 - 1$), 127 ($= 2^7 - 1$), and 8191 ($= 2^{13} - 1$). Note that $2^{11} - 1 = 2047 = (23)(89)$ is *not* a Mersenne prime.

In 1644, Mersenne authored *Cogitata Physica-Mathematica* in which he claimed that $2^p - 1$ was a prime number for p equal to 2, 3, 5, 7, 13, 17, 19, 31, 67, 127, and 257 and a composite number for all other prime numbers p with $p < 257$. Work completed in 1947 revealed that Mersenne made five mistakes: $2^p - 1$ is a prime number for p equal to 61, 89, and 107 (not included in Mersenne's list) and is a composite number for p equal to 67 and 257 (included in Mersenne's list). There are currently 38 known Mersenne primes. The largest known Mersenne prime is of vintage 1999 and was discovered by N. Hajratwala, G. Woltman, and S. Kurowski; it is $2^{6972593} - 1$, a number containing 2098960 digits. (The observant reader will recall that this Mersenne prime is also the largest known prime number!) Curiously enough, the Mersenne primes have *not* been discovered in increasing order. For example, the 31st known Mersenne prime, $2^{110503} - 1$, was discovered three years after the larger 30th known Mersenne prime, $2^{216091} - 1$. It may be that other Mersenne primes lie in the gaps formed by known Mersenne primes. In any event, most mathematicians believe that there are infinitely many Mersenne primes, and so we state the following conjecture.

Conjecture 3: There are infinitely many Mersenne primes.

The second form for prime numbers is named after French mathematician Pierre de Fermat.

Definition 8: Any prime number expressible in the form $2^{2^n} + 1$ with $n \in \mathbf{Z}$ and $n \geq 0$ is said to be a *Fermat prime.*

Example 10:

The first five Fermat primes are 3 ($= 2^{2^0} + 1$), 5 ($2^{2^1} + 1$), 17 ($= 2^{2^2} + 1$), 257 ($= 2^{2^3} + 1$), and 65537 ($= 2^{2^4} + 1$).

Fermat conjectured in 1640 that any number expressible in the form $2^{2^n} + 1$ with $n \in \mathbf{Z}$ and $n \geq 0$ is prime. The conjecture was disproved in 1732 by Euler, who proved that $641 \mid 2^{2^5} + 1$ and hence $2^{2^5} + 1$ is *not* a Fermat prime.

Currently, only five Fermat primes are known, namely those prime numbers given in Example 10 above. Many mathematicians believe that there are no Fermat primes other than these five; thus the following conjecture.

Conjecture 4: There are exactly five Fermat primes.

The final form for prime numbers that we examine here are those prime numbers that are expressible as one more than a perfect square. The conjecture below was made in 1922 by Hardy and Littlewood:

Conjecture 5: There are infinitely many prime numbers expressible in the form $n^2 + 1$ where n is a positive integer.

If you are interested, there are several examples of primes of the form in Conjecture 5.

We make one final remark here. In view of Conjecture 5 above, it may be tempting to conjecture that there are infinitely many prime numbers expressible in the form $n^2 - 1$ where n is a positive integer. This conjecture, however, is easily seen to be false. Note first that $n = 1$ does not give a prime number when substituted in the desired form, while $n = 2$ does. Now note that

$$n^2 - 1 = (n - 1)(n + 1)$$

Inasmuch as the product $(n - 1)(n + 1)$ gives a nontrivial factorization of $n^2 - 1$ if $n > 2$, we have the fact that $n^2 - 1$ is a prime number if and only if $n = 2$. Something as simple as factoring expressions can be a powerful tool in number theory (see also Exercise 9 of Section 1). Remember this tool!

Exercise Set 1.2

16. Determine whether the following positive integers are prime or composite by using the primality test motivated by Proposition 1.7.
(a) 127
(b) 129
(c) 131
(d) 133
(e) 137
(f) 139

17. Use the sieve of Eratosthenes to find all prime numbers less than 200.

18. *(a)* Find 13 consecutive composite positive integers.
(b) Find the least occurrence of 13 consecutive composite positive integers. (*Hint*: Use Table 3 in Appendix E.)

19. Find all twin primes less than 200.

20. *(a)* Find $\pi(10)$, $\pi(100)$, and $\pi(200)$.
(b) Compute $\frac{\pi(x)\ln x}{x}$ for $x = 10$, $x = 100$, and $x = 200$ and compare the obtained values with those in Table 1.1.

21. Verify Goldbach's Conjecture (Conjecture 2) for the following even integers.
(a) 30
(b) 98

(c) 114
(d) 222

22. Prove that 2 is the only even prime number.

23. Prove or disprove the following conjecture, which is similar to Conjecture 1.
Conjecture: There are infinitely many prime numbers p for which $p + 2$ and $p + 4$ are also prime numbers.

24. Prove that every integer greater than 11 can be expressed as the sum of two composite numbers.

25. *(a)* Prove that all odd prime numbers can be expressed as the difference of squares of two successive integers.
(b) Prove that no prime number can be expressed as the difference of two fourth powers of integers. (*Hint*: Use the factorization tool discussed in the final paragraph of this section.)

26. Prove or disprove the following statements.
(a) If p is a prime number, then $2^p - 1$ is a prime number.
(b) If $2^p - 1$ is a prime number, then p is a prime number. (*Hint*: Consider the contrapositive of the statement.)

27. Let a and n be positive integers with $n \neq 1$. Prove that, if $a^n - 1$ is a prime number, then $a = 2$ and n is a prime number. Conclude that the only prime numbers of the form $a^n - 1$ with $n \neq 1$ are Mersenne primes.

28. Let a and n be positive integers with $a > 1$. Prove that, if $a^n + 1$ is a prime number, then a is even and n is a power of 2.

29. Let n be a positive integer with $n \neq 1$. Prove that, if $n^2 + 1$ is a prime number, then $n^2 + 1$ is expressible in the form $4k + 1$ with $k \in \mathbf{Z}$.

30. Prove or disprove the following conjecture, which is similar to Conjecture 5.
Conjecture: There are infinitely many prime numbers expressible in the form $n^3 + 1$ where n is a positive integer.

31. Prove or disprove the following conjecture.
Conjecture: If n is a positive integer, then $n^2 - n + 41$ is a prime number.

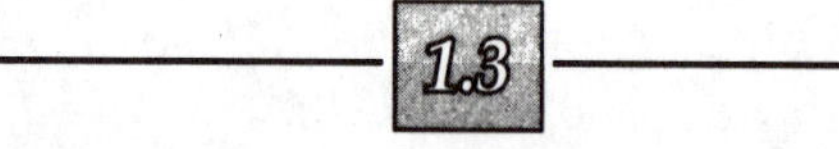

Greatest Common Divisors

Given two integers a and b, not both zero, consider the set S of integers that divide both a and b. The set S is necessarily nonempty (since $\pm 1 \in S$) and finite (since zero is the only integer that has an infinite number of divisors and at least one of a and b is nonzero). So it makes sense to speak of the greatest element of S. Note that such an element is necessarily positive.

Definition 9: Let $a, b \in \mathbf{Z}$ with a and b not both zero. The *greatest common divisor* of a and b, denoted (a, b), is the greatest positive integer d such that $d \mid a$ and $d \mid b$. If $(a, b) = 1$, then a and b are said to be *relatively prime*.

Note that $(0, 0)$ is undefined. (Why?) Furthermore, it is easy to see that if $(a, b) = d$, then

$$(-a, b) = (a, -b) = (-a, -b) = d$$

So, in Example 11, we restrict our discussion to the computation of the greatest common divisor of two nonnegative integers.

Example 11:

(a) The divisors of 24 are ± 1, ± 2, ± 3, ± 4, ± 6, ± 8, ± 12, and ± 24. The divisors of 60 are ± 1, ± 2, ± 3, ± 4, ± 5, ± 6, ± 10, ± 12, ± 15, ± 20, ± 30, and ± 60. The common divisors of 24 and 60 are ± 1, ± 2, ± 3, ± 4, ± 6, and ± 12. So the greatest common divisor of 24 and 60 is 12, which is denoted $(24, 60) = 12$.

(b) Every integer is a divisor of zero. So the common divisors of 24 and 0 are precisely the divisors of 24; from (a) above, the greatest common divisor of 24 and 0 is 24, which is denoted $(24, 0) = 24$. In general, if $a \in \mathbf{Z}$ with $a \neq 0$, we have $(a, 0) = |a|$.

(c) The divisors of 35 are ± 1, ± 5, ± 7, and ± 35. From (a) above, the common divisors of 24 and 35 are ± 1. So the greatest common divisor of 24 and 35 is 1, which is denoted $(24, 35) = 1$. In other words, 24 and 35 are relatively prime.

One crucial fact to remember when solving theoretical problems involving greatest common divisors is that $(a, b) = d$ implies $d \mid a$ and $d \mid b$; this observation in conjunction with Definition 1 or Proposition 1.2 or both is an extremely effective theoretical tool. At this point, it may be particularly instructive to attempt the solution of Exercises 36 and 38 with the above facts in mind.

We now prove two properties of greatest common divisors, which are recorded as propositions.

Proposition 1.10: Let $a, b \in \mathbf{Z}$ with $(a, b) = d$. Then $(a/d, b/d) = 1$.

Proof: Let $(a/d, b/d) = d'$. Then $d' \mid a/d$ and $d' \mid b/d$ so there exist $e, f \in \mathbf{Z}$ such that $a/d = d'e$ and $b/d = d'f$. So $a = d'de$ and $b = d'df$; consequently, we have $d'd \mid a$ and $d'd \mid b$. This implies that $d'd$ is a common divisor of a and b and, since d is the greatest common divisor of a and b, we have $d' = 1$, from which comes the desired result. ■

Proposition 1.11: Let $a, b \in \mathbf{Z}$ with a and b not both zero. Then

$$(a, b) = \min\{ma + nb : m, n \in \mathbf{Z}, ma + nb > 0\}$$

The set-theoretic function "min" produces the minimum element of a set. (Similarly, the set-theoretic function "max," to be used later, produces the maximum element of a set.) So Proposition 1.11 says that the greatest common divisor of two integers is the least positive number that is expressible as an integral linear combination of the integers.

Proof: (of Proposition 1.11) Note that $\{ma + nb: m, n \in \mathbf{Z}, ma + nb > 0\} \neq \varnothing$ since, without loss of generality, $a \neq 0$ and then either $1a + 0b > 0$ or $-1a + 0b > 0$. So $\min\{ma + nb: m, n \in \mathbf{Z},\ ma + nb > 0\}$ exists by the well-ordering property; let

$$d = \min\{ma + nb: m, n \in \mathbf{Z}, ma + nb > 0\} = m'a + n'b$$

We first show that $d \mid a$ and $d \mid b$. By the division algorithm, there exist $q, r \in \mathbf{Z}$ such that

$$a = dq + r, \qquad 0 \leq r < d$$

Now

$$r = a - dq = a - (m'a + n'b)q = (1 - qm')a - qn'b$$

and we have that r is an integral linear combination of a and b. Since $0 \leq r < d$ and d is the minimum positive integral linear combination of a and b, we have $r = 0$, from which $a = dq + r$ implies that $a = dq$. So $d \mid a$. Similarly, $d \mid b$. It remains to show that d is the *greatest* common divisor of a and b. Let c be any common divisor of a and b so that $c \mid a$ and $c \mid b$. Then $c \mid m'a + n'b = d$ by Proposition 1.2, from which $c \leq d$. ■

Proposition 1.11 above gives another important theoretical tool when dealing with greatest common divisors, namely, that $(a, b) = d$ implies that d may be expressed as an integral linear combination of a and b. In fact, the converse of this fact is true if $d = 1$, namely, that if 1 is expressible as an integral linear combination of two integers a and b, then $(a, b) = 1$. (Why is this true?) We will use the tool of expressing the greatest common divisor of two integers as an integral linear combination of these integers in forthcoming chapters. Note further that Proposition 1.11 may be used to show that a common divisor of two integers is not only less than the greatest common divisor but also *divides* the greatest common divisor. (Do this!) This fact is also frequently useful in establishing theoretical results.

The concept of greatest common divisor can be extended to more than two integers.

Definition 10: Let $a_1, a_2, \ldots, a_n \in \mathbf{Z}$ with $a_1, a_2, \ldots, a_n$ not all zero. The *greatest common divisor* of $a_1, a_2, \ldots, a_n$, denoted $(a_1, a_2, \ldots, a_n)$, is the greatest integer d such that $d \mid a_i$, $i = 1, 2, \ldots, n$. If $(a_1, a_2, \ldots, a_n) = 1$, then $a_1, a_2, \ldots, a_n$ are said to be *relatively prime*. If $(a_i, a_j) = 1$ for all pairs i, j with $i \neq j$, then $a_1, a_2, \ldots, a_n$ are said to be *pairwise relatively prime.*

Example 12:

The divisors of 24 are ±1, ±2, ±3, ±4, ±6, ±8, ±12, and ±24. The divisors of 60 are ±1, ±2, ±3, ±4, ±5, ±6, ±10, ±12, ±15, ±20, ±30, and ±60. The divisors of 30 are ±1, ±2, ±3, ±5, ±6, ±10, ±15, and ±30. The common divisors of 24, 60, and 30 are ±1, ±2, ±3, and ±6. So the greatest common divisor of 24, 60, and 30 is 6, which is denoted $(24, 60, 30) = 6$. An alternate method for computing the greatest common divisor of more than two integers is investigated in Exercise 51.

Pairwise relatively prime integers are relatively prime; it is *not* the case that relatively prime integers are necessarily pairwise relatively prime as Example 13 illustrates.

Example 13:

Since $(24, 60, 49) = 1$ (verify this!), we have that 24, 60, and 49 are relatively prime. Since $(24, 60) = 12 \neq 1$ by Example 11(a), we have that 24, 60, and 49 are *not* pairwise relatively prime.

Exercise Set 1.3

32. Find the greatest common divisors below.
(a) $(21, 28)$
(b) $(32, 56)$
(c) $(58, 63)$
(d) $(0, 113)$
(e) $(111, 129)$
(f) $(120, 165)$

33. Let $a \in \mathbf{Z}$ with $a > 0$. Find the greatest common divisors below.
(a) (a, a^n) where n is a positive integer
(b) $(a, a + 1)$
(c) $(a, a + 2)$
(d) $(3a + 5, 7a + 12)$

34. Find the greatest common divisors below.
(a) $(18, 36, 63)$
(b) $(30, 42, 70)$
(c) $(0, 51, 0)$
(d) $(35, 55, 77)$
(e) $(36, 42, 54, 78)$
(f) $(35, 63, 70, 98)$

35. Find four integers that are relatively prime (when taken together) but such that no two of the integers are relatively prime when taken separately.

36. *(a)* Do there exist integers x and y such that $x + y = 100$ and $(x, y) = 8$? Why or why not?
(b) Prove that there exist infinitely many pairs of integers x and y such that $x + y = 87$ and $(x, y) = 3$.

37. Let $a, b \in \mathbf{Z}$ with a and b not both zero and let c be a nonzero integer. Prove that $(ca, cb) = |c|\,(a, b)$.

38. Let a and b be relatively prime integers. Prove that $(a + b, a - b)$ is either 1 or 2.

39. Let a and b be relatively prime integers. Find all values of $(a + 2b, 2a + b)$.

40. Let $a, b \in \mathbf{Z}$ with $(a, 4) = 2$ and $(b, 4) = 2$. Find $(a + b, 4)$ and prove that your answer is correct.

41. Let $a, b, c \in \mathbf{Z}$ with $(a, b) = 1$ and $c \mid a + b$. Prove that $(a, c) = 1$ and $(b, c) = 1$.

42. *(a)* Let $a, b, c \in \mathbf{Z}$ with $(a, b) = (a, c) = 1$. Prove that $(a, bc) = 1$.

(b) Let $a, b_1, b_2, \ldots, b_n \in \mathbf{Z}$ with $(a, b_1) = (a, b_2) = \cdots = (a, b_n) = 1$. Prove that

$$(a, b_1 b_2 \cdots b_n) = 1$$

43. (a) Let $a, b, c \in \mathbf{Z}$ with $(a, b) = 1$. Prove that if $a \mid c$ and $b \mid c$, then $ab \mid c$.
    (b) Provide a counterexample to show why the statement of part (a) does not hold if $(a, b) \neq 1$.
    (c) Let $a, a_2, \ldots, a_n, c \in \mathbf{Z}$ with $a_1, a_2, \ldots, a_n$ pairwise relatively prime. Prove that if $a_i \mid c$ for each i, then $a_1 a_2 \cdots a_n \mid c$.
44. (a) Let $a, b, c \in \mathbf{Z}$ with $(a, b) = 1$ and $a \mid bc$. Prove that $a \mid c$.
    (b) Provide a counterexample to show why the statement of part (a) does not hold if $(a, b) \neq 1$.
45. Let a, b, c, and d be positive integers. If $b \neq d$ and $(a, b) = (c, d) = 1$, prove that $\frac{a}{b} + \frac{c}{d} \notin \mathbf{Z}$.
46. Let a, b, c, and d be integers with b and d positive and $(a, b) = (c, d) = 1$. A mistake often made when first encountering fractions is to assume that $\frac{a}{b} + \frac{c}{d} = \frac{a+c}{b+d}$. Find all solutions of this equation.
47. (a) Let $a, b \in \mathbf{Z}$ and let m be a nonnegative integer. Prove that $(a, b) = 1$ if and only if $(a^m, b) = 1$.
    (b) Let $a, b \in \mathbf{Z}$ and let m and n be nonnegative integers. Prove that $(a, b) = 1$ if and only if $(a^m, b^n) = 1$.
48. Let $a, b \in \mathbf{Z}$. Prove that $(a, b) = 1$ if and only if $(a + b, ab) = 1$.
49. Prove that in any eight composite positive integers not exceeding 360, at least two are not relatively prime.
50. Prove that every integer greater than 6 can be expressed as the sum of two relatively prime integers greater than 1.
51. Let $a_1, a_2, \ldots, a_n \in \mathbf{Z}$ with $a_1 \neq 0$. Prove that

$$(a_1, a_2, a_3, \ldots, a_n) = ((a_1, a_2), a_3, \ldots, a_n)$$

(This method can be used generally to compute the greatest common divisor of more than two integers.) Use this method to compute the greatest common divisor of each set of integers in Exercise 34.

52. Let $a_1, a_2, \ldots, a_n \in \mathbf{Z}$ with $a_1 \neq 0$ and let c be a nonzero integer. Prove that

$$(ca_1, ca_2, ca_3, \ldots, ca_n) = |c|\,(a_1, a_2, a_3, \ldots, a_n).$$

53. Let $n \in \mathbf{Z}$. Prove that the integers $6n - 1$, $6n + 1$, $6n + 2$, $6n + 3$, and $6n + 5$ are pairwise relatively prime.

1.4

The Euclidean Algorithm

Our current method for finding the greatest common divisor of two integers is to list all divisors of each integer, find those divisors common to the two lists, and then choose the greatest such common divisor. Surely this method becomes unwieldy for large integers! (For example, what is the greatest common divisor of 803 and 154?) Is there a better method for computing the greatest common divisor of two integers? Yes, fortunately. Before discussing this method, we need a preliminary lemma.

Lemma 1.12: If $a, b \in \mathbf{Z}$, $a \geq b > 0$, and $a = bq + r$ with $q, r \in \mathbf{Z}$, then $(a, b) = (b, r)$.

Proof: Let c be a common divisor of a and b. Then $c \mid a$ and $c \mid b$ imply $c \mid a - qb$ by Proposition 1.2, from which $c \mid r$; we then have that c is a common divisor of b and r. Now let c be a common divisor of b and r so that $c \mid b$ and $c \mid r$. Then $c \mid qb + r$ by Proposition 1.2, from which $c \mid a$; we then have that c is a common divisor of a and b. So the common divisors of a and b are the same as the common divisors of b and r from which $(a, b) = (b, r)$. ■

The suggested method for computing the greatest common divisor of two positive integers is called the Euclidean algorithm after Euclid, who describes the algorithm in Book VII of his *Elements*. The Euclidean algorithm repeatedly uses the division algorithm (Theorem 1.4) to generate quotients and remainders from smaller and smaller pairs of positive integers. Eventually in this process, a remainder of zero is encountered, which terminates the algorithm; the greatest common divisor of the two integers is the remainder encountered just prior to the zero remainder. All of this is contained in the following theorem. As you read the theorem, note that the only parts requiring proof are the last three statements.

Theorem 1.13: (The Euclidean Algorithm) Let $a, b \in \mathbf{Z}$ with $a \geq b > 0$. By the division algorithm, there exist $q_1, r_1 \in \mathbf{Z}$ such that

$$a = bq_1 + r_1, \qquad 0 \leq r_1 < b$$

If $r_1 > 0$, there exist (by the division algorithm) $q_2, r_2 \in \mathbf{Z}$ such that

$$b = r_1 q_2 + r_2, \qquad 0 \leq r_2 < r_1$$

If $r_2 > 0$, there exist (by the division algorithm) $q_3, r_3 \in \mathbf{Z}$ such that

$$r_1 = r_2 q_3 + r_3, \qquad 0 \leq r_3 < r_2$$

Continue this process. Then $r_n = 0$ for some n. If $n > 1$, then $(a, b) = r_{n-1}$. If $n = 1$, then $(a, b) = b$.

Proof: Note that $r_1 > r_2 > r_3 > \ldots$. If $r_n \neq 0$ for all n, then $r_1, r_2, r_3, \ldots$ is an infinite, strictly decreasing sequence of positive integers, which is impossible. So $r_n = 0$ for some n. Now, if $n > 1$, repeated applications of Lemma 1.12 give

$$(a, b) = (b, r_1) = (r_1, r_2) = (r_2, r_3) = \cdots = (r_{n-1}, r_n) = (r_{n-1}, 0) = r_{n-1}$$

as desired. If $n = 1$, the desired statement is obvious. ■

Note that the Euclidean algorithm finds the greatest common divisor in the case of two *positive* integers, a and b (with $a \geq b$). In view of the remarks immediately after Definition 9 and the general remark in (b) of Example 11, all other computations are either reducible to this case or trivial. We now illustrate the use of the Euclidean algorithm with an example.

Example 14:

Find (803, 154) by using the Euclidean algorithm.
The notation of Theorem 1.13 is used throughout. Here $a = 803$ and $b = 154$. By the division algorithm,

$$803 = 154 \cdot 5 + 33 \tag{2}$$

Since $r_1 = 33 > 0$ apply the division algorithm to $b = 154$ and $r_1 = 33$ to obtain

$$154 = 33 \cdot 4 + 22 \tag{3}$$

Since $r_2 = 22 > 0$, apply the division algorithm to $r_1 = 33$ and $r_2 = 22$ to obtain

$$33 = 22 \cdot 1 + 11 \tag{4}$$

Since $r_3 = 11 > 0$, we apply the division algorithm to $r_2 = 22$ and $r_3 = 11$ to obtain

$$22 = 11 \cdot 2 + 0$$

Since $r_4 = 0$, the Euclidean algorithm terminates and

$$(803, 154) = r_3 = 11$$

Recall that, by Proposition 1.11, the greatest common divisor of two integers is expressible as an integral linear combination of the two integers. (As already noted, this expression will be used in forthcoming chapters.) The Euclidean algorithm provides a systematic procedure for obtaining such a linear combination, as the following example illustrates.

Example 15:

Express (803, 154) as an integral linear combination of 803 and 154.
Essentially, work through the steps of Example 14 backward. We have

$$\begin{aligned}
(803, 154) = 11 &= 33 - 22 \quad [\text{by } (4)] \\
&= 33 - (154 - 33 \cdot 4) \quad [\text{by } (3)] \\
&= 33 \cdot 5 - 154 \\
&= (803 - 154 \cdot 5)5 - 154 \quad [\text{by } (2)] \\
&= 803 \cdot 5 - 154 \cdot 26 \\
&= 5 \cdot 803 + (-26)154
\end{aligned}$$

This is an expression of (803, 154) as an integral linear combination of 803 and 154 as desired. Two remarks are in order here. First, by Proposition 1.11, the greatest common divisor of two integers is the least positive number expressible as an integral linear combination of the integers. So *no* integral linear combination of 803 and 154 can yield any of the integers 10, 9, ..., 1. Second, the expression of 11 as an integral linear combination of 803 and 154 above is *not* unique. For example, the reader may verify that $11 = 19\cdot 803 - 99\cdot 154$ is another such expression. In fact, there are infinitely many expressions of 11 as an integral linear combination of 803 and 154. We will have more to say on this issue in Chapter 6.

A useful programming project related to Examples 14 and 15 appears as Student Project 1.

Exercise Set 1.4

54. Use the Euclidean algorithm (Theorem 1.13) to find the greatest common divisors below. Express each greatest common divisor as an integral linear combination of the original integers.
(a) $(37, 60)$
(b) $(78, 708)$
(c) $(441, 1155)$
(d) $(793, 3172)$
(e) $(2059, 2581)$
(f) $(25174, 42722)$

55. Prove that 7 has no expression as an integral linear combination of 18209 and 19043.

56. Find two rational numbers with denominators 11 and 13, respectively, and a sum of $\frac{7}{143}$.

57. Use Exercise 51 and the Euclidean algorithm to find the greatest common divisors below. Express each greatest common divisor as an integral linear combination of the original integers.
(a) $(221, 247, 323)$
(b) $(210, 294, 490, 735)$

58. [The following exercise presents an algorithm for computing the greatest common divisor of two positive integers analogous to the Euclidean algorithm. This new algorithm is based on the absolute least remainder algorithm given in part (c) of Exercise 15.] Let $a, b \in \mathbf{Z}$ with $a \geq b > 0$. By the absolute least remainder algorithm, there exist $q_1, r_1 \in \mathbf{Z}$ such that

$$a = bq_1 + r_1, \qquad -\frac{|b|}{2} < r_1 \leq \frac{|b|}{2}$$

If $r_1 \neq 0$, there exist (by the absolute least remainder algorithm) $q_2, r_2 \in \mathbf{Z}$ such that

$$b = r_1 q_2 + r_2, \qquad -\frac{|r_1|}{2} < r_2 \leq \frac{|r_1|}{2}$$

If $r_2 \neq 0$, there exist (by the absolute least remainder algorithm) $q_3, r_3 \in \mathbf{Z}$ such that

$$r_1 = r_2 q_3 + r_3, \qquad -\frac{|r_2|}{2} < r_3 \leq \frac{|r_2|}{2}$$

Continue this process.
(a) Prove that $r_n = 0$ for some n. If $n > 1$, prove that $(a, b) = |r_{n-1}|$.
(b) Use the new algorithm above to find $(204, 228)$ and $(233, 377)$.

1.5

The Fundamental Theorem of Arithmetic

The Fundamental Theorem of Arithmetic (the essence of which appears as Proposition 14 in Book IX of Euclid's *Elements*) is our first big theorem of number theory, but certainly not our last! This theorem guarantees that any integer greater than 1 can be decomposed into a product of prime numbers; furthermore, this decomposition is unique except for the order in which the prime numbers are listed. (For example, the decomposition of 12 into $2 \cdot 2 \cdot 3$ is the same as the decomposition of 12 into $2 \cdot 3 \cdot 2$ except for the order in which the two 2's and one 3 are listed.) We first prove an important preliminary lemma.

Lemma 1.14: (Euclid) Let $a, b, p \in \mathbf{Z}$ with p prime. If $p \mid ab$, then $p \mid a$ or $p \mid b$.

Proof: Assume that $p \nmid a$. Then $(a, p) = 1$. We must show that $p \mid b$. By Proposition 1.11, there exist $m, n \in \mathbf{Z}$ such that $ma + np = 1$. Also, $p \mid ab$ implies $ab = pc$ for some $c \in \mathbf{Z}$. Now multiplying both sides of $ma + np = 1$ by b, we have $mab + npb = b$; $ab = pc$ then implies that $mpc + npb = b$ or $p(mc + nb) = b$. So we have $p \mid b$ as desired. ■

Note that Lemma 1.14 does not hold if p is composite. (Your solution to Exercise 2 in Section 1 should provide a counterexample.) In view of this, the criterion in Lemma 1.14 could be used to *define* a prime number as follows:

> An integer $p > 1$ is said to be *prime* if whenever a and b are integers with $p \mid ab$, then $p \mid a$ or $p \mid b$.

This alternate definition of *prime* is useful in more general mathematical settings (see, for example, Definition 3.3 on page 136 of Hungerford, 1974); in the system of integers, we have opted instead for the more traditional definition given in Definition 5.

Lemma 1.14 above can be generalized. We state this generalization as a corollary in which we use the powerful proof technique called *mathematical induction.* Mathematical induction will be used frequently in this book, especially in Chapter 7. (You may have already used mathematical induction a few times in the preceding exercises.) See Appendix A for a discussion of mathematical induction.

Corollary 1.15: Let $a_1, a_2, \ldots, a_n, p \in \mathbf{Z}$ with p prime. If $p \mid a_1a_2 \cdots a_n$, then $p \mid a_i$ for some i.

Proof: We use induction on n. The statement for $n = 1$ is obvious. The statement for $n = 2$ is Lemma 1.14. Assume that $k \geq 2$ and that the desired statement is true for $n = k$ so that $p \mid a_1a_2 \cdots a_k$ implies that $p \mid a_i$ for some i (with $1 \leq i \leq k$). We must show that $p \mid a_1a_2 \cdots a_{k+1}$ implies that $p \mid a_i$ for some i (with $1 \le i \le k+1$) so that the desired statement holds for $n = k+1$. Now $p \mid a_1 a_2 \cdots a_{k+1}$ implies $p \mid (a_1 a_2 \cdots a_k) a_{k+1}$; Lemma 1.14 then implies $p \mid a_1 a_2 \cdots a_k$ or $p \mid a_{k+1}$. If $p \mid a_{k+1}$, then the desired statement holds for $n = k+1$. If $p \nmid a_{k+1}$, then $p \mid a_1 a_2 \cdots a_k$, which implies that $p \mid a_i$ for some i (with $1 \le i \le k$) by the induction hypothesis, and the desired statement holds for $n = k+1$, which completes the proof. ■

We now state and prove the Fundamental Theorem of Arithmetic.

Theorem 1.16: (Fundamental Theorem of Arithmetic) Every integer greater than 1 can be expressed in the form $p_1^{a_1} p_2^{a_2} \cdots p_n^{a_n}$ with $p_1, p_2, \ldots, p_n$ distinct prime numbers and $a_1, a_2, \ldots, a_n$ positive integers. This form is said to be the *prime factorization* of the integer. This prime factorization is unique except for the arrangement of the $p_i^{a_i}$.

Proof: Assume, by way of contradiction, that k is an integer greater than 1 that does not have an expression as in the statement of the theorem. Without loss of generality, we may assume that k is the least such integer. Now k cannot be prime because it would then be of the desired form. So k is composite and $k = ab$ with $1 < a < k$ and $1 < b < k$. But then a and b are of the desired form due to the minimality of k, from which it follows that k is of the desired form, a contradiction. So every integer greater than 1 has an expression of the desired form. We still must show the uniqueness of such an expression. Assume that k has two such expressions, say

$$k = p_1^{a_1} p_2^{a_2} \cdots p_n^{a_n} = q_1^{b_1} q_2^{b_2} \cdots q_m^{b_m}$$

with $p_1, p_2, \ldots, p_n$ distinct prime numbers; $q_1, q_2, \ldots, q_m$ distinct prime numbers; and $a_1, a_2, \ldots, a_n$, $b_1, b_2, \ldots, b_m$ positive integers. Without loss of generality, we may assume that $p_1 < p_2 < \cdots < p_n$ and $q_1 < q_2 < \cdots < q_m$. We must show that

$$n = m$$

$$p_i = q_i, \quad i = 1, 2, \ldots, n$$

and

$$a_i = b_i, \quad i = 1, 2, \ldots, n$$

Now, given a p_i, we have $p_i \mid q_1^{b_1} q_2^{b_2} \cdots q_m^{b_m}$, which implies that $p_i \mid q_j$ for some j by Corollary 1.15. So $p_i = q_j$ for some j. Similarly, given a q_j, we have $q_j = p_i$ for some i. So $n = m$; by the ordering of the p_i's and q_j's, we have $p_i = q_i$, $i = 1, 2, \ldots, n$. Consequently,

$$k = p_1^{a_1} p_2^{a_2} \cdots p_n^{a_n} = p_1^{b_1} p_2^{b_2} \cdots p_n^{b_n}$$

Now assume, by way of contradiction, that $a_i \ne b_i$ for some i. Without loss of generality, we may assume that $a_i < b_i$. Then

$$p_i^{b_i} \mid p_1^{a_1} p_2^{a_2} \cdots p_n^{a_n}$$

which implies that

$$p_i^{b_i - a_i} \mid p_1^{a_1} p_2^{a_2} \cdots p_{i-1}^{a_{i-1}} p_{i+1}^{a_{i+1}} \cdots p_n^{a_n}$$

Since $b_i - a_i > 0$, we have that

$$p_i \mid p_1^{a_1} p_2^{a_2} \cdots p_{i-1}^{a_{i-1}} p_{i+1}^{a_{i+1}} \cdots p_n^{a_n}$$

from which $p_i \mid p_j$ for some $j \neq i$, by Corollary 1.15. This is a contradiction. So $a_i = b_i$, $i = 1, 2, \ldots, n$. ■

Example 16:

Find the prime factorization of 756.
We have

$$\begin{aligned} 756 &= 2 \cdot 378 \\ &= 2 \cdot 2 \cdot 189 \\ &= 2 \cdot 2 \cdot 3 \cdot 63 \\ &= 2 \cdot 2 \cdot 3 \cdot 7 \cdot 9 \\ &= 2 \cdot 2 \cdot 3 \cdot 7 \cdot 3 \cdot 3 \\ &= 2^2 3^3 7 \end{aligned}$$

Because of our familiarity with the integers, we may tend to take the Fundamental Theorem of Arithmetic for granted. Note, however, that number systems exist for which unique factorization does *not* hold. Such a system is investigated more fully in Exercise 77. The presence of the Fundamental Theorem of Arithmetic for the integers makes the integers into what is called a *unique factorization domain* (*UFD*). UFDs are important number systems in a more advanced branch of number theory called *algebraic number theory.* The moral here is to appreciate the Fundamental Theorem of Arithmetic for the extremely nice property that it is!

A concept that is parallel to the greatest common divisor of two integers is the least common multiple of the integers, which we define now.

Definition 11: Let $a, b \in \mathbf{Z}$ with $a, b > 0$. The *least common multiple* of a and b, denoted $[a, b]$, is the least positive integer m such that $a \mid m$ and $b \mid m$.

Given two positive integers a and b, note that $ab > 0$, $a \mid ab$, and $b \mid ab$; in other words, ab is always a positive common multiple of a and b. Hence the set of positive common multiples of two positive integers is always nonempty; consequently, the *least* common multiple of these integers always exists by the well-ordering property of the integers (see the Introduction).

Example 17:

(a) The positive multiples of 6 are $6, 12, 18, 24, 30, 36, 42, 48, \ldots$. The positive multiples of 7 are 7, 14, 21, 28, 35, 42, 49, 54, ... It is a fact (perhaps not obvious) that the positive *common* multiples of 6 and 7 (the multiples common to the two infinite lists above) are $42, 84, 126, 168, 210, \ldots$. (Convince yourself of this!) Obviously from these common multiples of 6 and 7, the least common multiple of 6 and 7 is 42, which is denoted $[6, 7] = 42$.

(b) The positive multiples of 8 are $8, 16, 24, 32, 40, 48, 56, 64, \ldots$. From the multiples of 6 in (a) above, it is obvious that the least common multiple of 6 and 8 is 24, which is denoted $[6, 8] = 24$. (As an exercise, find the first five positive common multiples of 6 and 8.)

The Fundamental Theorem of Arithmetic may be used to compute greatest common divisors and least common multiples. We illustrate such computations with an example.

Example 18:

Find $(756, 2205)$ and $[756, 2205]$ by using prime factorization.
By Example 16, we have $756 = 2^2 3^3 7$. The reader may easily verify that $2205 = 3^2 5 \cdot 7^2$. Now

$$756 = 2^2 3^3 5^0 7^1$$

and

$$2205 = 2^0 3^2 5^1 7^2$$

For $(756, 2205)$, we compare the exponents appearing on like prime numbers and choose the minimum exponent appearing in each comparison [since $(756, 2205)$ must divide both 756 and 2205]. So

$$(756, 2205) = 2^0 3^2 5^0 7 = 63$$

Similarly, for $[756, 2205]$, we compare the exponents appearing on like prime numbers and choose the maximum exponent appearing in each comparison (since both 756 and 2205 must divide $[756, 2205]$). So

$$[756, 2205] = 2^2 3^3 5^1 7^2 = 26460$$

Example 18 motivates the following proposition.

Proposition 1.17: Let $a, b \in \mathbf{Z}$ with $a, b > 1$. Write $a = p_1^{a_1} p_2^{a_2} \cdots p_n^{a_n}$ and $b = p_1^{b_1} p_2^{b_2} \cdots p_n^{b_n}$ where $p_1, p_2, \ldots, p_n$ are distinct prime numbers and $a_1, a_2, \ldots, a_n, b_1, b_2, \ldots, b_n$ are nonnegative integers (possibly zero). Then

$$(a, b) = p_1^{\min\{a_1, b_1\}} p_2^{\min\{a_2, b_2\}} \cdots p_n^{\min\{a_n, b_n\}}$$

and

$$[a, b] = p_1^{\max\{a_1, b_1\}} p_2^{\max\{a_2, b_2\}} \cdots p_n^{\max\{a_n, b_n\}}$$

Proof: The proof is obvious from the Fundamental Theorem of Arithmetic (Theorem 1.16) and the definitions of (a, b) and $[a, b]$. ■

It is instructive to pause here. Proposition 1.17 seemingly gives us a beautiful way to compute greatest common divisors and least common multiples of positive integers. Has our Euclidean algorithm (for computation of the greatest common divisor) been rendered obsolete? The answer is a resounding *no.* While Proposition 1.17 is useful as a theoretical result, it has little practical use except for computations involving greatest common divisors and least common multiples of relatively small integers. Proposition 1.17 requires the computation of prime factorizations. Unfortunately, prime factorizations of large integers are generally difficult to obtain. [The interested reader may wish to consult Bressoud (1990) in this regard.] The power inherent in the Euclidean algorithm is that greatest common divisors of integers may be computed *without obtaining the prime factorizations of the integers.* But what about least common multiples? Our goal here is to establish a simple relationship between the least common multiple and the greatest common divisor of two positive integers; the desired result is a practical method for computing least common multiples. We first need an easy lemma.

Lemma 1.18: Let $x, y \in \mathbf{R}$. Then $\max\{x, y\} + \min\{x, y\} = x + y$.

Proof: If $x < y$, then $\max\{x, y\} = y$ and $\min\{x, y\} = x$, and the desired result follows. The cases are similar for $x = y$ and $x > y$. ■

The relationship between the greatest common divisor and the least common multiple of two positive integers is now given by the following theorem.

Theorem 1.19: Let $a, b \in \mathbf{Z}$ with $a, b > 0$. Then $(a, b)[a, b] = ab$.

Proof: If $a, b > 1$, write $a = p_1^{a_1} p_2^{a_2} \cdots p_n^{a_n}$ and $b = p_1^{b_1} p_2^{b_2} \cdots p_n^{b_n}$ as in Proposition 1.17. Then

$$\begin{aligned}(a, b)[a, b] &= \{p_1^{\min\{a_1,b_1\}} p_2^{\min\{a_2,b_2\}} \cdots p_n^{\min\{a_n,b_n\}}\}\{p_1^{\max\{a_1,b_1\}} p_2^{\max\{a_2,b_2\}} \cdots p_n^{\max\{a_n,b_n\}}\} \\ &= p_1^{\min\{a_1,b_1\}+\max\{a_1,b_1\}} p_2^{\min\{a_2,b_2\}+\max\{a_2,b_2\}} \cdots p_n^{\min\{a_n,b_n\}+\max\{a_n,b_n\}} \\ &= p_1^{a_1+b_1} p_2^{a_2+b_2} \cdots p_n^{a_n+b_n} \quad \text{(by Lemma 1.18)} \\ &= p_1^{a_1} p_2^{a_2} \cdots p_n^{a_n} p_1^{b_1} p_2^{b_2} \cdots p_n^{b_n} \\ &= ab\end{aligned}$$

as desired. The cases $a = 1$ and $b > 1$, $a > 1$ and $b = 1$, and $a = b = 1$ are easily checked and are left as exercises. ■

So, to find the least common multiple of two positive integers without using prime factorization, one finds the greatest common divisor of the two integers by using the Euclidean algorithm and then Theorem 1.19 to obtain the least common multiple. Again, this method is usually much easier than using Proposition 1.17, which requires the prime factorizations of the integers! The reader is invited to pause here to compute $[803, 154]$ using $(803, 154) = 11$ from Example 14.

Note in Example 17 that the least common multiple of 6 and 7 was the product of 6 and 7, while the least common multiple of 6 and 8 was *not* the product of 6 and 8. (Is the least common multiple of 803 and 154 the product of 803 and 154?) Under what circumstances is the least common multiple of two positive integers simply the product of the two integers? In view of Theorem 1.19, the answer is not surprising, as the next corollary shows.

Corollary 1.20: Let $a, b \in \mathbf{Z}$ with $a, b > 0$. Then $[a, b] = ab$ if and only if $(a, b) = 1$.

Proof: The (extremely easy) proof is left as an exercise for the reader. ■

We conclude this chapter with another illustration of the power of the

Biography

Peter Gustav Lejeune Dirichlet (1805–1859)

P. G. L. Dirichlet was a student of Gauss and Gauss's successor at Göttingen University. Born in Germany, Dirichlet was fluent in both German and French and served as a mathematical liasion between Germany and France. Along with his close friend Jacobi (see the latter's biography in Chapter 4), Dirichlet aided in the shift of mathematical activity from France to Germany in the nineteenth century. Dirichlet authored *Vorlesungen über Zahlentheorie,* a work essentially devoted to the further understanding of the many treasures in Gauss's profound *Disquisitiones Arithmeticae.* In addition, he analyzed the convergence of Fourier series and proved the theorem on prime numbers in arithmetic progressions below. Dirichlet frequently relied on the principles of real and complex analysis in his proofs; as such, Dirichlet set the stage for that area of mathematics known as *analytic number theory.*

Fundamental Theorem of Arithmetic. We begin by stating a famous theorem of number theory from P. G. L. Dirichlet. This theorem generalizes Euclid's theorem on the infinitude of the prime numbers (Theorem 1.6). Unfortunately, the proof of this theorem is beyond the scope of this book.

Theorem 1.21: (Dirichlet's Theorem on Prime Numbers in Arithmetic Progressions) Let $a, b \in \mathbf{Z}$ with $a, b > 0$ and $(a, b) = 1$. Then the arithmetic progression

$$a, a + b, a + 2b, \ldots, a + nb, \ldots$$

contains infinitely many prime numbers.

Note that Theorem 1.6 (the infinitude of the prime numbers) follows from

Theorem 1.21 above by putting $a = b = 1$. Dirichlet's Theorem, however, may be used to establish many other interesting facts, one of which is investigated in Exercise 84. The next proposition is a special case of Dirichlet's Theorem (with $a = 3$ and $b = 4$).

Proposition 1.22: There are infinitely many prime numbers expressible in the form $4n + 3$ where n is a nonnegative integer.

The Fundamental Theorem of Arithmetic allows us to prove the special case of Dirichlet's Theorem given in Proposition 1.22 above. We need a preliminary lemma.

Lemma 1.23: Let $a, b \in \mathbf{Z}$. If a and b are expressible in the form $4n + 1$ where n is an integer, then ab is also expressible in that form.

Proof: Let $a = 4n_1 + 1$ and $b = 4n_2 + 1$ with $n_1, n_2 \in \mathbf{Z}$. Then

$$\begin{aligned} ab &= (4n_1 + 1)(4n_2 + 1) = 16n_1n_2 + 4n_1 + 4n_2 + 1 \\ &= 4(4n_1n_2 + n_1 + n_2) + 1 \\ &= 4n + 1 \end{aligned}$$

where $n = 4n_1n_2 + n_1 + n_2 \in \mathbf{Z}$. ■

We now prove Proposition 1.22.

Proof: (of Proposition 1.22) Assume, by way of contradiction, that there are only finitely many prime numbers of the form $4n + 3$ where n is a nonnegative integer, say $p_0 = 3$, $p_1, p_2, \ldots, p_r$. (Here the prime numbers $p_1, p_2, \ldots, p_r$ are assumed to be distinct from 3.) Consider the number $N = 4p_1p_2 \cdots p_r + 3$. The prime factorization of N must contain a prime number p of the desired form or it would otherwise contain only prime numbers of the form $4n + 1$ where n is a positive integer (see Exercise 85) and hence be of the form $4n + 1$ where n is a positive integer by Lemma 1.23. So $p = p_0 = 3$ or $p = p_i$, $i = 1, 2, \ldots, r$.

Case I: $p = p_0 = 3$

Then $3 \mid N$ implies $3 \mid N - 3$ by Proposition 1.2, and we have that $3 \mid 4p_1p_2 \cdots p_r$. By Corollary 1.15, we now have $3 \mid 4$ or $3 \mid p_i$, $i = 1, 2, \ldots, r$, a contradiction.

Case II: $p = p_i$, $i = 1, 2, \ldots, r$

Then $p_i \mid N$ implies $p_i \mid N - 4p_1p_2 \cdots p_r$ by Proposition 1.2, and we have that $p_i \mid 3$, a contradiction.

So there are infinitely many prime numbers of the desired form. ■

Note the ingenuity in the construction of the number N above, because it is precisely consideration of this number that results in the desired contradictions. (Where have we seen such ingenuity before?) More important, Proposition 1.22 was billed as an application of the Fundamental Theorem of Arithmetic; where was the Fundamental Theorem of Arithmetic used in the proof of Proposition 1.22 above?

The proof of Proposition 1.22 cannot be extended to prove Dirichlet's Theorem in general, since the analogue of Lemma 1.23 is not true in general (see Exercise 86). The analogue of Lemma 1.23 is true in other special cases of Dirichlet's Theorem, one of which is investigated in Exercise 87.

Exercise Set 1.5

59. Find the prime factorization of each integer below.
(a) 51
(b) 87
(c) 361
(d) 367
(e) 422
(f) 945
(g) 1001
(h) 6292

60. Find the least common multiples below.
(a) $[6, 9]$
(b) $[7, 9]$
(c) $[13, 91]$
(d) $[24, 60]$
(e) $[100, 105]$
(f) $[101, 1111]$

61. Find the greatest common divisor and the least common multiple of each pair of integers below.
(a) $2^2 \cdot 3^3 \cdot 5 \cdot 7,\ 2^2 \cdot 3^2 \cdot 5 \cdot 7^2$
(b) $2^2 \cdot 5^2 \cdot 7^3 \cdot 11^2,\ 3 \cdot 5 \cdot 11 \cdot 13 \cdot 17$
(c) $2^2 \cdot 5^7 \cdot 11^{13},\ 3^2 \cdot 7^5 \cdot 13^{11}$
(d) $3 \cdot 17 \cdot 19^2 \cdot 23,\ 5 \cdot 7^2 \cdot 11 \cdot 19 \cdot 29$

62. Find five integers that are relatively prime (when taken together) but such that no two of the integers are relatively prime when taken separately.

63. Find each of the least common multiples below by using the Euclidean algorithm and Theorem 1.19.
(a) $[221, 323]$
(b) $[257, 419]$
(c) $[313, 1252]$
(d) $[1911, 9702]$

64. Find all pairs of positive integers a and b such that $(a, b) = 12$ and $[a, b] = 360$.

65. ***Definition:*** Let $a_1, a_2, \ldots, a_n \in \mathbf{Z}$ with $a_1, a_2, \ldots, a_n$ nonzero. The *least common multiple* of $a_1, a_2, \ldots, a_n$, denoted $[a_1, a_2, \ldots, a_n]$, is the least positive integer m such that $a_i \mid m$, $i = 1, 2, \ldots, n$.
Find the least common multiples below.

(*a*) $[10, 12, 15]$
(*b*) $[9, 13, 16]$
(*c*) $[6, 7, 8, 9]$

66. Why is 1 excluded as a prime number?
67. (*a*) Let $n \in \mathbf{Z}$ with $n > 1$. Prove that n is a perfect square if and only if all exponents in the prime factorization of n are even.
    (*b*) Let $n \in \mathbf{Z}$ with $n > 0$. Prove that n is the product of a perfect square and (possibly zero) distinct prime numbers.
68. Let $n \in \mathbf{Z}$ with $n > 0$. Prove that there exist $k, m \in \mathbf{Z}$ with m odd such that $n = 2^k m$.
69. ***Definition:*** Let $n \in \mathbf{Z}$ with $n > 1$. Then n is said to be *powerful* if all exponents in the prime factorization of n are at least 2.
    Prove that a powerful number is the product of a perfect square and a perfect cube.
70. Prove or disprove the following statements.
    (*a*) If $a, b \in \mathbf{Z}$, $a, b > 0$, and $a^2 \mid b^3$, then $a \mid b$.
    (*b*) If $a, b \in \mathbf{Z}$, $a, b > 0$, and $a^2 \mid b^2$, then $a \mid b$.
    (*c*) If $a \in \mathbf{Z}$, $a > 0$, p is a prime number, and $p^4 \mid a^3$, then $p^2 \mid a$.
71. Let n, a, and b be positive integers such that $ab = n^2$. If $(a, b) = 1$, prove that there exist positive integers c and d such that $a = c^2$ and $b = d^2$.
72. ***Definition:*** Let n and a be positive integers and let p be a prime number. Then p^a is said to *exactly divide n*, denoted $p^a \parallel n$, if $p^a \mid n$ and $p^{a+1} \nmid n$.
    Assume that $p^a \parallel m$ and $p^b \parallel n$.
    (*a*) What power of p exactly divides $m + n$? Prove your assertion.
    (*b*) What power of p exactly divides mn? Prove your assertion.
    (*c*) What power of p exactly divides m^n? Prove your assertion.
73. Let $n \in \mathbf{Z}$ with $n > 0$. Find the largest integer guaranteed to divide *all* products of n consecutive integers and prove your assertion. (*Hint*: Consider several small values of n and look for a pattern.)
74. Find all positive integers a and b such that $a^b = b^a$.
75. (*Note*: You may wish to review Exercise 14 before attempting this problem.)
    (*a*) Find the number of positive integers not exceeding 500 that are divisible by 2 and by 3.
    (*b*) Find the number of positive integers not exceeding 500 that are divisible by neither 2 nor 3.
    (*c*) Find the number of positive integers not exceeding 500 that are divisible by 2 but not by 3.
76. (*a*) Let $n \in \mathbf{Z}$ with $n > 1$, and let p be a prime number. If $p \mid n!$, prove that the exponent of p in the prime factorization of $n!$ is $[n/p] + [n/p^2] + [n/p^3] + \cdots$. (Note that this sum is finite, since $[n/p^m] = 0$ if $p^m > n$.)
    (*b*) Use part (a) above to find the prime factorization of $20!$.
    (*c*) Find the number of zeros with which the decimal representation of $100!$ terminates.
77. (The following exercise develops a number system that does not possess unique factorization and thus no Fundamental Theorem of Arithmetic.) Let $\mathbf{Z}[\sqrt{-10}]$ denote the set of all complex numbers of the form

$a + b\sqrt{-10}$ with $a, b \in \mathbf{Z}$ under the usual operations of addition and multiplication of complex numbers. Define an element $a + b\sqrt{-10}$ of $\mathbf{Z}[\sqrt{-10}]$ to be *irreducible* if $a + b\sqrt{-10}$ cannot be expressed as a product of two elements of $\mathbf{Z}[\sqrt{-10}]$ except as the trivial factorizations

$$a + b\sqrt{-10} = (1)(a + b\sqrt{-10})$$

or

$$a + b\sqrt{-10} = (-1)(-a - b\sqrt{-10})$$

(a) Prove that 2 and 7 are irreducible elements in $\mathbf{Z}[\sqrt{-10}]$.
(b) Prove that $2 + \sqrt{-10}$ and $2 - \sqrt{-10}$ are irreducible elements in $\mathbf{Z}[\sqrt{-10}]$. [*Hint*: For $2 + \sqrt{-10}$, assume that $2 + \sqrt{-10} = (a + b\sqrt{-10})(c + d\sqrt{-10})$ with $a, b, c, d \in \mathbf{Z}$. Then $2 - \sqrt{-10} = (a - b\sqrt{-10})(c - d\sqrt{-10})$ (why?). Now multiply the left-hand sides and the right-hand sides of the two equations and prove that the factorization of $2 + \sqrt{-10}$ must be a trivial factorization.]
(c) Prove that $\mathbf{Z}[\sqrt{-10}]$ does not possess unique factorization into irreducible elements. [*Hint*: In $\mathbf{Z}[\sqrt{-10}]$, we have $14 = 2 \cdot 7 = (2 + \sqrt{-10})(2 - \sqrt{-10})$.]

78. Let $n \in \mathbf{Z}$ with $n > 1$. Prove that $1 + \frac{1}{2} + \frac{1}{3} + \cdots + \frac{1}{n} \notin \mathbf{Z}$.

79. Let a and b be positive integers.
(a) Prove that $(a, b) \mid [a, b]$.
(b) Find and prove a necessary and sufficient condition that $(a, b) = [a, b]$.

80. Let a, b, and c be positive integers. Prove that $[ca, cb] = c[a, b]$.

81. Let $a_1, a_2, \ldots, a_n$ be positive integers. Prove that

$$[a_1, a_2, a_3, \ldots, a_n] = [[a_1, a_2], a_3, \ldots, a_n].$$

(This method can be used generally to compute the least common multiple of more than two integers.) Use this method to compute the least common multiple of each set of integers in Exercise 65.

82. Let $a_1, a_2, \ldots, a_n$, and c be positive integers. Prove that

$$[ca_1, ca_2, ca_3, \ldots, ca_n] = c[a_1, a_2, a_3, \ldots, a_n].$$

83. Let $a, b \in \mathbf{Z}$ with $a, b > 0$ and $(a, b) = 1$. Prove that the arithmetic progression

$$a, a + b, a + 2b, \ldots, a + nb, \ldots$$

contains infinitely many composite numbers.

84. Let $k \in \mathbf{Z}$ with $k > 0$. Prove that there are infinitely many prime numbers ending in k 1's.

85. *(a)* Prove that any integer is expressible in the form $4n$, $4n + 1$, $4n + 2$, or $4n + 3$ where n is an integer.
(b) Prove that any odd number is expressible in the form $4n + 1$ or $4n + 3$ where n is an integer.

86. Prove that the analogue of Lemma 1.23 is not true for numbers of the form $4n + 3$ where n is an integer.

87. *(a)* Let $a, b \in \mathbf{Z}$. Prove that if a and b are expressible in the form $6n + 1$, where n is an integer, then ab is also expressible in that form.
(b) Prove that there are infinitely many prime numbers of the form $6n + 5$

where n is an integer. (*Hint*: Assume, by way of contradiction, that there are only finitely many prime numbers of the desired form and ingeniously construct a number N that will eventually lead to a contradiction. In other words, parallel the proof of Proposition 1.22.)

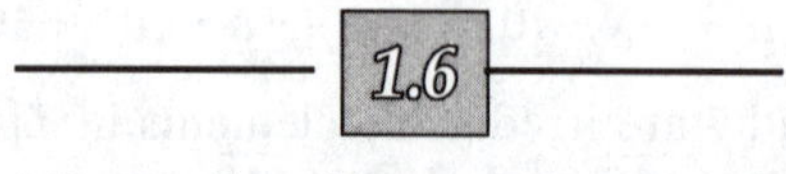

Concluding Remarks

The importance of divisibility and factorization in elementary number theory cannot be overestimated; the concepts and ideas presented in this chapter will appear again and again in succeeding chapters. The repercussions of the Fundamental Theorem of Arithmetic will be monumental. (Moral: Learn these concepts and ideas *now*!) The ultimate simplicity of these concepts has its disadvantages: Many of the open problems in elementary number theory are easily stated but maddeningly difficult to prove. (Go back and look at the conjectures in this chapter. How many of these conjectures would you have guessed still require proofs or counterexamples?) The simple nature of elementary number theory seemingly does not give us adequate techniques for attacking these questions; most mathematicians focus on more advanced algebraic and analytic techniques in their attempts to find solutions to these problems. It is perhaps unexpected (but certainly fascinating!) that the integers apparently contain so many secrets waiting to be discovered.

Student Projects

1. (Programming project for calculator or computer)
   (a) Given positive integers a and b, compute (a, b) by using the Euclidean algorithm.
   (b) Given positive integers a and b, express (a, b) as an integral linear combination of a and b.
2. Prove or disprove the following conjectures.
   Conjecture: If n is a nonnegative integer, then $n^2 - 79n + 1601$ is a prime number.
   Conjecture: If n is a nonnegative integer, then $n^2 + n + 41$ is a prime number.
   [A discussion of prime-producing polynomials of the form $n^2 + n + c$ (as in the second conjecture above) may be found in Daniel Fendel, "Prime-producing Polynomials and Principal Ideal Domains," *Mathematics Magazine, 58* (1985), 204–210.]
3. *(a)* Let p_n denote the nth prime number. Prove that the infinite series $\sum_{n=1}^{\infty} \frac{1}{p_n}$ diverges. (A short proof of this fact can be found in Apostol, 1976).
   (b) Do some research to find the behavior of the series $\sum_{p} \frac{1}{p}$, where p ranges over all twin primes.

4. Two integers m and n between 2 and 100 inclusive are chosen. The sum of the two integers is given to one mathematician, Sam, and the product of the two integers is given to another mathematician, Prudence. Suppose that, after some thought, the two mathematicians exchange the following dialogue:

   Prudence: "I don't know your sum, Sam."
   Sam: "I knew that you didn't, Prudence."
   Prudence: "Now I know your sum."
   Sam: "And now I know your product."

   What are m and n?

   [A discussion of the problem above and similar problems may be found in John O. Kiltinen & Peter B. Young, "Goldbach, Lemoine, and a Know/Don't Know Problem," *Mathematics Magazine, 58* (1985), 195–203.]
5. Answer the problem posed by Archimedes Andrews at the conclusion of the following article: Barry A. Cipra, "Archimedes Andrews and the Euclidean Time Bomb," *Mathematical Intelligencer, 9* (1987), 44–47.
6. For positive integral n, consider the function

$$h(n) = \begin{cases} \dfrac{n}{2}, & \text{if } n \text{ is even} \\ 3n + 1, & \text{if } n \text{ is odd} \end{cases}$$

   Given n, one may iterate the function $h(n)$ to obtain a sequence of positive integers $\{n, h(n), h(h(n)), h(h(h(n))), \ldots\}$. For example, the sequence associated with the starting integer 6 is $\{6, 3, 10, 5, 16, 8, 4, 2, 1, \ldots\}$. By experimenting with various starting values of n, formulate a conjecture concerning the behavior of the associated sequences.

   [There is a famous unsolved problem concerning the function $h(n)$ that you may have formulated above as a conjecture. For further discussion of this problem, consult section 11.3 of the following book: Clifford A. Pickover, *Computers, Pattern, Chaos and Beauty* (New York: St. Martin's Press, 1990).]
7. Find the next term in the following sequence:

$$2, \quad 12, \quad 360, \quad 75600, \quad 174636000, \quad \ldots$$

   [This sequence, created by Paul Chernoff, is discussed in Chapter 66 of the following book: Clifford A. Pickover, *Mazes for the Mind: Computers and the Unexpected* (New York: St. Martin's Press, 1992).]

2

Congruences

Up to the beginning of the nineteenth century, number theory consisted mainly of a series of brilliant results with no connecting theme. This all changed in 1801 when Carl Friedrich Gauss published his great work, *Disquisitiones Arithmeticae,* at 24 years of age. In this work, which many mathematicians consider the greatest ever in the theory of numbers, Gauss developed the theory of congruences, the subject of this chapter. We will see that this theory has important ramifications throughout much of the remainder of this book.

There will be four big theorems involving congruences in this chapter, each in its own section. The Chinese Remainder Theorem (Section 2.3) will enable us to solve certain systems of congruences. Consideration of the congruence known as Wilson's Theorem (Section 2.4) will result in a necessary and sufficient condition for the primality of a positive integer. Finally, Fermat's Little Theorem (Section 2.5) and Euler's Theorem (Section 2.6) provide two congruences that will be useful again and again as computational and theoretical tools.

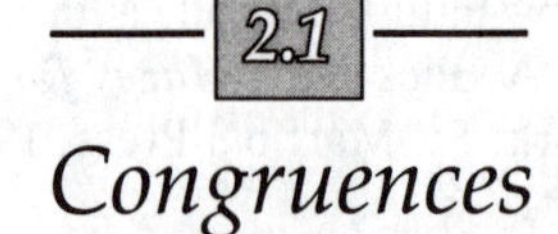

2.1 Congruences

The notion of congruence is defined in terms of the divisibility relation of Chapter 1. Two integers are related under congruence if their difference is evenly divisible by a third fixed positive integer as now specified.

Definition 1: Let $a, b, m \in \mathbf{Z}$ with $m > 0$. Then a is said to be *congruent to b modulo m,* denoted $a \equiv b \bmod m$, if $m \mid a - b$. If $a \equiv b \bmod m$, then m is said to be the *modulus* of the congruence. The notation $a \not\equiv b \bmod m$ means that a is not congruent to b modulo m; a is said to be *incongruent to b modulo m.*

Biography

Carl Friedrich Gauss (1777–1855)

Carl Friedrich Gauss was the greatest mathematician of the nineteenth century and is generally regarded as one of the three greatest mathematicians of all time, along with Archimedes and Newton. Born in Germany, he never once left his homeland. From 1807 until his death, Gauss was professor of mathematics and director of the observatory at Göttingen University. He married twice and fathered six children.

Gauss was a child prodigy. At the age of three, he corrected an arithmetic error in his father's bookkeeping. An often-told story illustrates his brilliance: In an arithmetic class when he was 10 years of age, he almost immediately solved a problem involving the sum of a large number of terms of a certain arithmetic progression. Rather than performing the straightforward (but extremely tedious) addition, Gauss essentially discovered the formula for such a sum on his own, a magnificent feat for a boy of 10. (Furthermore, he was presumably the only student in the class who obtained the correct answer!)

Gauss' doctoral dissertation, written at age 20, contained a proof of the Fundamental Theorem of Algebra. Unsuccessful attempts to prove this theorem had been made by the likes of Newton, Euler, and Lagrange. During his lifetime, Gauss would find three more proofs of the Fundamental Theorem of Algebra.

Gauss is best known for his *Disquisitiones Arithmeticae,* published in 1801; it is perhaps the greatest work ever in the theory of numbers. In it, Gauss develops the theory of congruences, the theory of quadratic residues, and the theory of binary and ternary quadratic forms, and he shows how this theory may be applied to the classical geometric problem of constructing a regular polygon with a prescribed number of sides. In addition, *Disquisitiones Arithmeticae* contains the first published proof of the beautiful law of quadratic reciprocity. (The law of quadratic reciprocity is discussed in Section 4.3.) In all, Gauss found eight different proofs of this law.

Note that the modulus in a congruence is defined to be a *positive* integer. Always assume this fact even though it may not be stated explicitly.

Example 1:

(a) Since $4 \mid 25 - 1$, we have $25 \equiv 1 \bmod 4$.
(b) Since $6 \mid 4 - 10$, we have $4 \equiv 10 \bmod 6$.
(c) Since $7 \mid 10 - (-4)$, we have $10 \equiv -4 \bmod 7$.
(d) Since $1 \mid a - b$ for all integers a and b, we have $a \equiv b \bmod 1$ for all integers a and b. So any two integers are congruent modulo 1.
(e) Since $5 \nmid -7 - 2$, we have $-7 \not\equiv 2 \bmod 5$.

The congruence relation on $\mathbf{Z}$ enjoys many (but not all!) of the properties satisfied by the usual relation of equality on $\mathbf{Z}$. In fact, Gauss chose the symbol $\equiv$ to emphasize this close relationship between the congruence relation and the relation of equality. In particular, the congruence relation is reflexive on $\mathbf{Z}$ ($a \equiv a \bmod m$ for all $a \in \mathbf{Z}$), symmetric ($a \equiv b \bmod m$ implies that $b \equiv a \bmod m$ for all $a, b \in \mathbf{Z}$), and transitive ($a \equiv b \bmod m$ and $b \equiv c \bmod m$ imply that $a \equiv c \bmod m$ for all $a, b, c \in \mathbf{Z}$), which makes the congruence relation an equivalence relation on $\mathbf{Z}$ as we now show. (See Appendix B for a brief discussion of equivalence relations.)

Proposition 2.1: Congruence modulo m is an equivalence relation on $\mathbf{Z}$.

Proof: Let $a \in \mathbf{Z}$. Since any integer divides 0, we have $m \mid 0$, from which $m \mid a - a$ and $a \equiv a \bmod m$. So congruence modulo m is reflexive on $\mathbf{Z}$. Let $a, b \in \mathbf{Z}$ and assume that $a \equiv b \bmod m$. Then $m \mid a - b$, and we have $m \mid (-1)(a - b)$ or, equivalently, $m \mid b - a$. So $b \equiv a \bmod m$ and congruence modulo m is symmetric. Let $a, b, c \in \mathbf{Z}$ and assume that $a \equiv b \bmod m$ and $b \equiv c \bmod m$. Then $m \mid a - b$ and $m \mid b - c$, and we have $m \mid (a - b) + (b - c)$ by Proposition 1.2 or, equivalently, $m \mid a - c$. So $a \equiv c \bmod m$, and congruence modulo m is transitive. Thus, congruence modulo m is an equivalence relation on $\mathbf{Z}$. ■

An immediate consequence of Proposition 2.1 is important enough to be highlighted separately:

Consequence 2.2: (of Proposition 2.1) $\mathbf{Z}$ is partitioned into equivalence classes under congruence modulo m.

We now undertake an analysis of the structure of the equivalence classes of the integers under congruence modulo m. Throughout our discussion, we use the notation $[x]$ to denote the equivalence class containing x. This notation will cause no confusion with the greatest integer function (Definition 3 of Chapter 1), since the meaning of the symbol $[\cdot]$ will either be stated explicitly or be clear from the context in which it is used. [We note here that the potential for ambiguous notation occurs quite frequently throughout mathematics (in fact, it will occur again in this text!); this potential is an unfortunate by-product of the mathematician's desire to keep notation as simple as possible. Be aware of such possible ambiguities and make sure that you completely understand the meanings of the symbols involved in any mathematical expression]. Our analysis begins with a motivational example.

Example 2:

The equivalence classes of $\mathbf{Z}$ under congruence modulo 4 are

$$\begin{aligned}[0] &= \{x \in \mathbf{Z}: x \equiv 0 \bmod 4\} \\ &= \{x \in \mathbf{Z}: 4 \mid x - 0\} \\ &= \{x \in \mathbf{Z}: 4 \mid x\} \\ &= \{x: x = 4n \text{ for some } n \in \mathbf{Z}\} \\ &= \{\ldots, -8, -4, 0, 4, 8, \ldots\},\end{aligned}$$

$$\begin{aligned}[1] &= \{x \in \mathbf{Z}: x \equiv 1 \bmod 4\} \\ &= \{x \in \mathbf{Z}: 4 \mid x - 1\} \\ &= \{x: x - 1 = 4n \text{ for some } n \in \mathbf{Z}\} \\ &= \{x: x = 4n + 1 \text{ for some } n \in \mathbf{Z}\} \\ &= \{\ldots, -7, -3, 1, 5, 9, \ldots\},\end{aligned}$$

$$\begin{aligned}[2] &= \{x \in \mathbf{Z}: x \equiv 2 \bmod 4\} \\ &= \{x \in \mathbf{Z}: 4 \mid x - 2\} \\ &= \{x: x - 2 = 4n \text{ for some } n \in \mathbf{Z}\} \\ &= \{x: x = 4n + 2 \text{ for some } n \in \mathbf{Z}\} \\ &= \{\ldots, -6, -2, 2, 6, 10, \ldots\},\end{aligned}$$

and

$$\begin{aligned}[3] &= \{x \in \mathbf{Z}: x \equiv 3 \bmod 4\} \\ &= \{x \in \mathbf{Z}: 4 \mid x - 3\} \\ &= \{x: x - 3 = 4n \text{ for some } n \in \mathbf{Z}\} \\ &= \{x: x = 4n + 3 \text{ for some } n \in \mathbf{Z}\} \\ &= \{\ldots, -5, -1, 3, 7, 11, \ldots\}.\end{aligned}$$

Consequently, $\mathbf{Z}$ is partitioned into the four equivalence classes

$$\{\ldots, -8, -4, 0, 4, 8, \ldots\}$$
$$\{\ldots, -7, -3, 1, 5, 9, \ldots\}$$
$$\{\ldots, -6, -2, 2, 6, 10, \ldots\}$$
$$\{\ldots, -5, -1, 3, 7, 11, \ldots\}$$

under congruence modulo 4.

A remark is in order here. In the computations of each equivalence class in Example 2 above, congruences were translated into equivalent divisibility statements that were used to obtain equations involving the unknown quantity. (For example, in the computation of the equivalence class [3], the congruence

$x \equiv 3 \bmod 4$ became $4 \mid x - 3$; this in turn was used to obtain $x = 4n + 3$.) This technique of translating congruences into equations can be quite powerful as a proof technique. In fact, we will use this technique again in Section 2.2.

From Example 2, we see that any integer is congruent modulo 4 to exactly one of 0, 1, 2, and 3. Equivalently, any integer is expressible in exactly one of the forms $4n$, $4n + 1$, $4n + 2$, and $4n + 3$, where n is an integer; this was Exercise 85(a) of Chapter 1. These observations lead to the following definition.

Definition 2: A set of integers such that every integer is congruent modulo m to exactly one integer of the set is said to be a *complete residue system modulo m.*

Example 3:

(a) By the remark prior to Definition 2, $\{0, 1, 2, 3\}$ is a complete residue system modulo 4.

(b) We show that $\{6, -11, 19, 1988\}$ is also a complete residue system modulo 4. In view of the complete system in (a) above, it suffices to show that the integers 6, -11, 19, and 1988 are congruent modulo 4 to 0, 1, 2, and 3, possibly in some permuted order. Indeed, we have $6 \equiv 2 \bmod 4$, $-11 \equiv 1 \bmod 4$, $19 \equiv 3 \bmod 4$, and $1988 \equiv 0 \bmod 4$ as desired.

Certainly the complete residue system in (a) of Example 3 is much more convenient for most purposes than that in (b) of Example 3. Such a convenient complete residue system exists for any modulus m as the following proposition shows.

Proposition 2.3: The set $\{0, 1, 2, \ldots, m - 1\}$ is a complete residue system modulo m.

Proof: We first show that every integer is congruent modulo m to at least one integer in $\{0, 1, 2, \ldots, m - 1\}$. Let $a \in \mathbf{Z}$. Then, by the division algorithm, there exist $q, r \in \mathbf{Z}$ such that

$$a = mq + r, \qquad 0 \leq r < m$$

Now $a = mq + r$ implies that $a - r = mq$. So $m \mid a - r$ or equivalently, $a \equiv r \bmod m$. Furthermore, $0 \leq r < m$ implies $r \in \{0, 1, 2, \ldots, m - 1\}$, so every integer is congruent modulo m to at least one integer in the desired set. We now show that every integer is congruent modulo m to at most one integer in $\{0, 1, 2, \ldots, m - 1\}$, which completes the proof. Let $a \in \mathbf{Z}$ and let $r_1, r_2 \in \{0, 1, 2, \ldots, m - 1\}$. Assume that $a \equiv r_1 \bmod m$ and $a \equiv r_2 \bmod m$. Then $r_1 \equiv r_2 \bmod m$ or, equivalently, $m \mid r_1 - r_2$. Now $r_1, r_2 \in \{0, 1, 2, \ldots, m - 1\}$ implies that

$$-(m - 1) \leq r_1 - r_2 \leq m - 1$$

Since $m \mid r_1 - r_2$, we have that $r_1 - r_2 = 0$ or $r_1 = r_2$. So every integer is congruent modulo m to at most one integer in $\{0, 1, 2, \ldots, m - 1\}$. ■

We pause to give the complete residue system of Proposition 2.3 a descriptive name.

Definition 3: The set $\{0, 1, 2, \ldots, m - 1\}$ of Proposition 2.3 is said to be the set of *least nonnegative residues modulo m.*

Given an integer, it follows from the proof of Proposition 2.3 that the least nonnegative residue modulo m congruent to the integer is simply the remainder when the integer is divided by m in accordance with the division algorithm. In other words, any integer is congruent modulo m to its remainder when divided by m in accordance with the division algorithm; this remainder r satisfies $0 \leq r < m$ and so is exactly one of $0, 1, 2, \ldots, m - 1$.

We now deal with some arithmetic properties of the congruence relation. These properties will be used shortly to obtain an arithmetic of equivalence classes under congruence modulo m.

Proposition 2.4: Let $a, b, c, d \in \mathbf{Z}$ such that $a \equiv b \bmod m$ and $c \equiv d \bmod m$.

(a) $(a + c) \equiv (b + d) \bmod m$

(b) $ac \equiv bd \bmod m$

Proof: Since $a \equiv b \bmod m$ and $c \equiv d \bmod m$, we have that $m \mid a - b$ and $m \mid c - d$. For (a), note that $m \mid (a - b) + (c - d)$; this is equivalent to $m \mid (a + c) - (b + d)$, from which

$$(a + c) \equiv (b + d) \bmod m$$

For (b), note first that $m \mid a - b$ implies $m \mid c(a - b)$ and that $m \mid c - d$ implies $m \mid b(c - d)$. So $m \mid c(a - b) + b(c - d)$; this is equivalent to $m \mid ac - bd$, from which

$$ac \equiv bd \bmod m \quad \blacksquare$$

Stated in equivalence class notation, (a) and (b) of Proposition 2.4 can be used to establish

(a)′ $[a] + [c] = [a + c]$

(b)′ $[a][c] = [ac]$

Here, (a)′ means that any element of the equivalence class containing a added to any element of the equivalence class containing c yields an element of the equivalence class containing $a + c$. In other words, the *addition of equivalence classes* under congruence modulo m is well defined. Similar observations apply to (b)′. (Exercise 12 investigates the possibility of defining subtraction and division of equivalence classes under congruence modulo m). Proposition 2.4 allows us to construct addition and multiplication tables for equivalence classes under congruence modulo m using any complete residue system modulo m. This construction is illustrated in Example 4. The traditional complete residue system used in congruence arithmetic is the set of least nonnegative residues modulo m given by Proposition 2.3.

Example 4:

Addition and multiplication tables for the equivalence classes under congruence modulo 4 on **Z** are given below:

+	[0]	[1]	[2]	[3]
[0]	[0]	[1]	[2]	[3]
[1]	[1]	[2]	[3]	[0]
[2]	[2]	[3]	[0]	[1]
[3]	[3]	[0]	[1]	[2]

·	[0]	[1]	[2]	[3]
[0]	[0]	[0]	[0]	[0]
[1]	[0]	[1]	[2]	[3]
[2]	[0]	[2]	[0]	[2]
[3]	[0]	[3]	[2]	[1]

The set of equivalence classes {[0], [1], [2], [3]} along with the operations of addition and multiplication above is usually denoted $\mathbf{Z}_4$ and is an example of an algebraic structure known as a *ring.* Addition and multiplication tables for the equivalence classes under congruence modulo 5 on **Z** are given below:

+	[0]	[1]	[2]	[3]	[4]
[0]	[0]	[1]	[2]	[3]	[4]
[1]	[1]	[2]	[3]	[4]	[0]
[2]	[2]	[3]	[4]	[0]	[1]
[3]	[3]	[4]	[0]	[1]	[2]
[4]	[4]	[0]	[1]	[2]	[3]

·	[0]	[1]	[2]	[3]	[4]
[0]	[0]	[0]	[0]	[0]	[0]
[1]	[0]	[1]	[2]	[3]	[4]
[2]	[0]	[2]	[4]	[1]	[3]
[3]	[0]	[3]	[1]	[4]	[2]
[4]	[0]	[4]	[3]	[2]	[1]

The set of equivalence classes {[0], [1], [2], [3], [4]} along with the operations of addition and multiplication above is usually denoted $\mathbf{Z}_5$ and is an example of an algebraic structure known as a *field.* Compare the operation tables of $\mathbf{Z}_4$ and $\mathbf{Z}_5$ above. Can you conjecture a reason why $\mathbf{Z}_4$ and $\mathbf{Z}_5$ would be considered different structures algebraically? (See also Exercise 7.)

We conclude this section with a caution concerning the congruence relation. Consider the congruence $6a \equiv 6b \bmod 3$, which is true for all integers a and b. (Convince yourself of this!) Because of our familiarity with algebra, it is a natural tendency to want to "cancel" the 6 from each side of the congruence leaving $a \equiv b \bmod 3$. But this is certainly false since it implies that every integer is congruent modulo 3 to every other integer! So *the cancellation law of multiplication does not hold in general for congruence modulo m.* This situation can be remedied, however, by suitably adjusting the modulus in accordance with the quantity canceled from each side. To wit:

Proposition 2.5: Let $a, b, c \in \mathbf{Z}$. Then $ca \equiv cb \bmod m$ if and only if $a \equiv b \bmod (m/(c, m))$.

Proof: ($\Rightarrow$) Assume that $ca \equiv cb \bmod m$ and let $d = (c, m)$. We have $m \mid ca - cb = c(a - b)$ and thus $m/d \mid (c/d)(a - b)$. By Proposition 1.10, we have $(m/d, c/d) = 1$ so that $m/d \mid a - b$ (why?) and $a \equiv b \bmod m/d$ as desired.

($\Leftarrow$) Let $d = (c, m)$ and assume that $a \equiv b \bmod m/d$. Then $m/d \mid a - b$ so that $m \mid da - db$. So $m \mid (c/d)(da - db) = ca - cb$ and $ca \equiv cb \bmod m$ as desired. ■

Note that Proposition 2.5 applied to the congruence $6a \equiv 6b \bmod 3$ above would yield $a \equiv b \bmod (3/(6,3))$ or $a \equiv b \bmod 1$, which is true for all integers a and b. A further use of Proposition 2.5 is illustrated in Example 6 of Section 2.2. Be careful when dealing with congruences; familiar properties from algebra may not hold in congruence arithmetic.

Exercise Set 2.1

1. Prove or disprove each statement below.
(a) $7 \equiv 5 \bmod 2$
(b) $8 \equiv 12 \bmod 3$
(c) $-6 \equiv 2 \bmod 4$
(d) $0 \equiv -5 \bmod 5$
(e) $59 \equiv 31 \bmod 6$
(f) $-23 \equiv 19 \bmod 7$

2. Find all positive integers m for which the following statements are true.
(a) $13 \equiv 5 \bmod m$
(b) $10 \equiv 9 \bmod m$
(c) $-7 \equiv 6 \bmod m$
(d) $100 \equiv -5 \bmod m$

3. *(a)* Prove or disprove that $\{-39, 72, -23, 50, -15, 63, -52\}$ is a complete residue system modulo 7.
(b) Find a complete residue system modulo 7 consisting entirely of even integers.
(c) Find a complete residue system modulo 7 consisting entirely of odd integers.

4. Find the least nonnegative residue modulo m of each integer n below.
(a) $n = -157$, $m = 11$
(b) $n = 442$, $m = 26$
(c) $n = -531$, $m = 89$
(d) $n = 26158$, $m = 103$
(e) $n = 16^{16}$, $m = 6$ [*Hint*: Since $16 \equiv -2 \bmod 6$, Proposition 2.4(b) implies that $16^{16} \equiv (-2)^{16} \bmod 6$.]
(f) $n = 3^{1000}$, $m = 7$ [*Hint*: $3^{1000} = (3^2)^{500} \equiv (2)^{500} \bmod 7$]

5. Find the least nonnegative residue modulo m of each integer n below.
(a) $n = 6!$, $m = 7$
(b) $n = 10!$, $m = 11$
(c) $n = 12!$, $m = 13$
(d) Formulate a conjecture based on the (rather limited) numerical evidence above.

6. Find the least nonnegative residue modulo m of each integer n below.
(a) $n = 2^6$, $m = 7$
(b) $n = 4^{10}$, $m = 11$
(c) $n = 3^{12}$, $m = 13$

(d) Formulate a conjecture based on the (rather limited) numerical evidence above.

7. (a) Formulate a conjecture as to why $\mathbf{Z}_4$ and $\mathbf{Z}_5$ in Example 4 are different structures algebraically. [*Hint*: Examine the nonzero entries of the corresponding multiplication tables. (A brief discussion of rings and fields can be found in Appendix C.)]
   (b) Construct addition and multiplication tables modulo 6. Is $\mathbf{Z}_6$ a field?
   (c) Construct addition and multiplication tables modulo 7. Is $\mathbf{Z}_7$ a field?
   (d) On the basis of work in the text, your results in parts (a) and (b) above, and further experimentation, complete the following conjecture:
   Conjecture: $\mathbf{Z}_n$ is a field if and only if n is ____________.
   (*Hint*: Experiment further!)
8. Let $a, b \in \mathbf{Z}$ such that $a \equiv b \bmod m$. If n is a positive integer such that $n \mid m$, prove that $a \equiv b \bmod n$.
9. Let $a, b \in \mathbf{Z}$ such that $a \equiv b \bmod m$. If c is a positive integer, prove that $ca \equiv cb \bmod cm$.
10. Let $a, b \in \mathbf{Z}$ such that $a \equiv b \bmod m$. If d is a positive integer with $d \mid a$, $d \mid b$, and $d \mid m$, prove that $\frac{a}{d} \equiv \frac{b}{d} \bmod \frac{m}{d}$.
11. Let $a, b \in \mathbf{Z}$ such that $a \equiv b \bmod m$. Prove that $(a, m) = (b, m)$.
12. Let $a, b, c, d \in \mathbf{Z}$ such that $a \equiv b \bmod m$ and $c \equiv d \bmod m$. Prove or disprove the following statements.
    (a) $(a - c) \equiv (b - d) \bmod m$ (from which it would follow that the subtraction of equivalence classes under congruence modulo m is well defined).
    (b) If $c \mid a$ and $d \mid b$, then $\frac{a}{c} \equiv \frac{b}{d} \bmod m$.
13. (a) Let a be an even integer. Prove that $a^2 \equiv 0 \bmod 4$.
    (b) Let a be an odd integer. Prove that $a^2 \equiv 1 \bmod 8$. Deduce that $a^2 \equiv 1 \bmod 4$.
    (c) Prove that if n is a positive integer such that $n \equiv 3 \bmod 4$, then n cannot be written as the sum of two squares of integers.
    (d) Prove or disprove the converse of the statement in part (c) above.
14. Let n be an odd integer not divisible by 3. Prove that $n^2 \equiv 1 \bmod 24$.
15. Let $a, b \in \mathbf{Z}$ with $a \equiv b \bmod m$. If n is a positive integer, prove that $a^n \equiv b^n \bmod m$.

(*Note*: Exercises 16–19 below establish divisibility tests. As a global hint for these exercises, you will find it helpful to use Exercise 15 above as well as the fact that any positive integer n can be written in expanded base 10 form as

$$n = a_m 10^m + a_{m-1} 10^{m-1} + \cdots + a_1 10^1 + a_0$$

where each a_i is one of the digits $0, 1, 2, \ldots, 9$.)

16. Prove that a positive integer n is divisible by 2 (respectively 5) if and only if its units digit is divisible by 2 (respectively 5). (*Hint*: For divisibility by 2, note that $10 \equiv 0 \bmod 2$, from which Proposition 2.4 implies that

$$a_m 10^m + a_{m-1} 10^{m-1} + \cdots + a_1 10^1 + a_0$$
$$\equiv a_m 0^m + a_{m-1} 0^{m-1} + \cdots + a_1 0^1 + a_0 \bmod 2)$$

17. Prove that a positive integer n is divisible by 3 (respectively 9) if and only if the sum of its digits is divisible by 3 (respectively 9).

18. Prove that a positive integer n is divisible by 11 if and only if the integer obtained by alternately adding and subtracting its digits beginning with adding the units digit and working to the left is divisible by 11.

19. Prove that a positive integer is divisible by 7 (respectively 11 and 13) if and only if the integer obtained by alternately adding and subtracting the three digit integers formed from the successive blocks of three decimal digits beginning with adding the block containing the units digit and working to the left is divisible by 7 (respectively 11 and 13). (*Hint*: $1001 = 7 \cdot 11 \cdot 13$.)

20. Ask a friend to think of any positive integer. Tell the friend to scramble the digits of this integer to obtain another integer and then subtract the smaller integer from the larger integer. If the difference consists of at least two digits and the friend tells you all but one of these digits including all zeros (if any), you can provide the missing digit. Explain. (*Hint*: Use Exercise 17.)

21. The following multiplication was correct; unfortunately, the printer inserted an x in place of a digit in the product:

$$(172195)(572167) = 985242\text{x}6565$$

Determine x without redoing the multiplication.

22. ***Definition:*** A *repunit* is a positive integer with decimal representation consisting entirely of 1's. A repunit consisting of n 1's is denoted R_n.

(a) Prove that no repunit is divisible by 2 or 5.

(b) Find a necessary and sufficient condition on n for the repunit R_n to be divisible by 3.

(c) Repeat part (b) above for 7, 9, 11, and 13.

(d) Find as many prime repunits as you can. (Only five prime repunits are known; all are given in the Hints and Answers section. If you find more than one, you are an extremely ambitious person.)

23. Let n be a positive integer. Prove that

$$1 + 2 + 3 + \cdots + (n - 1) \equiv 0 \bmod n$$

if and only if n is odd.

24. Let n be a positive integer. Prove that

$$1^2 + 2^2 + 3^2 + \cdots + (n - 1)^2 \equiv 0 \bmod n$$

if and only if $n \equiv \pm 1 \bmod 6$.

25. Let n be a positive integer. Prove that

$$1^3 + 2^3 + 3^3 + \cdots + (n - 1)^3 \equiv 0 \bmod n$$

if and only if $n \not\equiv 2 \bmod 4$.

26. Let $a, b \in \mathbf{Z}$ and let p be a prime number.

(a) Prove that $a^2 \equiv b^2 \bmod p$ implies that $a \equiv \pm b \bmod p$.

(b) Prove that $a^2 \equiv a \bmod p$ implies that $a \equiv 0 \bmod p$ or $a \equiv 1 \bmod p$.

27. Let $m \in \mathbf{Z}$ with $m > 0$. Prove that

$$\left\{-\frac{m-1}{2}, -\frac{m-3}{2}, \ldots, -2, -1, 0, 1, 2, \ldots, \frac{m-3}{2}, \frac{m-1}{2}\right\}$$

is a complete residue system modulo m if and only if m is odd.

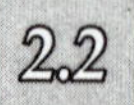

Linear Congruences in One Variable

We turn now to solving special types of congruences known as *linear congruences in one variable.*

Definition 4: Let $a, b \in \mathbf{Z}$. A congruence of the form $ax \equiv b \bmod m$ is said to be a *linear congruence in the variable x.*

The congruence is linear in the sense that the variable x occurs to the first power; more general congruences will be studied in later chapters. In a linear congruence in one variable, we are interested in finding all integer values of the variable for which the congruence is true. We illustrate with several examples.

Example 5:

(a) The congruence $2x \equiv 3 \bmod 4$ is a linear congruence in the variable x. Modulo 4, there are only four values that x can attain, namely, 0, 1, 2, and 3. (See Example 4.) A quick check shows that none of these values substituted for x in $2x \equiv 3 \bmod 4$ results in a true congruence; so the congruence $2x \equiv 3 \bmod 4$ is not solvable.

(b) The congruence $2x \equiv 3 \bmod 5$ is a linear congruence in the variable x. Modulo 5, there are only five values that x can attain, namely, 0, 1, 2, 3, and 4. (See Example 4.) A quick check shows that only $x = 4$ results in a true congruence; so the solution set for this congruence is 4 as well as any integer congruent to 4 modulo 5. In other words, the solution set for this congruence is $\{\ldots, -6, -1, 4, 9, \ldots\}$. Even though there are infinitely many solutions to $2x \equiv 3 \bmod 5$, there is exactly one incongruent solution to the congruence modulo 5, namely, 4.

(c) The congruence $2x \equiv 4 \bmod 6$ is a linear congruence in the variable x. It is left to the reader to show that this congruence has infinitely many solutions but exactly two incongruent solutions modulo 6, namely, 2 and 5.

(d) The congruence $3x \equiv 9 \bmod 6$ is a linear congruence in the variable x. It is left to the reader to show that this congruence has infinitely many solutions but exactly three incongruent solutions modulo 6, namely, 1, 3, and 5.

As motivated in (b) of Example 5, if one element of a congruence class is a solution of $ax \equiv b \bmod m$, then all elements of the congruence class are solutions of $ax \equiv b \bmod m$. So any linear congruence in one variable has either no solutions in $\mathbf{Z}$ or infinitely many solutions in $\mathbf{Z}$. When dealing with linear congruences in one variable, we are more interested in the number of *incongruent solutions modulo m in* $\mathbf{Z}$ rather than in the total number of solutions in $\mathbf{Z}$. As Example 5 shows, there seem to be many different possibilities for the numbers of incongruent solutions to a linear congruence in one variable. Is there some way to predict this number without actually trying each of the incongruent solutions one by one? Yes!

Theorem 2.6: Let $ax \equiv b \bmod m$ be a linear congruence in one variable and let $d = (a, m)$. If $d \nmid b$, then the congruence has no solutions in $\mathbf{Z}$; if $d \mid b$, then the congruence has exactly d incongruent solutions modulo m in $\mathbf{Z}$.

We make two preliminary remarks concerning the proof of Theorem 2.6. First, the proof will begin by utilizing the technique of translating a congruence into an equivalent equation that was used already in Example 2 of Section 2.1; you will see another illustration of the usefulness of this procedure. Second, establishing the existence of exactly d incongruent solutions modulo m if $d \mid b$ requires quite a bit of argument; accordingly, we prove this fact in four clearly outlined steps for easier comprehension. On with the proof!

Proof: (of Theorem 2.6) Note first that $ax \equiv b \bmod m$ if and only if $m \mid ax - b$ if and only if $ax - b = my$ for some $y \in \mathbf{Z}$ if and only if $ax - my = b$ for some $y \in \mathbf{Z}$. In other words, the linear congruence

$$ax \equiv b \bmod m$$

is solvable for x if and only if the equation

$$ax - my = b$$

is solvable for x and y. (What have we done here?) Now since $d \mid a$ and $d \mid m$, we have $d \mid ax - my$ by Proposition 1.2, which implies that $d \mid b$. So, if $d \nmid b$, the given linear congruence has no solutions. Assume that $d \mid b$. We now prove the second part of the theorem in four steps:

(a) We show that $ax \equiv b \bmod m$ has a solution, say x_0.

(b) Given any solution x_0, we show that $ax \equiv b \bmod m$ has infinitely many solutions in $\mathbf{Z}$ of a given form.

(c) Given any solution x_0, we show that every solution of $ax \equiv b \bmod m$ has the form in (b) [from which (b) and (c) combine to yield the solutions of $ax \equiv b \bmod m$ precisely given an initial particular solution x_0].

(d) We show that there are exactly d incongruent solutions modulo m in $\mathbf{Z}$ among the infinitely many solutions of (b).

We now prove (a) through (d).

For (a), by Proposition 1.11, there exist $r, s \in \mathbf{Z}$ such that

$$d = (a, m) = ar + ms$$

Furthermore, $d \mid b$ implies $b = de$ for some $e \in \mathbf{Z}$. Then

$$b = de = (ar + ms)e = a(re) + m(se)$$

So $x = re$ and $y = -se$ solve $ax - my = b$, from which $x_0 = re$ solves $ax \equiv b \bmod m$ (because of the connection established between $ax - my = b$ and $ax \equiv b \bmod m$ at the beginning of the proof).

For (b), let x_0 be any solution of $ax \equiv b \bmod m$. Let $n \in \mathbf{Z}$ and consider $x_0 + (\frac{m}{d})n$. Note that $x_0 + (\frac{m}{d})n$ is an integer, since $d \mid m$. Furthermore,

$$\begin{aligned} a\left(x_0 + \left(\frac{m}{d}\right)n\right) &= ax_0 + a\left(\frac{m}{d}\right)n \\ &= ax_0 + \left(\frac{a}{d}\right)mn \\ &\equiv b + 0 \bmod m \\ &\equiv b \bmod m \end{aligned}$$

So $x_0 + (\frac{m}{d})n$ is a solution of $ax \equiv b \bmod m$, and, given any solution x_0 of $ax \equiv b \bmod m$, the (infinitely many) values $x_0 + (\frac{m}{d})n, n \in \mathbf{Z}$, are also solutions.

For (c), let x_0 be any solution of $ax \equiv b \bmod m$. Then $ax_0 - my_0 = b$ for some $y_0 \in \mathbf{Z}$. (Why?) Now any other solution x of $ax \equiv b \bmod m$ implies the existence of $y \in \mathbf{Z}$ with $ax - my = b$, and we have

$$(ax - my) - (ax_0 - my_0) = b - b = 0$$

So,

$$a(x - x_0) = m(y - y_0)$$

Dividing both sides of this equation by d, we have

$$\left(\frac{a}{d}\right)(x - x_0) = \left(\frac{m}{d}\right)(y - y_0)$$

Now $\frac{m}{d} \mid (\frac{a}{d})(x - x_0)$; since $(\frac{a}{d}, \frac{m}{d}) = 1$ by Proposition 1.10, we have that $\frac{m}{d} \mid x - x_0$. Consequently $x - x_0 = (\frac{m}{d})n$ for some $n \in \mathbf{Z}$ or, equivalently, $x = x_0 + (\frac{m}{d})n$ for some $n \in \mathbf{Z}$. Hence, given a solution x_0 of $ax \equiv b \bmod m$, we have that every solution takes the form $x_0 + (\frac{m}{d})n$ with $n \in \mathbf{Z}$. Combining this result with the result of (b), we have that, given a solution x_0 of $ax \equiv b \bmod m$, all solutions of $ax \equiv b \bmod m$ are given precisely by $x_0 + (\frac{m}{d})n, n \in \mathbf{Z}$.

For (d), to determine how many incongruent solutions modulo m there are among the solutions $x_0 + (\frac{m}{d})n, n \in \mathbf{Z}$, we find a necessary and sufficient condition for the congruence modulo m of two such solutions. We have

$$x_0 + \left(\frac{m}{d}\right)n_1 \equiv x_0 + \left(\frac{m}{d}\right)n_2 \bmod m$$

if and only if

$$\left(\frac{m}{d}\right)n_1 \equiv \left(\frac{m}{d}\right)n_2 \bmod m$$

if and only if

$$m \,\Big|\, \left(\frac{m}{d}\right)(n_1 - n_2)$$

if and only if

$$d \mid n_1 - n_2 \quad \text{(why?)}$$

if and only if

$$n_1 \equiv n_2 \bmod d$$

So two solutions of the form $x_0 + (\frac{m}{d})n$, $n \in \mathbf{Z}$, are congruent modulo m if and only if the n-values of these two solutions are congruent modulo d. Thus, a complete set of incongruent solutions modulo m of $ax \equiv b \bmod m$ is obtained from $x_0 + (\frac{m}{d})n$, $n \in \mathbf{Z}$, by letting n range over a complete residue system modulo d ($\{0, 1, 2, \ldots, d - 1\}$ for example); inasmuch as there are d such residues, the proof is complete. ■

We note that the proof of Theorem 2.6 shows the existence of the claimed incongruent solutions modulo m of the linear congruence $ax \equiv b \bmod m$ by actually exhibiting these solutions as $x_0 + (\frac{m}{d})n$, $n = 0, 1, 2, \ldots, d - 1$ (for example). So we state the following porism.

Porism 2.7: Let $ax \equiv b \bmod m$ be a linear congruence in one variable and let $d = (a, m)$. If $d \mid b$, then d incongruent solutions modulo m of the congruence are given by

$$x_0 + \left(\frac{m}{d}\right)n, \qquad n = 0, 1, 2, \ldots, d - 1$$

where x_0 is any particular solution of the congruence.

We now illustrate the solution of a linear congruence in one variable with an example.

Example 6:

Find all incongruent solutions of the congruence $16x \equiv 8 \bmod 28$.

It is possible to solve the problem by substituting each of the values $0, 1, 2, \ldots, 27$ for the variable x and noting whether the given congruence is satisfied. Unfortunately, this procedure is lengthy and can be quite cumbersome. We instead illustrate a more appropriate method based on the content of Theorem 2.6 and Porism 2.7. We begin by using the Euclidean algorithm to find the d of Theorem 2.6; in this case, it would be $(16, 28)$:

$$28 = 16 \cdot 1 + 12$$

$$16 = 12 \cdot 1 + 4$$

$$12 = 4 \cdot 3 + 0$$

So $d = (16, 28) = 4$. Since $4 \mid 8$, the given congruence has four incongruent solutions modulo 28 by Theorem 2.6. To determine these four solutions by Porism 2.7, we must first find a particular solution x_0. Working backward through the Euclidean algorithm (as in Example 15 of Chapter 1) to express $(16, 28)$ as a linear combination of 16 and 28, we obtain

$$4 = 2 \cdot 16 + (-1) \cdot 28 \qquad \textbf{(1)}$$

By the proof of Theorem 2.6, the solvability of the congruence $16x \equiv 8 \bmod 28$ for x is equivalent to the solvability of the equation

$$16x - 28y = 8 \tag{2}$$

for x and y. Multiplying both sides of (1) by 2, we obtain

$$8 = 4 \cdot 16 + (-2) \cdot 28$$

When rewritten in the form of (2), this yields

$$16 \cdot 4 - 28 \cdot 2 = 8$$

So a particular solution of $16x - 28y = 8$ is given by $x_0 = 4$ and $y_0 = 2$, from which a particular solution of $16x \equiv 8 \bmod 28$ is given by $x_0 = 4$. Now, by Porism 2.7, all incongruent solutions of $16x \equiv 8 \bmod 28$ are given by

$$4 + \left(\frac{28}{4}\right)n, \qquad n = 0, 1, 2, 3$$

or, equivalently, 4, 11, 18, and 25. Proposition 2.5 can occasionally be invoked to give an easier solution procedure. The congruence $16x \equiv 8 \bmod 28$ is equivalent to the congruence $8 \cdot 2x \equiv 8 \cdot 1 \bmod 28$, which, by Proposition 2.5, is equivalent to the congruence $2x \equiv 1 \bmod \left(\frac{28}{(8, 28)}\right)$ or $2x \equiv 1 \bmod 7$, which has one incongruent solution modulo 7. Such a solution is given by $x = 4$ (by inspection!). The given congruence $16x \equiv 8 \bmod 28$ is now solved by those incongruent integers modulo 28 among the integers congruent to 4 modulo 7; such integers are given (for example) by 4, 11, 18, and 25.

The special linear congruence in one variable given by $ax \equiv 1 \bmod m$ is important enough to be considered separately.

Definition 5: Any solution of the linear congruence in one variable $ax \equiv 1 \bmod m$ is said to be a (*multiplicative*) *inverse of a modulo m.*

The terminology above is chosen to coincide with our usual idea of multiplicative inverses in the theory of equations. In the theory of equations, the multiplicative inverse of a nonzero integer is defined as that number whose product with the integer is equal to 1, the multiplicative identity. (Multiplicative inverses in this context are termed *reciprocals*; note that reciprocals of nonzero integers are themselves integers only in two cases — which two?) In the theory of congruences, the multiplicative inverse modulo m of a nonzero integer is defined as that *integer,* if it exists, whose product with the integer is *congruent to* 1 (modulo m).

For what integers a do inverses modulo m exist? Equivalently, under what condition(s) is the congruence $ax \equiv 1 \bmod m$ solvable? How many inverses modulo m can an integer have? Since the congruence $ax \equiv 1 \bmod m$ is simply a special case of the congruence $ax \equiv b \bmod m$, Theorem 2.6 yields the answer to these questions; it is recorded as the following corollary.

Corollary 2.8: (to Theorem 2.6) The linear congruence in one variable $ax \equiv 1 \bmod m$ has a solution if and only if $(a, m) = 1$; if $(a, m) = 1$, then $ax \equiv 1 \bmod m$ has exactly one incongruent solution modulo m. ■

Corollary 2.8 establishes that an inverse of an integer modulo m is unique modulo m if it exists. For this reason, we speak hereafter of *the* inverse of an integer modulo m. We now find such an inverse.

Example 7:

Find the inverse of 5 modulo 16.

We must solve the linear congruence $5x \equiv 1 \bmod 16$; since $(5, 16) = 1$, such a solution exists by Corollary 2.8 and is unique modulo 16. We leave it to the reader to employ the method of Example 6 or use inspection to show that $x = 13$ is the inverse of 5 modulo 16.

The existence of the inverse of an integer modulo m can be used to solve the (general) linear congruence in one variable $ax \equiv b \bmod m$, as the following example illustrates.

Example 8:

Find all incongruent solutions of the congruence $5x \equiv 12 \bmod 16$.

Since $(5, 16) = 1$, the given congruence has a unique solution modulo 16 by Theorem 2.6. We find this unique solution by multiplying both sides of the congruence by the inverse of 5 modulo 16. By Example 7, we have that 13 is the inverse of 5 modulo 16; multiplying both sides of $5x \equiv 12 \bmod 16$ by 13, we have

$$13 \cdot 5x \equiv 13 \cdot 12 \bmod 16$$

Since 13 is the inverse of 5 modulo 16, we have that $13 \cdot 5 \equiv 1 \bmod 16$, and the congruence above becomes

$$x \equiv 13 \cdot 12 \bmod 16$$

So $x \equiv 156 \bmod 16$ and $x = 156$ is the desired unique solution modulo 16. The least nonnegative solution modulo 16 can be obtained by either reducing 156 modulo 16 directly or by noting that $13 \equiv -3 \bmod 16$ and $12 \equiv -4 \bmod 16$, from which

$$\begin{aligned} x &\equiv 13 \cdot 12 \bmod 16 \text{ (from above)} \\ &\equiv (-3)(-4) \bmod 16 \\ &\equiv 12 \bmod 16 \end{aligned}$$

So $x = 12$ is the unique incongruent solution modulo 16.

The method of Example 8 above is more important theoretically than practically. Solving a linear congruence $ax \equiv b \bmod m$ with $(a, m) = 1$ by finding the inverse of a modulo m usually requires more work than simply solving the congruence directly by the technique of Example 6 (unless, of course, we somehow know this inverse beforehand, as in Example 8 above). We will return to inverses modulo m in Section 2.4.

Exercise Set 2.2

28. Find all least nonnegative incongruent solutions of the following congruences.

(a) $9x \equiv 21 \bmod 30$
(b) $18x \equiv 12 \bmod 28$
(c) $18x \equiv 15 \bmod 27$
(d) $12x \equiv 16 \bmod 32$
(e) $623x \equiv 511 \bmod 679$
(f) $481x \equiv 627 \bmod 703$

29. Find the inverse modulo m of each integer n below.

(a) $n = 5, m = 26$
(b) $n = 8, m = 35$
(c) $n = 40, m = 81$
(d) $n = 51, m = 99$
(e) $n = 101, m = 103$
(f) $n = 1333, m = 1517$

30. Find all least nonnegative solutions of each of the congruences (in two variables) below.

(a) $2x + 3y \equiv 4 \bmod 7$ (*Hint*: Write the congruence as $2x \equiv 4 - 3y \bmod 7$ and solve the linear congruences in one variable obtained by successively setting y equal to 0, 1, 2, 3, 4, 5, and 6.)
(b) $4x + 2y \equiv 6 \bmod 8$
(c) $3x + 6y \equiv 2 \bmod 9$
(d) $8x + 2y \equiv 4 \bmod 10$

31. Generalize Theorem 2.6 by proving the following result:
Let $a_1x_1 + a_2x_2 + \cdots + a_nx_n \equiv b \bmod m$ be a linear congruence in the n variables $x_1, x_2, \ldots, x_n$ and let $d = (a_1, a_2, \ldots, a_n, m)$. If $d \nmid b$, then the congruence has no solutions in $\mathbf{Z}$; if $d \mid b$, then the congruence has exactly dm^{n-1} incongruent solutions modulo m in $\mathbf{Z}$. (*Hint*: Use mathematical induction.)

32. Let a' be the inverse of a modulo m and let b' be the inverse of b modulo m. Prove that $a'b'$ is the inverse of ab modulo m.

2.3

The Chinese Remainder Theorem

We begin with a motivational example.

Example 9:

Find a positive integer having a remainder of 2 when divided by 3, a remainder of 1 when divided by 4, and a remainder of 3 when divided by 5.

Let x be the desired positive integer. Since we want x to have a remainder of 2 when divided by 3, we must have

$$x \equiv 2 \bmod 3$$

Similarly, since we want x to have a remainder of 1 when divided by 4 and a remainder of 3 when divided by 5, we have (respectively)

$$x \equiv 1 \bmod 4$$

$$x \equiv 3 \bmod 5$$

So we wish to find a positive integer x such that

$$x \equiv 2 \bmod 3$$

$$x \equiv 1 \bmod 4$$

$$x \equiv 3 \bmod 5$$

Note that this is a *system* of linear congruences in one variable. We temporarily suspend consideration of this system, ultimately giving the solution after the proof of Theorem 2.9. Can you find such a positive integer?

Problems such as that presented in Example 9 arose quite frequently in ancient Chinese puzzles (see also Student Project 5). The theory behind the solution of such systems of linear congruences in one variable and similar systems is given by the following theorem whose name reflects its Chinese heritage.

Theorem 2.9: (Chinese Remainder Theorem) Let $m_1, m_2, \ldots, m_n$ be pairwise relatively prime positive integers and let $b_1, b_2, \ldots, b_n$ be any integers. Then the system of linear congruences in one variable given by

$$x \equiv b_1 \bmod m_1$$

$$x \equiv b_2 \bmod m_2$$

$$\vdots$$

$$x \equiv b_n \bmod m_n$$

has a unique solution modulo $m_1 m_2 \cdots m_n$.

Proof: We first construct a solution to the given system of linear congruences in one variable. Let $M = m_1 m_2 \cdots m_n$ and, for $i = 1, 2, \ldots, n$, let $M_i = M/m_i$. Now $(M_i, m_i) = 1$ for each i. (Why?) So $M_i x_i \equiv 1 \bmod m_i$ has a solution for each i by Corollary 2.8. Form

$$x = b_1 M_1 x_1 + b_2 M_2 x_2 + \cdots + b_n M_n x_n$$

Note that x is a solution of the desired system since, for $i = 1, 2, \ldots, n$,

$$\begin{aligned} x &= b_1M_1x_1 + b_2M_2x_2 + \cdots + b_iM_ix_i + \cdots + b_nM_nx_n \\ &\equiv 0 + 0 + \cdots + b_i + \cdots + 0 \bmod m_i \\ &\equiv b_i \bmod m_i \end{aligned}$$

It remains to show the uniqueness of the solution modulo M. Let x' be another solution to the given system of linear congruences in one variable. Then, for all i, we have that $x' \equiv b_i \bmod m_i$; since $x \equiv b_i \bmod m_i$ for all i, we have that $x \equiv x' \bmod m_i$ for all i, or, equivalently, $m_i \mid x - x'$ for all i. Then $M \mid x - x'$ (why?), from which $x \equiv x' \bmod M$. The proof is complete. ■

Note that the proof of the Chinese Remainder Theorem shows the existence and uniqueness of the claimed solution modulo M by actually *constructing* this solution. Such a proof is said to be *constructive*; the advantage of constructive proofs is that they yield a procedure or algorithm for obtaining the desired quantity. We now use the procedure motivated in the proof of Theorem 2.9 to solve the system of linear congruences in one variable of Example 9.

Example 9 (Continued):

Recall that the system of linear congruences to be solved is given by

$$\begin{aligned} x &\equiv 2 \bmod 3 \\ x &\equiv 1 \bmod 4 \\ x &\equiv 3 \bmod 5 \end{aligned}$$

Using the notation in the proof of Theorem 2.9, we have

$$M = (3)(4)(5) = 60$$

$$M_1 = \frac{M}{m_1} = \frac{60}{3} = 20$$

$$M_2 = \frac{M}{m_2} = \frac{60}{4} = 15$$

$$M_3 = \frac{M}{m_3} = \frac{60}{5} = 12$$

We now solve the linear congruences $M_ix_i \equiv 1 \bmod m_i$, $i = 1, 2, 3$. (Each congruence has been solved below by inspection; the technique of Example 6 would be used in general.) $M_1x_1 \equiv 1 \bmod m_1$ becomes $20x_1 \equiv 1 \bmod 3$ or, equivalently, $2x_1 \equiv 1 \bmod 3$, from which $x_1 \equiv 2 \bmod 3$. $M_2x_2 \equiv 1 \bmod m_2$ becomes $15x_2 \equiv 1 \bmod 4$ or, equivalently, $3x_2 \equiv 1 \bmod 4$, from which $x_2 \equiv 3 \bmod 4$. Finally, $M_3x_3 \equiv 1 \bmod m_3$ becomes $12x_3 \equiv 1 \bmod 5$ or, equivalently, $2x_3 \equiv 1 \bmod 5$, from which $x_3 \equiv 3 \bmod 5$. So we put

$$x_1 = 2$$
$$x_2 = 3$$
$$x_3 = 3$$

We finally compute the desired solution x of the system:

$$\begin{aligned} x &= b_1M_1x_1 + b_2M_2x_2 + b_3M_3x_3 \\ &= (2)(20)(2) + (1)(15)(3) + (3)(12)(3) \\ &= 233 \\ &\equiv 53 \bmod 60 \end{aligned}$$

Any positive integer congruent to 53 modulo 60 solves the given system of linear congruences; in particular, the minimum positive solution to the system is 53. (Please check it!)

The Chinese Remainder Theorem solves an extremely specialized system of linear congruences in one variable: The moduli must be pairwise relatively prime and the coefficient of x in each congruence must be 1. Exercise 34 and Exercises 38 through 41 investigate the solvability of systems of linear congruences in one variable more general than those handled by the Chinese Remainder Theorem.

Exercise Set 2.3

33. Find the least nonnegative solution of each system of congruences below.

(a) $x \equiv 3 \bmod 4$
$x \equiv 2 \bmod 5$

(b) $x \equiv 2 \bmod 5$
$x \equiv 3 \bmod 7$
$x \equiv 1 \bmod 8$

(c) $x \equiv 2 \bmod 3$
$x \equiv 1 \bmod 5$
$x \equiv 4 \bmod 7$

(d) $x \equiv 1 \bmod 7$
$x \equiv 4 \bmod 6$
$x \equiv 3 \bmod 5$

(e) $x \equiv 1 \bmod 2$
$x \equiv 2 \bmod 3$
$x \equiv 4 \bmod 5$
$x \equiv 6 \bmod 7$

(f) $x \equiv 1 \bmod 3$
$x \equiv 2 \bmod 4$
$x \equiv 3 \bmod 5$
$x \equiv 4 \bmod 7$

34. Solve each system of congruences below.

(a) $2x \equiv 1 \bmod 3$
$3x \equiv 2 \bmod 5$
$5x \equiv 4 \bmod 7$ (*Hint*: Use the technique of Example 8 to rewrite each congruence of the system in the form $x \equiv a \bmod b$.)

(b) $3x \equiv 2 \bmod 4$
$4x \equiv 1 \bmod 5$
$6x \equiv 3 \bmod 9$

(c) $5x \equiv 3 \bmod 7$
$2x \equiv 4 \bmod 8$
$3x \equiv 6 \bmod 9$

35. (Brahmagupta, circa 600 A.D.) There are n eggs in a basket. If eggs are removed from the basket 2, 3, 4, 5, and 6 at a time, there remain 1, 2, 3, 4, and 5 eggs in the basket, respectively. If eggs are removed from the basket 7 at a time, no eggs remain in the basket. What is the smallest value of n for which the above scenario could occur?

36. Fifteen pirates steal a sack of gold coins. Each coin is of equal denomination. The pirates attempt to divide the coins evenly but find that two coins are left over. A quarrel over these two coins erupts and one pirate is killed. Another attempt is made to divide the coins evenly but this time one coin is left over. Another quarrel erupts and another pirate is killed. A third attempt to divide the coins evenly succeeds. Find the smallest number of gold coins that could have been in the sack initially.

37. Three vegetarians are shipwrecked on an island on which there is nothing to eat except coconuts. They soon notice that monkeys are eating a large number of the coconuts so, to protect their food supply, they gather all of the coconuts into one big pile. That night, one of the vegetarians gets up and, not trusting the other vegetarians, attempts to take his fair share of coconuts. He divides the coconuts into three equal piles, finds one coconut left over which he tosses to the monkeys, hides one of the piles, combines the remaining two piles into one pile, and returns to bed. The second vegetarian then gets up and attempts to take his fair share of coconuts. He divides the pile of coconuts into three equal piles, finds one coconut left over (which he tosses to the monkeys), hides one of the piles, combines the remaining two piles into one pile, and returns to bed. The third vegetarian then goes through exactly the same procedure. In the morning, the three vegetarians approach the substantially diminished pile of coconuts and, upon attempting to divide the pile fairly, find themselves presenting the monkeys with their fourth and final coconut. What is the smallest number of coconuts that could have been in the pile initially?

38. Prove that the system of linear congruences in one variable given by

$$x \equiv b_1 \bmod m_1$$

$$x \equiv b_2 \bmod m_2$$

is solvable if and only if $(m_1, m_2) \mid b_1 - b_2$. In this case, prove that the solution is unique modulo $[m_1, m_2]$.

39. Solve each system of congruences below.

(a) $x \equiv 3 \bmod 4$
$x \equiv 1 \bmod 6$

(b) $x \equiv 2 \bmod 6$
$x \equiv 8 \bmod 9$

(c) $x \equiv 2 \bmod 4$
$x \equiv 4 \bmod 8$

40. Prove that the system of linear congruences in one variable given by

$$x \equiv b_1 \bmod m_1$$
$$x \equiv b_2 \bmod m_2$$
$$\vdots$$
$$x \equiv b_n \bmod m_n$$

is solvable if and only if $(m_i, m_j) \mid b_i - b_j$ for all $i \neq j$. In this case, prove that the solution is unique modulo $[m_1, m_2, \ldots, m_n]$.

41. Solve each system of congruences below.

(a) $x \equiv 2 \bmod 6$
$x \equiv 1 \bmod 7$
$x \equiv 5 \bmod 8$

(b) $x \equiv 3 \bmod 6$
$x \equiv 7 \bmod 10$
$x \equiv 12 \bmod 15$

(c) $x \equiv 3 \bmod 4$
$x \equiv 5 \bmod 10$
$x \equiv 11 \bmod 12$
$x \equiv 5 \bmod 15$

Wilson's Theorem

This section deals with an important congruence in elementary number theory that is known as Wilson's Theorem. If p is a prime number, Wilson's Theorem states that $(p - 1)! \equiv -1 \bmod p$ [or, equivalently, $(p - 1)! \equiv p - 1 \bmod p$]. The theorem is named for John Wilson, who conjectured the result on numerical evidence but could not provide a proof. (Perhaps you conjectured this result in Exercise 5 of Section 2.1.) The theorem was first proved by Joseph Louis Lagrange in 1771.

Before we prove Wilson's Theorem, we need a preliminary lemma.

Lemma 2.10: Let p be a prime number and let $a \in \mathbf{Z}$. Then a is its own inverse modulo p if and only if $a \equiv \pm 1 \bmod p$.

Biography

Joseph Louis Lagrange (1736–1813)

Joseph Louis Lagrange was one of the two greatest mathematicians of the eighteenth century (the other was Euler). Born in Italy, Lagrange was the youngest of 11 children and the only child to survive beyond infancy. As is the case with many great mathematicians, Lagrange's mathematical work covered many diverse areas. He attempted a rigorization of the calculus, had several noteworthy achievements in differential equations, and aided in the development of the calculus of variations. A major theorem of group theory in abstract algebra bears his name. He described a method for approximating the real roots of an equation using continued fractions. Two of Lagrange's most notable achievements were number theoretic: He published the first proof of Wilson's Theorem and proved the difficult "four-square theorem" that will be discussed in Section 6.5.

Proof: ($\Rightarrow$) Assume that a is its own inverse modulo p. Then $a^2 \equiv 1 \bmod p$. So $p \mid a^2 - 1$ or, equivalently, $p \mid (a-1)(a+1)$. By Lemma 1.14, we have that $p \mid a - 1$ or $p \mid a + 1$; so $a \equiv 1 \bmod p$ or $a \equiv -1 \bmod p$. Thus, $a \equiv \pm 1 \bmod p$ as desired.

($\Leftarrow$) Assume that $a \equiv \pm 1 \bmod p$. Then $a^2 \equiv 1 \bmod p$, and a is its own inverse modulo p, as desired. ■

Before formally stating and proving Wilson's Theorem, the idea behind the proof is illustrated with an example. Study this example carefully; the proof of Wilson's Theorem will be much easier if you remember this example.

Example 10:

We use the idea behind the proof of Wilson's Theorem to establish that $(11 - 1)! \equiv -1 \bmod 11$ or, equivalently, $10! \equiv -1 \bmod 11$. We have $10! = 1 \cdot 2 \cdot 3 \cdot 4 \cdot 5 \cdot 6 \cdot 7 \cdot 8 \cdot 9 \cdot 10$. By Lemma 2.10, the only integers between 1 and 10 inclusive that are their own inverses modulo 11 are 1 (which is congruent to 1 modulo 11) and 10 (which is congruent to -1 modulo 11). We now take each integer from 2 to 9 inclusive and pair it with its inverse: 2 is the inverse of 6 modulo 11 (and vice versa), 3 is the inverse of 4 modulo 11 (and vice versa), 5 is the inverse of 9 modulo 11 (and vice versa), and 7 is the inverse of 8 modulo 11 (and vice versa). So the product of each such pair is congruent to 1 modulo 11, from which

$$\begin{aligned}10! &= 1 \cdot 2 \cdot 3 \cdot 4 \cdot 5 \cdot 6 \cdot 7 \cdot 8 \cdot 9 \cdot 10 \\ &= 1 \cdot (2 \cdot 6) \cdot (3 \cdot 4) \cdot (5 \cdot 9) \cdot (7 \cdot 8) \cdot 10 \\ &\equiv 1 \cdot (1) \cdot (1) \cdot (1) \cdot (1) \cdot (-1) \bmod 11 \\ &\equiv -1 \bmod 11\end{aligned}$$

as desired.

We now formally state Wilson's Theorem and use the technique illustrated in Example 10 to prove it.

Theorem 2.11: (Wilson's Theorem) Let p be a prime number. Then

$$(p - 1)! \equiv -1 \bmod p$$

Proof: The statement of the theorem is easily checked for $p = 2$ and $p = 3$. Assume that $p > 3$. Each $a \in \mathbf{Z}$ with $1 \leq a \leq p - 1$ has an inverse modulo p by Corollary 2.8. Without loss of generality, we assume that the inverse of such an a, say a', satisfies $1 \leq a' \leq p - 1$. Now, by Lemma 2.10, the only integers that are their own inverses modulo p are those integers congruent to ± 1 modulo p; so, the only integers a with $1 \leq a \leq p - 1$ that are their own inverses modulo p are 1 (which obviously is congruent to 1 modulo p) and $p - 1$ (which is congruent to -1 modulo p). Therefore, for each integer a with $2 \leq a \leq p - 2$, there exists a different integer a' with $2 \leq a' \leq p - 2$ such that $aa' \equiv 1 \bmod p$. Accordingly, group the numbers $2, 3, \ldots, p - 2$ into pairs a, a' such that $aa' \equiv 1 \bmod p$. The product of the left-hand sides of the $\frac{p-3}{2}$ congruences so obtained is $(p - 2)!$ after rearrangement; the right-hand sides of these congruences obviously multiply to 1. Hence

$$(p - 2)! \equiv 1 \bmod p$$

Multiplying both sides of this congruence by $p - 1$, we obtain

$$(p - 2)!(p - 1) \equiv (p - 1) \bmod p$$

or

$$(p - 1)! \equiv -1 \bmod p$$

as desired. ■

We will use Wilson's Theorem in Section 2.5 as well as in Chapter 4.

The converse of Wilson's Theorem is also true:

Proposition 2.12: Let $n \in \mathbf{Z}$ with $n > 1$. If $(n-1)! \equiv -1 \bmod n$, then n is a prime number.

Proof: Let $n = ab$ where $a, b \in \mathbf{Z}$ with $1 \leq a < n$. To prove that n is a prime number, it suffices to show that $a = 1$ so that any such factorization of n is trivial (consisting of 1 and n only). Since $1 \leq a < n$, we have $a \mid (n-1)!$. Now $(n-1)! \equiv -1 \bmod n$ implies $n \mid (n-1)! + 1$; since $a \mid n$, we have $a \mid (n-1)! + 1$ by Proposition 1.1. So $a \mid (n-1)! + 1 - (n-1)!$ by Proposition 1.2, from which $a \mid 1$. So $a = 1$, which completes the proof. ■

Note that Proposition 2.12 gives a criterion for determining whether a given positive integer is prime. In other words, Proposition 2.12 is a primality test! (What other primality test have we encountered?) The primality test of Proposition 2.12 is more important theoretically than practically—indeed, as n increases, $(n-1)!$ becomes prohibitively cumbersome to compute since the computation of $(n-1)!$ requires at least $n-3$ multiplications modulo n. To see this, consider a small example with $n = 5$. We have $(5-1)! = 2 \cdot 3 \cdot 4$; the computation of $2 \cdot 3 \cdot 4$ requires at least two (or $5-3$) multiplications modulo 5. (Try it!) These considerations generalize to arbitrary n. While the small example is not too prohibitive, consider the task of computing 3000! to determine the primality of 3001; one would have to perform at least 2998 multiplications modulo 3001. This is certainly prohibitive!

Wilson's Theorem (Theorem 2.11) and its converse (Proposition 2.12) taken together provide a necessary and sufficient condition (respectively) for primality; the prime numbers are said to be *characterized* by the condition $(n-1)! \equiv -1 \bmod n$. The word *characterize* will be used frequently in this book. Keep in mind that characterizing a certain property of an object (in this case, the primality of an integer greater than 1) requires a necessary and sufficient condition for the object to have that property.

We conclude this section by mentioning that Wilson's name is associated with a class of prime numbers (putting Wilson in the same "prime" company as Mersenne and Fermat). A prime number p is said to be a *Wilson prime* if $(p-1)! \equiv -1 \bmod p^2$. Alas, only three Wilson primes are known, namely 5, 13, and 563. In addition, it has been verified that no Wilson prime exists below $5 \cdot 10^8$, but whether the number of Wilson primes is finite or infinite has yet to be established.

Exercise Set 2.4

42. Use Wilson's Theorem (Theorem 2.11) to find the least nonnegative residue modulo m of each integer n below.

(a) $n = 30!$, $m = 31$

(b) $n = 88!$, $m = 89$

(c) $n = 21!$, $m = 23$ [*Hint*: $22(21!) \equiv -1 \bmod 23$ by Wilson's Theorem.]

(d) $n = 64!$, $m = 67$

(e) $n = 31!/22!$, $m = 11$

(f) $n = 65!/51!$, $m = 17$

43. *(a)* Prove that if p is an odd prime number, then $2(p-3)! \equiv -1 \bmod p$.
(b) Find the least nonnegative residue of $2(100!)$ modulo 103.

44. Let $n \in \mathbf{Z}$ with $n > 1$. Prove that n is a prime number if and only if $(n-2)! \equiv 1 \bmod n$.

45. Let n be a composite integer greater than 4. Prove that $(n-1)! \equiv 0 \bmod n$.

46. Let p be a prime number. Prove that the numerator of $1 + \frac{1}{2} + \frac{1}{3} + \cdots + \frac{1}{p-1}$ (when expressed as a single fraction) is divisible by p.

47. Let p be an odd prime number.
(a) Prove that $\left(\left(\frac{p-1}{2}\right)!\right)^2 \equiv (-1)^{(p+1)/2} \bmod p$.
(b) If $p \equiv 1 \bmod 4$, prove that $\left(\frac{p-1}{2}\right)!$ is a solution of the quadratic congruence $x^2 \equiv -1 \bmod p$.
(c) If $p \equiv 3 \bmod 4$, prove that $\left(\frac{p-1}{2}\right)!$ is a solution of the quadratic congruence $x^2 \equiv 1 \bmod p$.

48. Let p be an odd prime number. Prove that

$$1^2 3^2 5^2 \cdots (p-4)^2 (p-2)^2 \equiv (-1)^{(p+1)/2} \bmod p$$

49. Let p be a prime number congruent to 3 modulo 4. Prove that $\left(\frac{p-1}{2}\right)! \equiv \pm 1 \bmod p$.

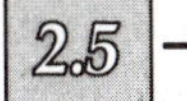

Fermat's Little Theorem; Pseudoprime Numbers

Another important congruence in elementary number theory is known as Fermat's Little Theorem. In 1640, Pierre de Fermat revealed what was to become known as his Little Theorem in a letter to a friend. Fermat did not provide a proof of the theorem; in fact, the first published proof came from Euler in 1736.

Theorem 2.13: (Fermat's Little Theorem) Let p be a prime number and let $a \in \mathbf{Z}$. If $p \nmid a$, then

$$a^{p-1} \equiv 1 \bmod p$$

(Did you conjecture this result in Exercise 6 of Section 2.1?)

Proof: Consider the $p-1$ integers given by $a, 2a, 3a, \ldots, (p-1)a$. Note that $p \nmid ia$, $i = 1, 2, \ldots, p-1$. (If $p \mid ia$ for some i, then $p \mid a$ or $p \mid i$ by Lemma 1.14; since $p \nmid a$, we have $p \mid i$, which is impossible.) Note also that no two of the given $p-1$ integers are congruent modulo p. (Since $p \nmid a$, the inverse of a modulo p, say a', exists by Corollary 2.8. So, if $ia \equiv ja \bmod p$ with $i \neq j$, then $iaa' \equiv jaa' \bmod p$, from which $i \equiv j \bmod p$, which is impossible.) So the least nonnegative residues modulo p of the integers $a, 2a, 3a, \ldots,$

$(p-1)a$, taken in some order, must be $1, 2, 3, \ldots, p-1$. Then

$$(a)(2a)(3a)\cdots((p-1)a) \equiv (1)(2)(3)\cdots(p-1) \bmod p$$

or, equivalently,

$$a^{p-1}(p-1)! \equiv (p-1)! \bmod p$$

By Wilson's Theorem, the congruence above becomes $-a^{p-1} \equiv -1 \bmod p$, which is equivalent to $a^{p-1} \equiv 1 \bmod p$, as desired. ■

Exercise 50 leads the reader through the proof of Fermat's Little Theorem above with a concrete example; we suggest that you solve this exercise now.

Fermat's Little Theorem has several easy corollaries.

Corollary 2.14: Let p be a prime number and let $a \in \mathbf{Z}$. If $p \nmid a$, then a^{p-2} is the inverse of a modulo p.

Proof: By Fermat's Little Theorem, we have $a^{p-1} \equiv 1 \bmod p$. Consequently, $aa^{p-2} \equiv 1 \bmod p$, and a^{p-2} is the inverse of a modulo p as desired. ■

In view of Example 8 (please review this example now!), Corollary 2.14 is useful in solving linear congruences in one variable of the form $ax \equiv b \bmod p$ where p is a prime number with $p \nmid a$ (see Exercise 53).

Corollary 2.15: Let p be a prime number and let $a \in \mathbf{Z}$. Then $a^p \equiv a \bmod p$.

Proof: If $p \nmid a$, the corollary follows from Fermat's Little Theorem on multiplication of both sides of the congruence by a. If $p \mid a$, then $a^p \equiv 0 \bmod p$ and $a \equiv 0 \bmod p$, and the desired congruence is obvious. ■

Corollary 2.16: Let p be a prime number. Then $2^p \equiv 2 \bmod p$.

Proof: Put $a = 2$ in the statement of Corollary 2.15. ■

Is the converse of Corollary 2.16 true? In other words, if $n \in \mathbf{Z}$, $n > 1$, and $2^n \equiv 2 \bmod n$, then is n necessarily a prime number? If so, we would have another necessary and sufficient condition for an integer greater than 1 to be prime [joining the one given by Wilson's Theorem (Theorem 2.11) and its converse (Proposition 2.12) of Section 2.4] and so have another characterization of primality. Chinese mathematicians more than 2000 years ago claimed such a characterization. Unfortunately, this characterization is disproved by the following counterexample.

Example 11:

$2^{341} \equiv 2 \bmod 341$, but 341 is not a prime number.

Proof: We have $341 = (11)(31)$; so 341 is not a prime number. To prove $2^{341} \equiv 2 \bmod 341$, it suffices to show $2^{341} \equiv 2 \bmod 11$ and $2^{341} \equiv 2 \bmod 31$,

Biography

Pierre de Fermat (1601–1665)

Pierre de Fermat, a Frenchman, was the founding father of the modern theory of numbers. A lawyer by profession, Fermat pursued mathematics as a sideline and has the distinction of being perhaps the most famous amateur mathematician of all time. Fermat published very little during his lifetime but considerably influenced his contemporaries through personal correspondence. Many of his unproved theorems were later shown to be correct. His "little theorem" was stated in a letter to Frénicle de Bessy in 1640 and proved by Euler in 1736; the fact that a prime number congruent to 1 modulo 4 is expressible as the sum of two squares of integers was stated in a letter to Mersenne in 1640 and proved by Euler in 1754; his assertion that every positive integer is expressible as the sum of four squares of integers was finally proved by Lagrange in 1770. Not all of Fermat's claims were true, however. As noted in Section 1.2, his claim that $2^{2^n} + 1$ is a prime number for every nonnegative integer n was disproved by Euler in 1732 when he showed that $2^{2^5} + 1$ was not prime.

Fermat is responsible for perhaps the most famous problem in number theory—his "last theorem" (see Section 6.4). In addition, he is credited with the discovery of the method of descent (also discussed in Section 6.4), a powerful proof technique for establishing that certain relations connecting positive integers are impossible. It is believed that Fermat made many of his discoveries using this method.

since $(11, 31) = 1$. We have

$$\begin{aligned} 2^{341} &\equiv (2^{10})^{34}2 \bmod 11 \\ &\equiv (1)^{34}2 \bmod 11 \quad \text{(by Fermat's Little Theorem)} \\ &\equiv 2 \bmod 11 \end{aligned}$$

and

$$\begin{aligned} 2^{341} &\equiv (2^{30})^{11}2^{11} \bmod 31 \\ &\equiv 1^{11}2^{11} \bmod 31 \quad \text{(by Fermat's Little Theorem)} \\ &\equiv 2^{11} \bmod 31 \\ &\equiv (2^5)^2 2 \bmod 31 \\ &\equiv (1)^2 2 \bmod 31 \\ &\equiv 2 \bmod 31 \end{aligned}$$

as desired. ■

The technique used in Example 11 is important enough to be highlighted as a fundamental strategy. To establish that $2^{341} \equiv 2 \bmod 341$, it suffices to establish the congruence of 2^{341} and 2 modulo each of the prime factors of 341. We chose this latter course of action (even though it involves solving two problems instead of one) since the two associated congruences have prime moduli and Fermat's Little Theorem can be used to reduce the number 2^{341} quickly. There are two morals here. First, splitting a problem into subproblems as in Example 11 may result in less work than attacking the original problem directly. Second, Fermat's Little Theorem can be very useful when you work with congruences involving large exponents! (See, for example, Exercise 51.)

Example 11 motivates the following definition.

Definition 6: Let n be a composite integer. If $2^n \equiv 2 \bmod n$, then n is said to be *pseudoprime.*

By Example 11, we have that 341 is a pseudoprime number; it is, moreover, the first pseudoprime number and was discovered in 1819. The next two pseudoprime numbers are 561 and 645 (see Exercise 54). There are infinitely many pseudoprime numbers; in fact, there are infinitely many odd pseudoprime numbers (a fact whose proof is motivated in Exercise 62) and infinitely many even pseudoprime numbers (see Beeger, 1951). Even pseudoprime numbers are somewhat more difficult to find than odd pseudoprime numbers; the first such even number is 161038 (see Exercise 55), which was discovered in 1950. Pseudoprime numbers are much rarer than prime numbers. For example, there are only 2057 pseudoprime numbers less than 10^8, but 5761455 prime numbers less than 10^8.

Exercise Set 2.5

50. Prove that $9^{10} \equiv 1 \bmod 11$ by following the steps in the proof of Fermat's Little Theorem (Theorem 2.13).

51. Using Fermat's Little Theorem, find the least nonnegative residue modulo m of each integer n below. (*Hint*: Review the procedure used in the two subproblems of Example 11.)

(a) $n = 29^{202}$, $m = 13$
(b) $n = 71^{71}$, $m = 17$
(c) $n = 3^{1000000}$, $m = 19$
(d) $n = 99^{999999}$, $m = 23$

52. Prove that 11 divides $456^{654} + 123^{321}$.

53. Use Fermat's Little Theorem to find all incongruent solutions of each congruence below.
(a) $9x \equiv 21 \bmod 23$ (*Hint*: Use Corollary 2.14; multiply both sides of the congruence by the inverse of 9 modulo 23.)
(b) $11x \equiv 15 \bmod 29$

54. *(a)* Prove that $561 = (3)(11)(17)$ is pseudoprime.
(b) Prove that $645 = (3)(5)(43)$ is pseudoprime.

55. Prove that $161038 = (2)(73)(1103)$ is pseudoprime.

56. ***Definition:*** Let b be a positive integer and let n be a composite integer. If $b^n \equiv b \bmod n$, then n is said to be *pseudoprime to the base b.* Note that pseudoprime numbers (with no base specified) are pseudoprime to the base 2.
(a) Find all integers that are pseudoprime to the base 1.
(b) Prove that 91 is pseudoprime to the base 3.
(c) Prove that 25 is pseudoprime to the base 7.

57. Let n be an integer. Prove each congruence below.
(a) $n^{21} \equiv n \bmod 30$
(b) $n^7 \equiv n \bmod 42$
(c) $n^{13} \equiv n \bmod 2730$

58. Let a and b be integers not divisible by the prime number p.
(a) If $a^p \equiv b^p \bmod p$, prove that $a \equiv b \bmod p$.
(b) If $a^p \equiv b^p \bmod p$, prove that $a^p \equiv b^p \bmod p^2$.

59. Let p and q be distinct prime numbers. Prove that $p^{q-1} + q^{p-1} \equiv 1 \bmod pq$.

60. Let p and q be distinct odd prime numbers with $p - 1 \mid q - 1$. If $a \in \mathbf{Z}$ with $(a, pq) = 1$, prove that $a^{q-1} \equiv 1 \bmod pq$.

61. Let $a \in \mathbf{Z}$ and let p be a prime number. Prove that $p \mid a^p + (p-1)!a$ and $p \mid (p-1)!a^p + a$.

62. (The following exercise proves that there are infinitely many odd pseudoprime numbers.)
(a) Let m and n be positive integers such that $m \mid n$. Prove that $2^m - 1 \mid 2^n - 1$.
(b) Prove that n is an odd pseudoprime number if and only if $2^{n-1} \equiv 1 \bmod n$.
(c) Prove that if n is an odd pseudoprime number, then $m = 2^n - 1$ is an odd pseudoprime number. [*Hint*: First show that m is composite by using part (a) above. Then show that m is an odd pseudoprime number by using part (b).]
(d) Prove that there are infinitely many odd pseudoprime numbers.

63. Prove that there are infinitely many pseudoprime numbers to an arbitrary base b. (See Exercise 56.)

64. Let p be a prime number. Prove that if $2^p - 1$ is a composite number, then $2^p - 1$ is a pseudoprime number.

65. ***Definition:*** Let n be a composite number. Then n is said to be an *absolute pseudoprime number* or a *Carmichael number* if $b^n \equiv b \bmod n$ for all integers b.

(a) Prove that a composite number n is a Carmichael number if and only if $b^{n-1} \equiv 1 \bmod n$ for all integers b with $(b, n) = 1$.

(b) Prove that 561 is a Carmichael number. [In fact, 561 is the smallest Carmichael number. R. C. Carmichael conjectured in 1912 that there were infinitely many Carmichael numbers; this was proved in 1992 by R. Alford, A. Granville, and C. Pomerance. For further discussion of this result, see Granville (1992).]

(c) Prove that the pseudoprime number 341 is not a Carmichael number.

(d) Prove that all Carmichael numbers are odd.

(e) If n is a product of distinct prime numbers, say $n = p_1 p_2 \cdots p_m$, and $p_i - 1 \mid n - 1$ for all i, prove that n is a Carmichael number. (In fact, the converse of this result is also true; see Student Project 4 in Chapter 5. See also Student Project 5 in Chapter 5.)

(f) Let m be a positive integer such that $6m + 1$, $12m + 1$, and $18m + 1$ are prime numbers. Prove that $n = (6m + 1)(12m + 1)(18m + 1)$ is a Carmichael number.

(g) Prove that 1729, 10585, and 294409 are Carmichael numbers. [*Hint*: Use parts (e) and (f) above.]

2.6

Euler's Theorem

Fermat's Little Theorem can be thought of as providing an answer to the following question: If p is a prime number and a is any integer with $(a, p) = 1$ (or $p \nmid a$), then what power of a is guaranteed to be congruent to 1 modulo p? The answer given by Fermat's Little Theorem is, of course, $p - 1$. In 1760, Leonhard Euler was able to generalize Fermat's Little Theorem. Euler's Theorem provides an answer to the same question for an arbitrary modulus m. In other words, Euler's Theorem answers the following question: If m is any positive integer and a is any integer with $(a, m) = 1$, then what power of a is guaranteed to be congruent to 1 modulo m? So Fermat's Little Theorem is a special case of Euler's Theorem. In addition, Euler's Theorem will be crucial in Chapter 5. Before stating Euler's theorem, we need a preliminary definition, which was also introduced by Euler.

Definition 7: Let $n \in \mathbf{Z}$ with $n > 0$. The *Euler phi-function,* denoted $\phi(n)$, is the function defined by

$$\phi(n) = |\{x \in \mathbf{Z}: 1 \le x \le n; (x, n) = 1\}|$$

In other words, $\phi(n)$ is the number of positive integers less than or equal to n that are relatively prime to n. Note that n will be relatively prime to itself if and only if $n = 1$.

Example 12:

(a) $\phi(12) = 4$ since there are four positive integers less than or equal to 12 that are relatively prime to 12, namely, 1, 5, 7, and 11.

(b) $\phi(14) = 6$ since there are six positive integers less than or equal to 14 that are relatively prime to 14, namely, 1, 3, 5, 9, 11, and 13.

(c) If p is a prime number, then *all* positive integers less than p are relatively prime to p. Inasmuch as there are $p - 1$ such numbers, we have $\phi(p) = p - 1$.

Values of $\phi(n)$ for integers n with $1 \leq n \leq 100$ are given in Table 2 of Appendix E. We will return to a more detailed study of $\phi(n)$ and related functions in Chapter 3.

We now state Euler's Theorem (and you will see immediately why we need the perhaps unmotivated definition above).

Theorem 2.17: (Euler's Theorem) Let $a, m \in \mathbf{Z}$ with $m > 0$. If $(a, m) = 1$, then

$$a^{\phi(m)} \equiv 1 \bmod m$$

Note immediately that Fermat's Little Theorem follows from Euler's Theorem by taking m to be a prime number [from which $\phi(m) = m - 1$ by Example 12(c)]. Although the proof of Euler's theorem below is a bit harder than the proof of Fermat's Little Theorem (since the former theorem is more general), the proof below is still very similar to the proof of Fermat's Little Theorem. We suggest that you become familiar again with the proof of Fermat's Little Theorem (by solving Exercise 50 again, for example) before reading the proof of Euler's Theorem below.

Proof: (of Theorem 2.17) Let $r_1, r_2, \ldots, r_{\phi(m)}$ be the $\phi(m)$ distinct positive integers not exceeding m such that $(r_i, m) = 1$, $i = 1, 2, \ldots, \phi(m)$. Consider the $\phi(m)$ integers given by $r_1a, r_2a, \ldots, r_{\phi(m)}a$. Note that $(r_ia, m) = 1$, $i = 1, 2, \ldots, \phi(m)$. [If $(r_ia, m) > 1$ for some i, then there is a prime divisor p of (r_ia, m) by Lemma 1.5 from which $p \mid r_ia$ and $p \mid m$. Now $p \mid r_ia$ implies $p \mid r_i$ or $p \mid a$ by Lemma 1.14; so either we have $p \mid r_i$ and $p \mid m$ or we have $p \mid a$ and $p \mid m$. But $p \mid r_i$ and $p \mid m$ is impossible since $(r_i, m) = 1$, and $p \mid a$ and $p \mid m$ is impossible since $(a, m) = 1$.] Note also that no two of r_1a, $r_2a, \ldots, r_{\phi(m)}a$ are congruent modulo m. [Since $(a, m) = 1$, the inverse of a modulo m, say a', exists by Corollary 2.8. Hence, if $r_ia \equiv r_ja \bmod m$ with $i \neq j$, then $r_iaa' \equiv r_jaa' \bmod m$, from which $r_i \equiv r_j \bmod \mathrm{m}$, which is impossible.] So the least nonnegative residues modulo m of the integers $r_1a, r_2a, \ldots, r_{\phi(m)}a$, taken in some order, must be $r_1, r_2, \ldots, r_{\phi(m)}$. Then

$$(r_1a)(r_2a) \cdots (r_{\phi(m)}a) \equiv r_1r_2 \cdots r_{\phi(m)} \bmod m$$

or, equivalently,

$$a^{\phi(m)}r_1r_2 \cdots r_{\phi(m)} \equiv r_1r_2 \cdots r_{\phi(m)} \bmod m$$

Biography

Leonhard Euler (1707–1783)

Leonhard Euler was a Swiss mathematician who has the honor of being the most prolific mathematician in history. Euler wrote more than 800 papers; it has been estimated that his collected works (currently being compiled by the Swiss Society of Natural Science) will fill more than 100 volumes. It is not surprising that one finds Euler's name in all areas of mathematics: Euler lines (in geometry), Euler's method (in algebra), Euler equations (in differential equations), and Euler's phi-function, Euler's Theorem, Euler's Criterion (see Section 4.2), and Euler's constant (in number theory), to name a few.

Euler lost the sight in his right eye in 1735 and was totally blind during the last 17 years of his life. Such an obstacle would seem to be insurmountable to an ordinary mathematician, but Euler's mind seemed to be sharpened by the lack of sight. He had a remarkable memory — he knew countless mathematical formulas by heart as well as many poems, including the entire *Aeneid*. In addition, Euler was gifted at mental computation. It is said that he could perform mathematical calculations in his head that other capable mathematicians had difficulty with on paper. Approximately half of his works were written during his last years of total blindness.

Euler's mathematical achievements did not come at the expense of other interests. He married twice and fathered 13 children. Euler also had an avid interest in fields outside mathematics, including theology, literature, oriental languages, and all areas of natural science.

Now $m \mid a^{\phi(m)} r_1 r_2 \cdots r_{\phi(m)} - r_1 r_2 \cdots r_{\phi(m)}$ implies that $m \mid r_1 r_2 \cdots r_{\phi(m)} \times (a^{\phi(m)} - 1)$; since $(r_1 r_2 \cdots r_{\phi(m)}, m) = 1$, we have that $m \mid a^{\phi(m)} - 1$ and $a^{\phi(m)} \equiv 1 \bmod m$, as desired. ■

Exercise 67 leads the reader through the proof of Euler's Theorem with a concrete example; we suggest that you solve this exercise now.

The proof of Euler's Theorem motivates the following definition.

Definition 8: Let m be a positive integer. A set of $\phi(m)$ integers such that each element of the set is relatively prime to m and no two elements of the set are congruent modulo m is said to be a *reduced residue system modulo m.*

Example 13:

(a) By Example 12(a), a reduced residue system modulo 12 is given by the set $\{1, 5, 7, 11\}$.

(b) By Example 12(b), a reduced residue system modulo 14 is given by the set $\{1, 3, 5, 9, 11, 13\}$.

(c) By Example 12(c), a reduced residue system modulo a prime number p is given by the set $\{1, 2, \ldots, p - 1\}$.

Note that the proof of Euler's Theorem begins with a reduced residue system modulo m, namely, $\{r_1, r_2, \ldots, r_{\phi(m)}\}$, and uses this system to produce a new reduced residue system modulo m given by $\{r_1 a, r_2 a, \ldots, r_{\phi(m)} a\}$ where $(a, m) = 1$. So we have the following porism.

Porism 2.18: Let m be a positive integer and let $\{r_1, r_2, \ldots, r_{\phi(m)}\}$ be a reduced residue system modulo m. If a is an integer with $(a, m) = 1$, then the set $\{r_1 a, r_2 a, \ldots, r_{\phi(m)} a\}$ is a reduced residue system modulo m. ■

Example 14:

(a) By Example 13(a), the set $\{1, 5, 7, 11\}$ is a reduced residue system modulo 12. Since $(5, 12) = 1$, the set $\{1 \cdot 5, 5 \cdot 5, 7 \cdot 5, 11 \cdot 5\} = \{5, 25, 35, 55\}$ is also a reduced residue system modulo 12 by Porism 2.18. It is left to the reader to verify that 5, 25, 35, and 55 are congruent modulo 12 to 1, 5, 7, and 11 in some permuted order.

(b) The reader is invited to illustrate Porism 2.18 by using the reduced residue system modulo 14 given in Example 13(b).

Euler's Theorem has the following easy corollary.

Corollary 2.19: Let $a, m \in \mathbf{Z}$ with $m > 0$. If $(a, m) = 1$, then $a^{\phi(m)-1}$ is the inverse of a modulo m.

Proof: By Euler's Theorem, we have $a^{\phi(m)} \equiv 1 \bmod m$. Hence $aa^{\phi(m)-1} \equiv 1 \bmod m$, and $a^{\phi(m)-1}$ is the inverse of a modulo m, as desired. ■

In view of Example 8 (please review this example again if necessary), Corollary 2.19 is useful in solving linear congruences in one variable of the form $ax \equiv b \bmod m$ where $(a, m) = 1$ (see Exercise 70).

Exercise Set 2.6

66. Compute $\phi(n)$ for all integral n between 1 and 25 inclusive.

67. Prove that $9^8 \equiv 1 \bmod 16$ by following the steps in the proof of Euler's Theorem (Theorem 2.17).

68. Using Euler's Theorem, find the least nonnegative residue modulo m of each integer n below. (*Hint*: Parallel the procedure used in the two subproblems of Example 11.)
(a) $n = 29^{198}$, $m = 20$
(b) $n = 79^{79}$, $m = 9$
(c) $n = 3^{1000000}$, $m = 14$
(d) $n = 99^{999999}$, $m = 26$

69. Find a reduced residue system modulo each integer below.
(a) 15
(b) 18
(c) p (p a prime number)
(d) 2^n (n a positive integer)

70. Use Euler's Theorem to find all incongruent solutions of each congruence below.
(a) $9x \equiv 21 \bmod 25$ (*Hint*: Use Corollary 2.19; multiply both sides of the congruence by the inverse of 9 modulo 25.)
(b) $5x \equiv 21 \bmod 36$

71. *(a)* Let n be an integer not divisible by 3. Prove that $n^7 \equiv n \bmod 63$.
(b) Let n be an integer divisible by 9. Prove that $n^7 \equiv n \bmod 63$.

72. *(a)* Prove that if n is an integer relatively prime to 72, then $n^{12} \equiv 1 \bmod 72$.
(b) Find the largest integer m such that $n^{12} \equiv 1 \bmod m$ for all integers n relatively prime to m. (*Note*: By part (a) above, $m \geq 72$.)

73. Let m and n be relatively prime positive integers. Prove that $m^{\phi(n)} + n^{\phi(m)} \equiv 1 \bmod mn$.

74. Let p be an odd prime number. Prove that

$$\left\{\frac{-(p-1)}{2}, \frac{-(p-3)}{2}, \ldots, -2, -1, 1, 2, \ldots, \frac{p-3}{2}, \frac{p-1}{2}\right\}$$

is a reduced residue system modulo p.

75. Let m be a positive integer with $m \neq 2$. If $\{r_1, r_2, \ldots, r_{\phi(m)}\}$ is a reduced residue system modulo m, prove that

$$r_1 + r_2 + \cdots + r_{\phi(m)} \equiv 0 \bmod m$$

76. Let $a, m \in \mathbf{Z}$ with $m > 0$, $(a, m) = 1$, and $(a - 1, m) = 1$. Prove that

$$a^{\phi(m)-1} + a^{\phi(m)-2} + \cdots + a^2 + a + 1 \equiv 0 \bmod m$$

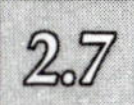

Concluding Remarks

The arithmetic associated with the relation of equality is familiar and customary. As a simple example, we know that $2x = 3$ if and only if $x = \frac{3}{2}$. In this chapter, we have learned that the notion of congruence gives us a new type of arithmetic, an arithmetic of the congruence relation known as *modular arithmetic.* The statement "$2x \equiv 3$ if and only if $x \equiv \frac{3}{2}$" makes no sense in modular arithmetic; on the one hand, no modulus for the congruence has been specified and, on the other hand, fractions such as $\frac{3}{2}$ have no meaning in this arithmetic (unless, of course, we *define* a meaning!). We can remedy this by specifying a modulus for the congruence (as in $2x \equiv 3 \bmod 4$, for example). We can now solve this linear congruence in one variable (just as we can solve linear equations). In a similar way we can solve special systems of linear congruences (just as we can solve systems of linear equations). The point made here is that this chapter has dealt with a *new* system of arithmetic as well as a discussion of the properties satisfied by the system. This system of modular arithmetic, along with its properties (including special congruences such as Wilson's Theorem and the like), will be crucial in forthcoming chapters.

Student Projects

1. (Programming project for calculator or computer)
Consider the following procedure for finding the least nonnegative residue of 456^{123} modulo 789.
Step 1. Express 123 as the sum of powers of two, namely,

$$\begin{aligned} 123 &= 2^0 + 2^1 + 2^3 + 2^4 + 2^5 + 2^6 \\ &= 1 + 2 + 8 + 16 + 32 + 64 \end{aligned}$$

Step 2. Compute the least nonnegative residue of 456^n modulo 789 where n ranges over all powers of 2 up to and including the maximal power of 2 of Step 1 as follows:

$$\begin{aligned} 456^1 &\equiv 456 \bmod 789 \\ 456^2 &\equiv 429 \bmod 789 \\ 456^4 &\equiv 429^2 \equiv 204 \bmod 789 \\ 456^8 &\equiv 204^2 \equiv 588 \bmod 789 \\ 456^{16} &\equiv 588^2 \equiv 162 \bmod 789 \\ 456^{32} &\equiv 162^2 \equiv 207 \bmod 789 \\ 456^{64} &\equiv 207^2 \equiv 243 \bmod 789 \end{aligned}$$

Step 3. Compute the desired least nonnegative residue of 456^{123} modulo 789 as follows:

$$\begin{aligned} 456^{123} &\equiv (456)(429)(588)(162)(207)(243) \bmod 789 \\ &\equiv (741)(588)(162)(207)(243) \bmod 789 \\ &\equiv (180)(162)(207)(243) \bmod 789 \\ &\equiv (756)(207)(243) \bmod 789 \\ &\equiv (270)(243) \bmod 789 \\ &\equiv 123 \bmod 789. \end{aligned}$$

Given positive integers a, b, and m, compute the least nonnegative residue of a^b modulo m by using the three-step procedure above.

2. Consider the date August 22, 1999. Let

$$M = 6$$

(the month, where March $= 1$, April $= 2, \ldots,$ February $= 12$),

$$N = 22$$

(the day of the month),

$$F = 19$$

(the first two digits in the year),

$$L = 99$$

(the last two digits in the year), and

$$Y = 0$$

($Y = 0$ for nonleap years and $Y = 1$ for leap years). Then the day of the week D for August 22, 1999 is given by

$$D \equiv [2.6M - 0.2] + N + [F/4] + L + [L/4] - 2F - (Y + 1)[M/11] \bmod 7$$

where Sunday $\equiv 0 \bmod 7$, Monday $\equiv 1 \bmod 7, \ldots,$ Saturday $\equiv 6 \bmod 7$, and $[\cdot]$ denotes the greatest integer function. In fact, the formula above is valid for *any* date after 1582. Do some research to find a justification of this result.

3. In 1828, N. H. Abel asked whether there existed an integer $x \neq \pm 1$ and a prime number p such that $x^{p-1} \equiv 1 \bmod p^2$. Investigate the solvability of this congruence. A discussion of the congruence above and related congruences may be found on pages 24–26 of Hua Loo Keng, *Introduction to Number Theory* (New York: Springer-Verlag, 1982).

4. For an odd prime number p, consider the prime numbers between 3 and p inclusive. Let M_p be the number of these prime numbers congruent to 1 modulo 4 and let N_p be the number of these prime numbers congruent to 3 modulo 4. Find as many values of p for which $M_p = N_p$ as you can. [There are infinitely many such values of p, as proved by Littlewood in 1914. For an interesting graphical depiction of this "congruence modulo 4 race," see the PrimePiMod4 exercise in Section 1.3 of Stan Wagon, *Mathematica*® *in Action* (New York: W. H. Freeman, 1991).]

5. Read the following essay for a historical discussion of the Chinese

Remainder Theorem: Li Wenlin and Yuan Xiangdong, "The Chinese Remainder Theorem," *Ancient China's Technology and Science* (1983), 99–110.

6. Let $n \in \mathbf{Z}$ with $n > 1$.
   (a) Prove that 2^n is never congruent to 1 modulo n.
   (b) When is 2^n congruent to 2 modulo n?
   (c) Is 2^n ever congruent to 3 modulo n?
   [The problem above appears as Example 13 in Richard K. Guy, "The Strong Law of Small Numbers," *American Mathematical Monthly, 95* (1988), 697–712.]
7. Let $m \in \mathbf{Z}$ with $m \geq 1$. Prove that the sequence

$$2, 2^2, 2^{2^2}, 2^{2^{2^2}}, \ldots$$

   is constant for all but finitely many terms when viewed modulo m. (This is Problem 3 of the Twentieth Annual U.S.A. Mathematical Olympiad. A solution appears in the June 1992 edition of *Mathematics Magazine.*)

3

Arithmetic Functions

The concept of a function is one of the most pervasive in all of mathematics. So it is not surprising that functions play a key role in number theory. We have already seen one example of such a function in Section 2.6, namely, the Euler phi-function. It is the purpose of this chapter to investigate the Euler phi-function and related number-theoretic functions more deeply.

A property called *multiplicativity* will allow us to obtain compact formulas for the evaluation of the most important functions of number theory. In addition, consideration of a certain function will result in a class of numbers known as *perfect numbers.* Dating back to the classical Greek period of mathematics, perfect numbers were assigned mystical and religious significance by the Pythagoreans. More fundamentally, perfect numbers yield several of the most intriguing unsolved problems in number theory today. We conclude the chapter with the Möbius Inversion Formula, a result of substantial significance in the theory of arithmetic functions.

3.1

Arithmetic Functions; Multiplicativity

The important functions in number theory are termed *arithmetic,* pronounced "ar′ ith-met′ ik." (Note the two accents and their locations.)

Definition 1: An *arithmetic function* is a function whose domain is the set of positive integers.

Here, *arithmetic* is an adjective (describing a type of function) rather than a noun (an area of elementary mathematics). The key arithmetic functions to be studied in this chapter are given in Example 1.

Biography

Pythagoras of Samos (572?–500? B.C.)

Pythagoras was born on the Aegean island of Samos. In southern Italy, he founded the famous Pythagorean school, which was a fraternity dedicated to the study of mathematics, philosophy, and natural science. The brotherhood of the Pythagoreans would dominate the first three centuries of classical Greek mathematics. The Pythagoreans believed that whole numbers completely described the universe. This belief led to the glorified study of arithmetic and number theory and is perhaps the reason for the assignment of mystical properties to certain numbers. (For example, the concepts of *amicable, perfect, deficient,* and *abundant numbers* discussed in this chapter may have originated with the Pythagoreans.) All discoveries within the Pythagorean brotherhood were attributed to Pythagoras himself, so it is difficult to say which results were actually his. Although the famous Pythagorean Theorem was known for more than a thousand years before the Pythagoreans, it is generally accepted that Pythagoras independently discovered the result and may have been the first to rigorously prove it. The related concept of Pythagorean triples is discussed in Section 6.3.

Example 1:

In what follows, n is a positive integer.

(a) The Euler phi-function $\phi(n)$ defined in Section 2.6 is an arithmetic function.

(b) Let $\nu(n)$ denote the number of positive divisors of n. [For example, $\nu(12) = 6$ since there are six positive divisors of 12, namely, 1, 2, 3, 4, 6, and

12.] Then $\nu(n)$ is an arithmetic function. We will study $\nu(n)$ more deeply in Section 3.3.

(c) Let $\sigma(n)$ denote the sum of the positive divisors of n. [For example, $\sigma(12) = 1 + 2 + 3 + 4 + 6 + 12 = 28$.] Then $\sigma(n)$ is an arithmetic function. We will study $\sigma(n)$ more deeply in Section 3.4.

Important properties possessed by certain arithmetic functions are multiplicativity and complete multiplicativity, which we define now.

Definition 2: An arithmetic function f is said to be *multiplicative* if $f(mn) = f(m)f(n)$ whenever m and n are relatively prime positive integers. An arithmetic function f is said to be *completely multiplicative* if $f(mn) = f(m)f(n)$ for all positive integers m and n.

It is clear that every completely multiplicative arithmetic function is multiplicative. For practice with multiplicativity and complete multiplicativity, you should solve Exercises 1, 2, and 5 now (consult the Hints and Answers section if you have difficulty). In addition, we will see several examples of multiplicative and completely multiplicative arithmetic functions in forthcoming sections and in the exercises of those sections.

There is a fundamental observation to be made concerning multiplicative arithmetic functions. Let f be a multiplicative arithmetic function. If n is a positive integer, then

$$n = p_1^{a_1}p_2^{a_2}\cdots p_r^{a_r}$$

with $p_1, p_2, \ldots, p_r$ distinct prime numbers and $a_1, a_2, \ldots, a_r$ nonnegative integers; since $(p_i^{a_i}, p_j^{a_j}) = 1$ if $i \neq j$, we have

$$f(n) = f(p_1^{a_1}p_2^{a_2}\cdots p_r^{a_r}) = f(p_1^{a_1})f(p_2^{a_2})\cdots f(p_r^{a_r})$$

In other words, *a multiplicative arithmetic function is completely determined by its values at powers of prime numbers.* This is an extremely nice fact that frequently allows us to obtain compact formulas for evaluating multiplicative arithmetic functions. We will exploit this fact shortly to obtain such formulas. An analogue of the preceding discussion establishes that a completely multiplicative arithmetic function is completely determined by its values at prime numbers (see Exercise 6).

We conclude this section with a theorem concerning multiplicative arithmetic functions that will be used in Sections 3.3 and 3.4. First, a bit of explanation concerning notation is in order. Throughout this book, the notation $\sum_{d \mid n, d>0}$ means "the summation over all distinct positive divisors d of n," so that, for example,

$$\sum_{d \mid 12, d>0} f(d) = f(1) + f(2) + f(3) + f(4) + f(6) + f(12)$$

You will notice that many results of this chapter, including the theorem below, utilize this notation.

Theorem 3.1: Let f be an arithmetic function and, for $n \in \mathbf{Z}$ with $n > 0$, let

$$F(n) = \sum_{d \mid n, d>0} f(d)$$

If f is multiplicative, then F is multiplicative.

Proof: Let m and n be relatively prime positive integers. To prove that F is multiplicative, we must show that $F(mn) = F(m)F(n)$. We have

$$F(mn) = \sum_{d \mid mn,\, d>0} f(d)$$

Since $(m, n) = 1$, each divisor $d > 0$ of mn can be written uniquely as $d_1 d_2$ where $d_1, d_2 > 0$, $d_1 \mid m$, $d_2 \mid n$, and $(d_1, d_2) = 1$ and each such product $d_1 d_2$ corresponds to a divisor d of mn (see Exercise 8). So we have

$$\begin{aligned} F(mn) &= \sum_{d_1 \mid m,\, d_2 \mid n,\, d_1, d_2>0} f(d_1 d_2) \\ &= \sum_{d_1 \mid m,\, d_2 \mid n,\, d_1, d_2>0} f(d_1) f(d_2) \quad \text{(since } f \text{ is multiplicative)} \\ &= \sum_{d_1 \mid m,\, d_1>0} f(d_1) \sum_{d_2 \mid n,\, d_2>0} f(d_2) \\ &= F(m)F(n) \end{aligned}$$

as desired. ■

You may find the proof of Theorem 3.1 confusing; we now provide an illustration of the idea behind this proof with a concrete m and n. Make sure that you see the parallels between the example below and the proof above.

Example 2:

Let $m = 3$ and $n = 4$ and let all notation be as in the proof of Theorem 3.1. We must show that $F(3 \cdot 4) = F(3)F(4)$. We have

$$\begin{aligned} F(3 \cdot 4) &= \sum_{d \mid 12,\, d>0} f(d) \\ &= f(1) + f(2) + f(3) + f(4) + f(6) + f(12) \\ &= f(1) + f(2) + f(4) + f(3) + f(6) + f(12) \\ &= f(1 \cdot 1) + f(1 \cdot 2) + f(1 \cdot 4) + f(3 \cdot 1) + f(3 \cdot 2) + f(3 \cdot 4) \\ &= f(1)f(1) + f(1)f(2) + f(1)f(4) + f(3)f(1) + f(3)f(2) + f(3)f(4) \\ &= [f(1) + f(3)][f(1) + f(2) + f(4)] \\ &= \sum_{d_1 \mid 3,\, d_1>0} f(d_1) \sum_{d_2 \mid 4,\, d_2>0} f(d_2) \\ &= F(3)F(4) \end{aligned}$$

as desired.

Exercise Set 3.1

1. Formulate conjectures on whether the arithmetic functions ϕ, ν, and σ are multiplicative by accumulating your own numerical evidence.
2. Formulate conjectures on whether the arithmetic functions ϕ, ν, and σ are completely multiplicative by accumulating your own numerical evidence.
3. Let f be a multiplicative arithmetic function. If there exists a positive integer n such that $f(n) \neq 0$ (so f is not identically zero), prove that $f(1) = 1$.
4. Let $n \in \mathbf{Z}$ with $n > 0$. Define an arithmetic function ρ by $\rho(1) = 1$ and $\rho(n) = 2^m$ where m is the number of distinct prime numbers in the prime factorization of n.
   (a) Prove that ρ is multiplicative but not completely multiplicative.
   (b) Let $n \in \mathbf{Z}$ with $n > 0$ and let

$$f(n) = \sum_{d \mid n,\, d>0} \rho(d)$$

   If $p_1^{a_1} p_2^{a_2} \cdots p_m^{a_m}$ is the prime factorization of n, find a formula for $f(n)$ in terms of this prime factorization. (*Hint:* Note that f is multiplicative by part (a) above and Theorem 3.1.)
5. Determine (with a proof or a counterexample) whether each of the arithmetic functions below is completely multiplicative, multiplicative, or both. In parts (d)–(f), k is a fixed real number.
   (a) $f(n) = 0$
   (b) $f(n) = 1$
   (c) $f(n) = 2$
   (d) $f(n) = n + k$
   (e) $f(n) = kn$
   (f) $f(n) = n^k$
   (g) $f(n) = n^n$
   (h) $f(n) = n!$
6. Prove that a completely multiplicative arithmetic function is completely determined by its values at prime numbers.
7. ***Definition:*** Let $n \in \mathbf{Z}$ with $n > 0$. The *Liouville λ-function,* denoted $\lambda(n)$, is

$$\lambda(n) = \begin{cases} 1, & \text{if } n = 1 \\ (-1)^k, & \text{if } n = p_1 p_2 \cdots p_k \text{ where } p_1, p_2, \ldots, p_k \\ & \text{are not necessarily distinct prime numbers} \end{cases}$$

   (a) Prove that λ is a completely multiplicative arithmetic function.
   (b) Let $F(n) = \sum_{d \mid n,\, d>0} \lambda(d)$. Prove that

$$F(n) = \begin{cases} 1, & \text{if } n \text{ is a perfect square} \\ 0, & \text{otherwise} \end{cases}$$

8. Let m and n be positive integers with $(m, n) = 1$. Prove that each divisor $d > 0$ of mn can be written uniquely as $d_1 d_2$ where $d_1, d_2 > 0$, $d_1 \mid m$,

$d_2 \mid n$, and $(d_1, d_2) = 1$ and each such product $d_1 d_2$ corresponds to a divisor d of mn.

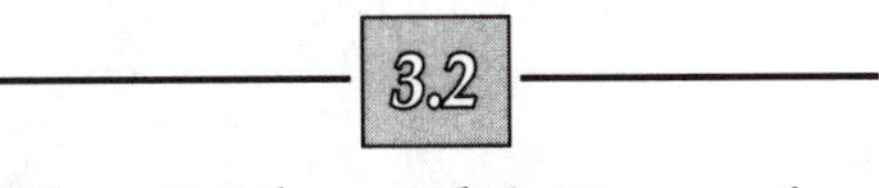

The Euler Phi-Function

We now return to the Euler phi-function $\phi(n)$. (You may wish to review Definition 7 in Chapter 2.) As stated before, values of $\phi(n)$ for integers n with $1 \leq n \leq 100$ are given in Table 2 of Appendix E. Our goal here is to prove that $\phi(n)$ is multiplicative and use this fact to obtain a nice formula for $\phi(n)$.

Theorem 3.2: The Euler phi-function $\phi(n)$ is multiplicative.

Proof: Let m and n be relatively prime positive integers. We must show that $\phi(mn) = \phi(m)\phi(n)$. Display the positive integers not exceeding mn in matrix form as follows:

$$\begin{array}{ccccc}
1 & m+1 & 2m+1 & \cdots & (n-1)m+1 \\
2 & m+2 & 2m+2 & \cdots & (n-1)m+2 \\
\vdots & \vdots & \vdots & \vdots & \vdots \\
i & m+i & 2m+i & \cdots & (n-1)m+i \\
\vdots & \vdots & \vdots & \vdots & \vdots \\
m & 2m & 3m & \cdots & nm
\end{array}$$

Consider the ith row of the matrix above. If $(m, i) = d > 1$, then no entry of the ith row is relatively prime to m and so not relatively prime to mn. [Every entry of the ith row is of the form $km + i$ where k is an integer; if $(m, i) = d > 1$, then $d \mid m$ and $d \mid i$ so that $d \mid km + i$ by Proposition 1.2.] Hence, to determine how many integers in the matrix are relatively prime to mn [which is $\phi(mn)$], it suffices to examine the ith row of the matrix if and only if $(m, i) = 1$; there are $\phi(m)$ such rows. Now, given such an ith row, we must determine how many integers in the row are relatively prime to mn. The entries of such a row are (as above) $i, m + i, 2m + i, \ldots, (n - 1)m + i$. Since $(m, i) = 1$, each of these n integers is relatively prime to m. Furthermore, these n integers form a complete system of residues modulo n (see Exercise 11) and so exactly $\phi(n)$ of these integers are relatively prime to n. Since these $\phi(n)$ integers are relatively prime to m, they are relatively prime to mn. In summary, then, exactly $\phi(m)$ rows in the matrix contain integers relatively prime to mn. Each such row contains exactly $\phi(n)$ integers relatively prime to mn. So $\phi(mn)$, the total number of integers in the matrix that are relatively prime to mn, is $\phi(m)\phi(n)$ as desired. ■

Did you conjecture the result of Theorem 3.2 above when you solved Exercise 1?

Now, since the Euler phi-function is multiplicative, it is completely determined by its values at powers of prime numbers. The next two theorems exploit this fact to obtain a formula for $\phi(n)$.

Theorem 3.3: Let p be a prime number and let $a \in \mathbf{Z}$ with $a > 0$. Then $\phi(p^a) = p^a - p^{a-1}$.

Proof: The total number of positive integers not exceeding p^a is clearly p^a. The positive integers not exceeding p^a that are *not* relatively prime to p^a are precisely the multiples of p given by $p, 2p, 3p, \ldots, (p^{a-1})p$. There are clearly p^{a-1} such multiples. So the total number of positive integers not exceeding p^a that are relatively prime to p^a is the difference of the above quantities, and $\phi(p^a) = p^a - p^{a-1}$, as desired. ■

Theorem 3.4: Let $n \in \mathbf{Z}$ with $n > 0$. Then

$$\phi(n) = n \prod_{p \mid n,\, p \text{ prime}} \left(1 - \frac{1}{p}\right)$$

The notation $\prod_{p \mid n,\, p \text{ prime}}$ means "the product over all distinct prime divisors p of n," so that, for example,

$$\prod_{p \mid 12,\, p \text{ prime}} \left(1 - \frac{1}{p}\right) = \left(1 - \frac{1}{2}\right)\left(1 - \frac{1}{3}\right)$$

Proof: (of Theorem 3.4) The desired result is clear for $n = 1$ (with the convention that the empty product is taken to be 1). Assume that $n > 1$ and that $n = p_1^{a_1}p_2^{a_2}\cdots p_r^{a_r}$ with $p_1, p_2, \ldots, p_r$ distinct prime numbers and $a_1, a_2, \ldots, a_r$ positive integers. Then

$$\begin{aligned}
\phi(n) &= \phi(p_1^{a_1}p_2^{a_2}\cdots p_r^{a_r}) \\
&= \phi(p_1^{a_1})\phi(p_2^{a_2})\cdots\phi(p_r^{a_r}) \quad \text{(by Theorem 3.2)} \\
&= (p_1^{a_1} - p_1^{a_1-1})(p_2^{a_2} - p_2^{a_2-1})\cdots(p_r^{a_r} - p_r^{a_r-1}) \quad \text{(by Theorem 3.3)} \\
&= p_1^{a_1}\left(1 - \frac{1}{p_1}\right)p_2^{a_2}\left(1 - \frac{1}{p_2}\right)\cdots p_r^{a_r}\left(1 - \frac{1}{p_r}\right) \\
&= p_1^{a_1}p_2^{a_2}\cdots p_r^{a_r}\left(1 - \frac{1}{p_1}\right)\left(1 - \frac{1}{p_2}\right)\cdots\left(1 - \frac{1}{p_r}\right) \\
&= n \prod_{p \mid n,\, p \text{ prime}} \left(1 - \frac{1}{p}\right) \quad ■
\end{aligned}$$

Theorem 3.4 gives the desired formula for $\phi(n)$. Note that this formula requires the distinct prime numbers in the prime factorization of n. We illustrate with an example.

Example 3:

We use the formula of Theorem 3.4 to compute $\phi(504)$. We have $504 = 2^3 3^2 7$, so the distinct prime numbers dividing 504 are 2, 3, and 7. So

$$\begin{aligned}\phi(504) &= 504\left(1 - \frac{1}{2}\right)\left(1 - \frac{1}{3}\right)\left(1 - \frac{1}{7}\right)\\ &= 504\left(\frac{1}{2}\right)\left(\frac{2}{3}\right)\left(\frac{6}{7}\right)\\ &= 144\end{aligned}$$

and there are exactly 144 positive integers not exceeding 504 that are relatively prime to 504. Are you impressed with the power of multiplicativity?

We end this section with a curious property of the Euler phi-function that was first noticed by Gauss.

Theorem 3.5: (Gauss) Let $n \in \mathbf{Z}$ with $n > 0$. Then

$$\sum_{d \mid n,\, d>0} \phi(d) = n$$

Proof: Let d be a positive divisor of n and put

$$S_d = \{m \in \mathbf{Z}: 1 \le m \le n, (m, n) = d\}$$

Now $(m, n) = d$ if and only if $\left(\frac{m}{d}, \frac{n}{d}\right) = 1$; hence the number of integers in S_d is equal to the number of positive integers not exceeding $\frac{n}{d}$ that are relatively prime to $\frac{n}{d}$. In other words, the number of integers in S_d is equal to $\phi\left(\frac{n}{d}\right)$. Since every integer from 1 to n inclusive is an element of one and only one S_d, we have

$$n = \sum_{d \mid n,\, d>0} \phi\left(\frac{n}{d}\right)$$

But as d runs through all positive divisors of n, we have that $\frac{n}{d}$ also runs through all positive divisors of n, and

$$n = \sum_{d \mid n,\, d>0} \phi\left(\frac{n}{d}\right) = \sum_{d \mid n,\, d>0} \phi(d)$$

as desired. ■

An illustrative concrete example of the proof of Theorem 3.5 is in order.

Example 4:

Let $n = 12$ and let all notation be as in the proof of Theorem 3.5. We must show that

$$\sum_{d \mid 12,\, d>0} \phi(d) = 12$$

We have

$$S_1 = \{m \in \mathbf{Z}: 1 \leq m \leq 12, (m, 12) = 1\} = \{1, 5, 7, 11\}$$
$$S_2 = \{m \in \mathbf{Z}: 1 \leq m \leq 12, (m, 12) = 2\} = \{2, 10\}$$
$$S_3 = \{m \in \mathbf{Z}: 1 \leq m \leq 12, (m, 12) = 3\} = \{3, 9\}$$
$$S_4 = \{m \in \mathbf{Z}: 1 \leq m \leq 12, (m, 12) = 4\} = \{4, 8\}$$
$$S_6 = \{m \in \mathbf{Z}: 1 \leq m \leq 12, (m, 12) = 6\} = \{6\}$$
$$S_{12} = \{m \in \mathbf{Z}: 1 \leq m \leq 12, (m, 12) = 12\} = \{12\}$$

The number of integers in each S_d above is $\phi(\frac{12}{d})$(verify this!). Since each integer from 1 to 12 inclusive is an element of one and only one S_d, we have

$$\begin{aligned} 12 &= \sum_{d \mid 12,\, d>0} \phi\left(\frac{12}{d}\right) \\ &= \phi(12) + \phi(6) + \phi(4) + \phi(3) + \phi(2) + \phi(1) \\ &= \sum_{d \mid 12,\, d>0} \phi(d) \end{aligned}$$

as desired.

An alternate proof of Theorem 3.5 using Theorem 3.1 is motivated in Exercise 27.

Exercise Set 3.2

9. Disprove (via counterexample) that the Euler phi-function $\phi(n)$ is completely multiplicative.

10. Find $\phi(n)$ for each value n below.

(a) p (p a prime number)
(b) 64
(c) 105
(d) 2592
(e) 4851
(f) 111111
(g) 15!
(h) a googol (a *googol* is a 1 followed by 100 zeros)

11. Complete the proof of Theorem 3.2 by proving that, if m, n, and i are positive integers with $(m, n) = (m, i) = 1$, then the integers i, $m + i$, $2m + i, \ldots, (n - 1)m + i$ form a complete system of residues modulo n.

12. Let $n \in \mathbf{Z}$ with $n > 1$. If $p_1^{a_1}p_2^{a_2} \cdots p_m^{a_m}$ is the prime factorization of n, prove that

$$\phi(n) = p_1^{a_1-1}p_2^{a_2-1}\cdots p_m^{a_m-1}\prod_{i=1}^{m}(p_i - 1)$$

13. Characterize those positive integers n for which each of the following properties holds. (Recall that a characterization requires a necessary and sufficient condition for the positive integer to have the desired property.)
(a) $\phi(n)$ is odd
(b) $\phi(n) = n - 1$
(c) $4 \mid \phi(n)$
(d) $\phi(n) = n/2$
(e) $\phi(n) \mid n$
(f) $\phi(n) = 2^k$ (k a nonnegative integer)

14. *(a)* Prove that there are infinitely many integers n for which $\phi(n) = \frac{n}{3}$.
(b) Prove that there are no integers n for which $\phi(n) = \frac{n}{4}$.

15. Let $k \in \mathbf{Z}$ with $k > 0$. Prove that the equation $\phi(n) = k$ has at most finitely many solutions. [Carmichael conjectured that the equation $\phi(n) = k$ cannot have exactly one solution, but this conjecture remains unproven.]

16. Let n be a positive integer.
(a) Prove that $\sqrt{n}/2 \leq \phi(n) \leq n$.
(b) If n is composite, prove that $\phi(n) \leq n - \sqrt{n}$.

17. Prove that if n is a positive integer, then

$$\phi(2n) = \begin{cases} \phi(n), & \text{if } n \text{ is odd} \\ 2\phi(n), & \text{if } n \text{ is even} \end{cases}$$

18. *(a)* Let n be a positive integer having k distinct odd prime divisors. Prove that $2^k \mid \phi(n)$.
(b) Prove or disprove that $2^k \parallel \phi(n)$ in part (a) above (see Exercise 72 of Chapter 1 for a definition of $\parallel$).

19. Let n and k be positive integers. Prove that $\phi(n^k) = n^{k-1}\phi(n)$.

20. *(a)* Let m and n be positive integers with $m \mid n$. Prove that $\phi(m) \mid \phi(n)$.
(b) Prove or disprove the converse of part (a) above.

21. *(a)* Let m and n be positive integers with $m \mid n$. Prove that $\phi(mn) = m\phi(n)$.
(b) Prove or disprove the converse of part (a) above.

22. *(a)* Let p and $p + 2$ be twin primes. Prove that $\phi(p + 2) = \phi(p) + 2$.
(b) Prove or disprove the converse of part (a) above.

23. Let $n \in \mathbf{Z}$ with $n > 1$. Prove that the sum of all positive integers k with $1 \leq k < n$ and $(k, n) = 1$ is $\frac{1}{2}n\phi(n)$.

24. The following problem is open:

Let $n \in \mathbf{Z}$ with $n > 1$. If $\phi(n) \mid n - 1$, then n is a prime number.

Prove the following weaker result:

Let $n \in \mathbf{Z}$ with $n > 1$. If $\phi(n) \mid n - 1$, then $n = p_1 p_2 \cdots p_k$ with $p_1, p_2, \ldots, p_k$ distinct prime numbers.

25. Let $n \in \mathbf{Z}$ with $n > 0$. Prove that the number of fractions $\frac{a}{b}$ in lowest terms with $0 < \frac{a}{b} \leq 1$ and $b \leq n$ is $\phi(1) + \phi(2) + \cdots + \phi(n)$.

26. Prove that if n is a positive integer, then

$$\sum_{d \mid n,\, d>0} (-1)^{n/d}\phi(d) = \begin{cases} -n, & \text{if } n \text{ is odd} \\ 0, & \text{if } n \text{ is even} \end{cases}$$

27. (The following exercise presents an alternate proof of Theorem 3.5.) Let $n \in \mathbf{Z}$ with $n > 0$ and let

$$f(n) = \sum_{d \mid n,\, d>0} \phi(d)$$

(*a*) Prove that f is a multiplicative arithmetic function. Deduce that f is completely determined by its values at powers of prime numbers. (*Hint*: Use Theorem 3.1.)

(*b*) Let p be a prime number and let $a \in \mathbf{Z}$ with $a \geq 0$. Prove that $f(p^a) = p^a$. Deduce that $f(n) = n$ for all $n \in \mathbf{Z}$ with $n > 0$ and so Theorem 3.5 is proven.

28. ***Definition:*** Let $n \in \mathbf{Z}$ with $n > 0$. If k is a nonnegative integer, then

$$\phi_k(n) = \sum_{1 \leq d \leq n,\, (d,\, n)=1} d^k$$

[Here, the summation is taken over all positive integers d not exceeding n that are relatively prime to n. Note also that $\phi_0(n) = \phi(n)$.] Let $n \in \mathbf{Z}$ with $n > 0$ and let k be a nonnegative integer. Prove that

$$\sum_{d \mid n,\, d>0} \frac{\phi_k(d)}{d^k} = \frac{1^k + 2^k + \cdots + n^k}{n^k}$$

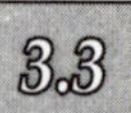

The Number of Positive Divisors Function

We begin by formally defining the arithmetic function $\nu(n)$ first introduced in Example 1(b).

Definition 3: Let $n \in \mathbf{Z}$ with $n > 0$. The *number of positive divisors function*, denoted $\nu(n)$, is the function defined by

$$\nu(n) = |\{d \in \mathbf{Z}: d > 0; d \mid n\}|$$

In other words, $\nu(n)$ is the number of positive divisors of n. [The notation here is chosen by this author for ease of remembrance: ν (lowercase Greek letter nu) represents the "number" of positive divisors. However, the number of positive divisors function is denoted variously by $d(n)$ and $\tau(n)$.]

Values of $\nu(n)$ for integers n with $1 \leq n \leq 100$ are given in Table 2 of Appendix E. We now subject $\nu(n)$ to the same analysis that we performed on $\phi(n)$ in Section 3.2, with the hope of obtaining a formula for $\nu(n)$.

Theorem 3.6: The number of positive divisors function $\nu(n)$ is multiplicative.

Proof: Note that

$$\nu(n) = \sum_{d \mid n,\, d>0} 1$$

Since the constant arithmetic function $f(d) = 1$ is multiplicative (verify this!), we have that $\nu(n)$ is multiplicative by Theorem 3.1. ■

Did you conjecture the result of Theorem 3.6 above when you solved Exercise 1?

You might be wondering why the proof of the multiplicativity of $\nu(n)$ in Theorem 3.6 above is so short compared to the proof of the multiplicativity of $\phi(n)$ in Theorem 3.2. The answer is that $\phi(n)$ has no representation as

$$\sum_{d \mid n,\, d>0} f(d)$$

where $f(d)$ is a multiplicative arithmetic function, so Theorem 3.1 cannot be invoked in the proof of the multiplicativity of $\phi(n)$. The proof of Theorem 3.6 used such a representation of $\nu(n)$ along with Theorem 3.1.

Since the number of positive divisors function is multiplicative, it is completely determined by its values at powers of prime numbers. The next two theorems exploit this fact to obtain a formula for $\nu(n)$.

Theorem 3.7: Let p be a prime number and let $a \in \mathbf{Z}$ with $a \geq 0$. Then $\nu(p^a) = a + 1$.

Proof: The positive divisors of p^a are precisely $1, p, p^2, \ldots, p^{a-1}, p^a$; there are $a + 1$ such divisors. ■

Theorem 3.8: Let $n = p_1^{a_1} p_2^{a_2} \cdots p_r^{a_r}$ with $p_1, p_2, \ldots, p_r$ distinct prime numbers and $a_1, a_2, \ldots, a_r$ nonnegative integers. Then

$$\nu(n) = \prod_{i=1}^{r} (a_i + 1)$$

Proof: We have

$$\begin{aligned}
\nu(n) &= \nu(p_1^{a_1} p_2^{a_2} \cdots p_r^{a_r}) \\
&= \nu(p_1^{a_1})\nu(p_2^{a_2}) \cdots \nu(p_r^{a_r}) \quad \text{(by Theorem 3.6)} \\
&= (a_1 + 1)(a_2 + 1) \cdots (a_r + 1) \quad \text{(by Theorem 3.7)} \\
&= \prod_{i=1}^{r} (a_i + 1) \; ■
\end{aligned}$$

Theorem 3.8 gives the desired formula for $\nu(n)$. Note that this formula requires the exponents of the distinct prime numbers in the prime factorization of n. We illustrate with an example.

Example 5:

We use the formula of Theorem 3.8 to compute $\nu(504)$. We have $504 = 2^3 3^2 7$, so the exponents of the distinct prime numbers dividing 504 are 3, 2, and 1. So

$$\begin{aligned}\nu(504) &= (3 + 1)(2 + 1)(1 + 1)\\ &= 24\end{aligned}$$

and there are exactly 24 positive divisors of 504.

Exercise Set 3.3

29. Disprove (via counterexample) that the number of positive divisors function $\nu(n)$ is completely multiplicative (but see Exercise 39 below).

30. Find $\nu(n)$ for each value n below.
(a) p (p a prime number)
(b) 64
(c) 105
(d) 2592
(e) 4851
(f) 111111
(g) 15!
(h) a googol (a *googol* is a 1 followed by 100 zeros)

31. Characterize those positive integers n for which each of the following properties holds.
(a) $\nu(n) = 1$
(b) $\nu(n) = 2$
(c) $\nu(n) = 3$
(d) $\nu(n) = 4$
(e) $\nu(n) = 15$
(f) $\nu(n) = 20$

32. Characterize those positive integers n for which $\nu(n)$ is odd.

33. There are n cells numbered $1, 2, \ldots, n$ in a linear cell block, each cell containing one inmate. The warden and the inmates agree to perform the following n-day experiment. All cell doors are originally locked. On the first day of the experiment, the warden turns the key in all cells. On the second day of the experiment, the warden turns the key in the cells numbered $2, 4, 6, 8, \ldots$. On the third day of the experiment, the warden turns the key in the cells numbered $3, 6, 9, 12, \ldots$. In general, on the kth day of the experiment ($1 \le k \le n$), the warden turns the key in the cells numbered k, $2k$, $3k$, $4k, \ldots$. The gentlemen's agreement is that no inmate will attempt to escape during the course of the experiment and that an inmate goes free at the conclusion of the experiment if and only if his cell door is unlocked at that time. Determine the cell numbers of those inmates who go free at the conclusion of the experiment. Explain.

34. Let $k \in \mathbf{Z}$ with $k > 1$. Prove that the equation $\nu(n) = k$ has infinitely many solutions.

35. Let $n = p_1 p_2 \cdots p_m$ where $p_1, p_2, \ldots, p_m$ are distinct prime numbers. Prove that $\nu(n) = 2^m$.

36. Let $n \in \mathbf{Z}$ with $n > 0$. Find a formula for the product of the positive divisors of n involving $\nu(n)$.

37. Let $n \in \mathbf{Z}$ with $n > 0$. Prove that $\nu(n) \leq 2\sqrt{n}$.

38. Let $n \in \mathbf{Z}$ with $n > 0$. Prove that $\nu(n) \leq \nu(2^n - 1)$.

39. Let m and n be positive integers. Prove that $\nu(mn) \leq \nu(m)\nu(n)$. [*Hint*: It suffices to prove the inequality for powers of prime numbers. (Why?)]

40. Let $n \in \mathbf{Z}$ with $n > 0$. Prove that

$$\left(\sum_{d \mid n,\, d>0} \nu(d)\right)^2 = \sum_{d \mid n,\, d>0} (\nu(d))^3$$

[*Hint*: It suffices to prove the equation above for powers of prime numbers. (Why?)]

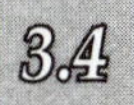

The Sum of Positive Divisors Function

We begin by formally defining the arithmetic function $\sigma(n)$ first introduced in Example 1(c).

Definition 4: Let $n \in \mathbf{Z}$ with $n > 0$. The *sum of positive divisors function*, denoted $\sigma(n)$, is the function defined by

$$\sigma(n) = \sum_{d \mid n,\, d>0} d$$

In other words, $\sigma(n)$ is the sum of the positive divisors of n. [As with $\nu(n)$, the notation is chosen for ease of remembrance: σ (lowercase Greek letter sigma) represents the "sum" of positive divisors.]

Values of $\sigma(n)$ for integers n with $1 \leq n \leq 100$ are given in Table 2 of Appendix E. We now subject $\sigma(n)$ to the same analysis that we performed on $\phi(n)$ in Section 3.2 and $\nu(n)$ in Section 3.3, with the hope of obtaining a formula for $\sigma(n)$.

Theorem 3.9: The sum of positive divisors function $\sigma(n)$ is multiplicative.

Proof: Since the arithmetic function $f(d) = d$ is multiplicative (verify this!), we have that $\sigma(n)$ is multiplicative by Theorem 3.1. ■

Did you conjecture the result of Theorem 3.9 above when you solved Exercise 1?

Since the sum of positive divisors function is multiplicative, it is completely determined by its values at powers of prime numbers. The next two theorems exploit this fact to obtain a formula for $\sigma(n)$.

Theorem 3.10: Let p be a prime number and let $a \in \mathbf{Z}$ with $a \geq 0$. Then

$$\sigma(p^a) = \frac{p^{a+1} - 1}{p - 1}$$

Proof: The positive divisors of p^a are precisely $1, p, p^2, \ldots, p^{a-1}, p^a$; the sum of these divisors is $(p^{a+1} - 1)/(p - 1)$ from the formula for the sum of the first $a + 1$ terms of a geometric series with ratio p. ■

Theorem 3.11: Let $n = p_1^{a_1} p_2^{a_2} \cdots p_r^{a_r}$ with $p_1, p_2, \ldots, p_r$ distinct prime numbers and $a_1, a_2, \ldots, a_r$ nonnegative integers. Then

$$\sigma(n) = \prod_{i=1}^{r} \frac{p_i^{a_i+1} - 1}{p_i - 1}$$

Proof: We have

$$\begin{aligned}
\sigma(n) &= \sigma(p_1^{a_1} p_2^{a_2} \cdots p_r^{a_r}) \\
&= \sigma(p_1^{a_1})\sigma(p_2^{a_2}) \cdots \sigma(p_r^{a_r}) \quad \text{(by Theorem 3.9)} \\
&= \left(\frac{p_1^{a_1+1} - 1}{p_1 - 1}\right)\left(\frac{p_2^{a_2+1} - 1}{p_2 - 1}\right) \cdots \left(\frac{p_r^{a_r+1} - 1}{p_r - 1}\right) \quad \text{(by Theorem 3.10)} \\
&= \prod_{i=1}^{r} \frac{p_i^{a_i+1} - 1}{p_i - 1} \quad ■
\end{aligned}$$

Theorem 3.11 gives the desired formula for $\sigma(n)$. Note that this formula requires the prime factorization of n. We illustrate with an example.

Example 6:

We use the formula of Theorem 3.11 to compute $\sigma(504)$. We have $504 = 2^3 3^2 7$. So

$$\begin{aligned}
\sigma(504) &= \left(\frac{2^4 - 1}{2 - 1}\right)\left(\frac{3^3 - 1}{3 - 1}\right)\left(\frac{7^2 - 1}{7 - 1}\right) \\
&= 1560
\end{aligned}$$

and 1560 is the sum of the positive divisors of 504.

Exercise Set 3.4

41. Disprove (via counterexample) that the sum of positive divisors function $\sigma(n)$ is completely multiplicative.

42. Find $\sigma(n)$ for each value n below.
 (a) p (p a prime number)
 (b) 64
 (c) 105
 (d) 2592
 (e) 4851
 (f) 111111
 (g) 15!
 (h) a googol (a *googol* is a 1 followed by 100 zeros)

43. ***Definition:*** Let m and n be distinct positive integers. Then m and n are said to be *amicable* if

$$\sigma(m) = \sigma(n) = m + n$$

Prove that 220 and 284 are amicable. (In fact, 220 and 284 form the smallest amicable pair of positive integers. It is believed that there are infinitely many amicable pairs of positive integers, but this possibility remains unproven.)

44. Characterize those positive integers n for which $\sigma(n)$ is odd.

45. Let $k \in \mathbf{Z}$ with $k > 0$. Prove that the equation $\sigma(n) = k$ has at most finitely many solutions.

46. Let $n \in \mathbf{Z}$ with $n > 0$. Prove that $n \leq \sigma(n) \leq n^2$.

47. Let $n \in \mathbf{Z}$ with $n > 0$. Prove that $\displaystyle\sum_{d \mid n,\, d>0} \frac{1}{d} = \frac{\sigma(n)}{n}$.

48. Let $n \in \mathbf{Z}$ with $n > 0$. Prove that $\displaystyle\frac{\sigma(n!)}{n!} \geq \sum_{i=1}^{n} \frac{1}{i}$.

49. Let $n \in \mathbf{Z}$ with $n > 0$. Prove that n is composite if and only if $\sigma(n) > n + \sqrt{n}$.

50. *(a)* Let p and $p + 2$ be twin primes. Prove that $\sigma(p + 2) = \sigma(p) + 2$.
 (b) Prove or disprove the converse of part (a) above. (*Hint*: Examine 434.)

51. ***Definition:*** Let $n \in \mathbf{Z}$ with $n > 0$. If k is a positive integer, then

$$\sigma_k(n) = \sum_{d \mid n,\, d>0} d^k$$

[Note that $\sigma_1(n) = \sigma(n)$.]
 (a) Find $\sigma_3(12)$ and $\sigma_4(8)$.
 (b) Prove that $\sigma_k(n)$ is multiplicative.
 (c) Let p be a prime number and let $a \in \mathbf{Z}$ with $a > 0$. Find a formula for $\sigma_k(p^a)$.
 (d) Let $n = p_1^{a_1} p_2^{a_2} \cdots p_r^{a_r}$ with $p_1, p_2, \ldots, p_r$ distinct prime numbers and $a_1, a_2, \ldots, a_r$ nonnegative integers. Use the results of parts (b) and (c) above to find a formula for $\sigma_k(n)$.

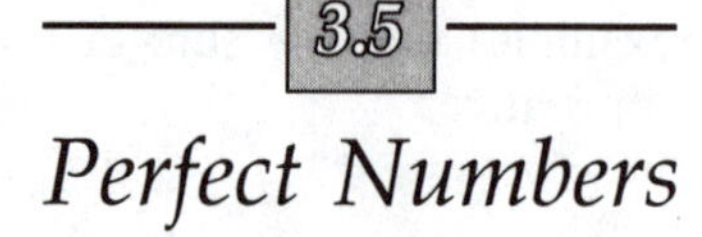

Perfect Numbers

Perfect numbers are positive integers that satisfy a special relationship involving the sum of positive divisors function $\sigma(n)$. We provide two equivalent definitions of perfect numbers.

Definition 5: Let $n \in \mathbf{Z}$ with $n > 0$. Then n is said to be a *perfect number* if $\sigma(n) = 2n$.

If $n \in \mathbf{Z}$ with $n > 0$, then $\sigma(n) - n$ is the sum of all positive divisors of n other than n itself. It is clear that $\sigma(n) = 2n$ if and only if $\sigma(n) - n = n$. Consequently, Definition 5 is equivalent to the following definition.

Definition 5′: Let $n \in \mathbf{Z}$ with $n > 0$. Then n is said to be a *perfect number* if $\sigma(n) - n = n$ or, equivalently, if the sum of all positive divisors of n other than n itself is equal to n.

We now illustrate both definitions with examples.

Example 7:

(a) 6 is a perfect number since $\sigma(6) = 12 = 2(6)$ (please verify!). Equivalently, the sum of all positive divisors of 6 other than 6 is 6.

(b) 28 is a perfect number since $\sigma(28) = 56 = 2(28)$ (please verify!). Equivalently, the sum of all positive divisors of 28 other than 28 is 28.

The concept of a perfect number dates back to classical Greek civilization when mystical qualities and religious significance were assigned to such numbers. Only the four smallest perfect numbers were known to the ancient Greeks. It is a fact that the perfect numbers 6 and 28 of Example 7 above are the two smallest perfect numbers. The (extremely dedicated) reader may wish to find the next two perfect numbers. (The answer is given in Example 8.) As an aid in your search, it should be stated that 6 and 28 are the only perfect numbers less than 400. Are you still game?

Even perfect numbers have a nice characterization as given by Theorem 3.12 below. In fact, there is a one-to-one correspondence between Mersenne primes and even perfect numbers, with the existence of one giving rise to the existence of the other. The "if" part of the characterization below was proved by Euclid in Book IX of his *Elements*. The "only if" part of this characterization would have to wait for Euler, who proved it approximately 2000 years later.

Theorem 3.12: Let $n \in \mathbf{Z}$ with $n > 0$. Then n is an even perfect number if and only if

$$n = 2^{p-1}(2^p - 1)$$

where $2^p - 1$ is a prime number. [Note that, since $2^p - 1$ is a prime number, then p is a prime number by Exercise 26(b) of Chapter 1 so that $2^p - 1$ is indeed a Mersenne prime.]

Proof: ($\Rightarrow$) (Euler) Assume that n is an even perfect number. Write $n = 2^a b$ with $a, b \in \mathbf{Z}$ and b odd. Then

$$\sigma(n) = \sigma(2^a b) = \sigma(2^a)\sigma(b) = \left(\frac{2^{a+1} - 1}{2 - 1}\right)\sigma(b) = (2^{a+1} - 1)\sigma(b)$$

Also, since n is perfect, we have

$$\sigma(n) = 2n = 2(2^a b) = 2^{a+1} b$$

So

$$(2^{a+1} - 1)\sigma(b) = 2^{a+1} b \qquad \textbf{(1)}$$

Now $2^{a+1} \mid (2^{a+1} - 1)\sigma(b)$; since 2^{a+1} and $2^{a+1} - 1$ are relatively prime, we have $2^{a+1} \mid \sigma(b)$ or

$$\sigma(b) = 2^{a+1} c \qquad \textbf{(2)}$$

for some $c \in \mathbf{Z}$. Substituting for $\sigma(b)$ in (1), we obtain

$$(2^{a+1} - 1)2^{a+1} c = 2^{a+1} b$$

or

$$(2^{a+1} - 1)c = b \qquad \textbf{(3)}$$

We now prove that $c = 1$. Assume by way of contradiction that $c > 1$. Then, by (3), b has at least three distinct positive divisors, namely, 1, b, and c. So $\sigma(b) \geq 1 + b + c$. But

$$\sigma(b) = 2^{a+1} c = (2^{a+1} - 1)c + c = b + c$$

This is the desired contradiction. So $c = 1$, from which (3) gives $b = 2^{a+1} - 1$ and (2) gives $\sigma(b) = 2^{a+1} = b + 1$. The latter equation implies that b is a prime number, from which

$$n = 2^a b = 2^a(2^{a+1} - 1)$$

where $2^{a+1} - 1$ is a prime number as desired.
($\Leftarrow$) (Euclid) Assume that $n = 2^{p-1}(2^p - 1)$ where p and $2^p - 1$ are prime

numbers. Then

$$\begin{aligned}\sigma(n) &= \sigma(2^{p-1}(2^p - 1)) \\ &= \sigma(2^{p-1})\sigma(2^p - 1) \\ &= \left(\frac{2^p - 1}{2 - 1}\right)2^p \\ &= (2^p - 1)2^p \\ &= 2^p(2^p - 1) \\ &= 2(2^{p-1}(2^p - 1)) \\ &= 2n \ \blacksquare\end{aligned}$$

We may now use the characterization of Theorem 3.12 above to generate even perfect numbers as illustrated by the example below.

Example 8:

The first five even perfect numbers are given by 6, 28, 496, 8128, and 33550336 [corresponding to p equal to 2, 3, 5, 7, and 13 in $2^{p-1}(2^p - 1)$, respectively; see Example 9 of Chapter 1]. The first four of these perfect numbers were known to the ancient Greeks.

There are currently 38 known even perfect numbers, one corresponding to each known Mersenne prime. Since the largest known Mersenne prime is $2^{6972593} - 1$, we have that the largest known even perfect number is $2^{6972592}(2^{6972593} - 1)$, a number containing in excess of 4000000 digits. Inasmuch as mathematicians believe that there are infinitely many Mersenne primes (see Conjecture 3 in Chapter 1), we state the following.

Conjecture 1: There are infinitely many perfect numbers.

All of this talk about *even* perfect numbers has probably caused some readers to ask some rather natural questions. What about *odd* perfect numbers? What are some examples of these numbers? Is there a characterization similar to Theorem 3.12 for odd perfect numbers? Curiously enough, no odd perfect numbers are known. In fact, the existence or nonexistence of odd perfect numbers is one of the more famous unsolved problems in number theory. It is known that no odd perfect number exists below 10^{300}. Furthermore, any odd perfect number must be divisible by at least eight distinct prime numbers with the largest prime number being greater than 1000000 and the next largest prime number being greater than 10000. Our frustration with odd perfect numbers leads us to the following conjecture with which we conclude this section.

Conjecture 2: Every perfect number is even.

Exercise Set 3.5

52. Find the next two even perfect numbers after 33550336.

53. Let n be a perfect number. Prove that $\sum_{d \mid n,\, d>0} \frac{1}{d} = 2$. (*Hint*: This exercise could be a corollary of an exercise in the previous section.)

54. Prove or disprove each of the following statements.

(a) The known even perfect numbers, from smallest to largest, alternately end in 6 and 8.

(b) Every even perfect number ends in either 6 or 8.

55. Let $n_1, n_2, \ldots, n_m$ be distinct even perfect numbers. Prove that

$$\phi(n_1 n_2 \cdots n_m) = 2^{m-1}\phi(n_1)\phi(n_2)\cdots\phi(n_m)$$

56. ***Definition:*** Let $n \in \mathbf{Z}$ with $n > 0$. Then n is said to be *deficient* if $\sigma(n) < 2n$ and n is said to be *abundant* if $\sigma(n) > 2n$. (Note that every positive integer is either deficient, perfect, or abundant.)

(a) Label each of the first 20 positive integers as deficient, perfect, or abundant.

(b) Prove that every power of a prime number is deficient. Conclude that there are infinitely many deficient numbers.

(c) Prove that the product of two distinct odd prime numbers is deficient.

(d) Let n be a positive integer. Prove that if $2^n - 1$ is composite, then $2^{n-1}(2^n - 1)$ is abundant. Conclude that there are infinitely many abundant numbers.

(e) Prove that any positive divisor of a deficient number or perfect number other than the number itself is deficient.

(f) Prove that any positive multiple of an abundant number or perfect number other than the number itself is abundant.

57. ***Definition:*** Let $n \in \mathbf{Z}$ with $n > 1$. Then n is said to be *almost perfect* if $\sigma(n) = 2n - 1$.
Let $k \in \mathbf{Z}$ with $k > 0$. Prove that any integer of the form 2^k is almost perfect. (*Note*: The only known almost perfect numbers are numbers of this form.)

58. ***Definition:*** Let $k, n \in \mathbf{Z}$ with $k, n > 1$. Then n is said to be *k-perfect* if $\sigma(n) = kn$. (Note that a 2-perfect number is perfect in the usual sense.)

(a) Prove that 120 is 3-perfect.

(b) Find all 3-perfect numbers of the form $2^k 3p$ where p is an odd prime number.

(c) Let $n \in \mathbf{Z}$ with $n > 1$ and let p be a prime number not dividing n. Prove that if n is p-perfect, then pn is $(p + 1)$-perfect.

59. ***Definition:*** Let $n \in \mathbf{Z}$ with $n > 0$. Then n is said to be *superperfect* if $\sigma(\sigma(n)) = 2n$.

(a) Prove that 16 is superperfect.

(b) Prove that if $2^p - 1$ is a Mersenne prime, then 2^{p-1} is superperfect.

(c) Prove that if 2^a is superperfect, then $2^{a+1} - 1$ is a Mersenne prime. (*Note*: This statement is simply the converse of the statement in part (b) above. It is a fact that all even superperfect numbers are of the form 2^{p-1} where $2^p - 1$ is a Mersenne prime. No odd superperfect numbers are known.)

60. ***Definition:*** Let $n \in \mathbf{Z}$ with $n > 0$. Then n is said to be *multiplicatively perfect* if

$$\prod_{d \mid n,\, d>0} d = n^2$$

Characterize all multiplicatively perfect integers.

61. Prove that if n is an odd perfect number, then $n = p^a m$ where p, a, and m are positive integers with p an odd prime number not dividing m, $p \equiv a \equiv 1 \bmod 4$, and m a perfect square. Conclude that any odd perfect number is congruent to 1 modulo 4.

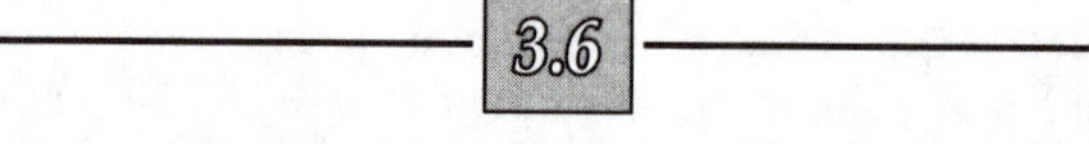

The Möbius Inversion Formula

The Möbius Inversion Formula can be used to obtain nontrivial identities among arithmetic functions from trivial identities among arithmetic functions. We first need to introduce a new arithmetic function that essentially gives the Möbius Inversion Formula its name.

Definition 6: Let $n \in \mathbf{Z}$ with $n > 0$. The *Möbius μ-function,* denoted $\mu(n)$, is

$$\mu(n) = \begin{cases} 1, & \text{if } n = 1 \\ 0, & \text{if } p^2 \mid n \text{ with } p \text{ prime} \\ (-1)^r, & \text{if } n = p_1 p_2 \cdots p_r \text{ with } p_1, p_2, \ldots, p_r \text{ distinct prime numbers} \end{cases}$$

Example 9:

(a) Since $504 = 2^3 3^2 7$, we have that two squares of prime numbers divide 504 (namely, 2^2 and 3^2); so $\mu(504) = 0$.

(b) Since $30 = 2 \cdot 3 \cdot 5$, we have that 30 is the product of three distinct prime numbers; so $\mu(30) = (-1)^3 = -1$.

Values of $\mu(n)$ for integers n with $1 \le n \le 100$ are given in Table 2 of Appendix E.

We now introduce two important properties of the Möbius μ-function.

Theorem 3.13: The Möbius μ-function $\mu(n)$ is multiplicative.

Proof: Let m and n be relatively prime positive integers. We must show that $\mu(mn) = \mu(m)\mu(n)$. If $m = 1$ or $n = 1$, the desired result is clear. Now m or n is divisible by the square of a prime number if and only if mn is divisible by the square of a prime number; in this case, both sides of the desired result reduce to zero. Assume that $m = p_1 p_2 \cdots p_r$ and $n = q_1 q_2 \cdots q_s$ with

$p_1, p_2, \ldots, p_r, q_1, q_2, \ldots, q_s$ distinct prime numbers. Then

$$\begin{aligned} \mu(mn) &= \mu(p_1 p_2 \cdots p_r q_1 q_2 \cdots q_s) \\ &= (-1)^{r+s} \\ &= (-1)^r(-1)^s \\ &= \mu(m)\mu(n) \blacksquare \end{aligned}$$

Proposition 3.14: Let $n \in \mathbf{Z}$ with $n > 0$. Then

$$\sum_{d \mid n,\, d>0} \mu(d) = \begin{cases} 1, & \text{if } n = 1 \\ 0, & \text{otherwise} \end{cases}$$

Proof: The arithmetic function

$$F(n) = \sum_{d \mid n,\, d>0} \mu(d)$$

is multiplicative by Theorem 3.13 and Theorem 3.1. So F is completely determined by its values at powers of prime numbers. Clearly, $F(1) = 1$ as desired. If p is a prime number and $a \in \mathbf{Z}$ with $a > 0$, then

$$\begin{aligned} F(p^a) &= \sum_{d \mid p^a,\, d>0} \mu(d) \\ &= \mu(1) + \mu(p) + \mu(p^2) + \cdots + \mu(p^{a-1}) + \mu(p^a) \\ &= 1 + (-1) + 0 + \cdots + 0 + 0 \\ &= 0 \end{aligned}$$

as desired. ■

A quick illustration perhaps clarifies the statement of Proposition 3.14.

Example 10:

Let $n = 12$ in the statement of Proposition 3.14. Then we have

$$\begin{aligned} \sum_{d \mid 12,\, d>0} \mu(d) &= \mu(1) + \mu(2) + \mu(3) + \mu(4) + \mu(6) + \mu(12) \\ &= 1 + (-1) + (-1) + 0 + 1 + 0 \\ &= 0 \end{aligned}$$

as desired.

We may now state and prove the Möbius Inversion Formula. Two prior remarks, however, are in order here. First, the Möbius Inversion Formula has a natural interpretation from the viewpoint of *convolutions*; this viewpoint is discussed in Exercise 69 and results in an almost trivial proof of the formula. (See also Student Project 7.) Second, the proof given below, while not relying on any advanced concepts, does require a bit of thought for full understanding.

In particular, the statements accompanied by "why?" are crucial steps to the proof; make sure that you carefully justify these steps. (A concrete example may help—construct your own!) On with the Möbius Inversion Formula!

Theorem 3.15: (Möbius Inversion Formula) Let f and g be arithmetic functions. Then

$$f(n) = \sum_{d \mid n,\, d>0} g(d)$$

if and only if

$$g(n) = \sum_{d \mid n,\, d>0} \mu(d) f\left(\frac{n}{d}\right) = \sum_{d \mid n,\, d>0} \mu\left(\frac{n}{d}\right) f(d)$$

Proof: We remark that the second equality on the right-hand side of the "if and only if" statement above is clear, since as d runs through all positive divisors of n, so does $\frac{n}{d}$.

($\Rightarrow$) Assume that

$$f(n) = \sum_{d \mid n,\, d>0} g(d)$$

Then

$$\sum_{d \mid n,\, d>0} \mu(d) f\left(\frac{n}{d}\right) = \sum_{d \mid n,\, d>0} \left(\mu(d) \sum_{c \mid n/d,\, c>0} g(c)\right)$$

$$= \sum_{c \mid n,\, c>0} \left(g(c) \sum_{d \mid n/c,\, d>0} \mu(d)\right) \quad \text{(why?)} \qquad \textbf{(4)}$$

Now, by Proposition 3.14, the summation inside the parentheses in (4) is 0 unless $\frac{n}{c} = 1$ or, equivalently, $c = n$; so the only contribution of the outer summation of (4) is when $c = n$, giving $g(n)$ as desired.

($\Leftarrow$) Assume that

$$g(n) = \sum_{d \mid n,\, d>0} \mu\left(\frac{n}{d}\right) f(d)$$

Then

$$\sum_{d \mid n,\, d>0} g(d) = \sum_{d \mid n,\, d>0} \left(\sum_{c \mid d,\, c>0} \mu\left(\frac{d}{c}\right) f(c)\right)$$

$$= \sum_{c \mid n,\, c>0} \left(f(c) \sum_{d=cm \mid n,\, d>0} \mu\left(\frac{d}{c}\right)\right) \quad \text{(why?)}$$

$$= \sum_{c \mid n,\, c>0} \left(f(c) \sum_{m \mid n/c,\, m>0} \mu(m)\right) \quad \text{(why?)} \qquad \textbf{(5)}$$

Now, by Proposition 3.14, the summation inside the parentheses in (5) is 0 unless $\frac{n}{c} = 1$ or, equivalently, $c = n$; so the only contribution of the outer summation of (5) is when $c = n$, giving $f(n)$ as desired. ■

As stated before, Theorem 3.15 can be used to obtain nontrivial identities among arithmetic functions from trivial identities among arithmetic functions. We illustrate with three examples.

Example 11:

Let $n \in \mathbf{Z}$ with $n > 0$ and let $g(n) = n$ for all such n. We have

$$g(n) = n = \sum_{d \mid n,\, d>0} \phi(d)$$

by Theorem 3.5. By the Möbius Inversion Formula, we obtain the nontrivial identity

$$\phi(n) = \sum_{d \mid n,\, d>0} \mu(d) g\left(\frac{n}{d}\right) = \sum_{d \mid n,\, d>0} \mu\left(\frac{n}{d}\right) g(d)$$

or, equivalently,

$$\phi(n) = \sum_{d \mid n,\, d>0} \mu(d) \frac{n}{d} = \sum_{d \mid n,\, d>0} \mu\left(\frac{n}{d}\right) d$$

Example 12:

Let $n \in \mathbf{Z}$ with $n > 0$. We have

$$\nu(n) = \sum_{d \mid n,\, d>0} 1 = \sum_{d \mid n,\, d>0} g(d)$$

where $g(d) = 1$ for $d \in \mathbf{Z}$ with $d > 0$. By the Möbius Inversion Formula, we obtain the nontrivial identity

$$1 = g(n) = \sum_{d \mid n,\, d>0} \mu(d) \nu\left(\frac{n}{d}\right) = \sum_{d \mid n,\, d>0} \mu\left(\frac{n}{d}\right) \nu(d)$$

Example 13:

Let $n \in \mathbf{Z}$ with $n > 0$. We have

$$\sigma(n) = \sum_{d \mid n,\, d>0} d = \sum_{d \mid n,\, d>0} g(d)$$

where $g(d) = d$ for $d \in \mathbf{Z}$ with $d > 0$. By the Möbius Inversion Formula, we obtain the nontrivial identity

$$n = g(n) = \sum_{d \mid n,\, d>0} \mu(d) \sigma\left(\frac{n}{d}\right) = \sum_{d \mid n,\, d>0} \mu\left(\frac{n}{d}\right) \sigma(d)$$

Exercise Set 3.6

62. Verify the nontrivial identities of Examples 11–13 with numerical examples.

63. Let f and g be arithmetic functions such that $f(n) = \sum_{d \mid n,\, d>0} g(d)$. Find an expression for $g(12)$ in terms of f.

64. Let $n \in \mathbf{Z}$ with $n > 0$ and let $\omega(n)$ denote the number of distinct prime numbers dividing n. Prove that

$$\sum_{d \mid n,\, d>0} |\mu(d)| = 2^{\omega(n)}$$

65. ***Definition:*** Let $n \in \mathbf{Z}$ with $n > 0$. *Von Mangoldt's function,* denoted $\Lambda(n)$, is

$$\Lambda(n) = \begin{cases} \ln p, & \text{if } n = p^a \text{ for some } a \in \mathbf{Z} \\ 0, & \text{otherwise} \end{cases}$$

Prove that

$$\Lambda(n) = -\sum_{d \mid n,\, d>0} \mu(d) \ln d$$

66. Let $n \in \mathbf{Z}$ with $n > 1$. If f is a multiplicative arithmetic function and $p_1^{a_1}p_2^{a_2} \cdots p_m^{a_m}$ is the prime factorization of n, prove that

$$\sum_{d \mid n,\, d>0} \mu(d)f(d) = \prod_{i=1}^{m} (1 - f(p_i))$$

67. Let $n \in \mathbf{Z}$ with $n > 1$. Use Exercise 66 above to obtain formulas for each sum below in terms of the prime factorization of n. [See Exercise 7 for a definition of the λ-function in part (d).]

(a) $\sum_{d \mid n,\, d>0} \mu(d)\phi(d)$

(b) $\sum_{d \mid n,\, d>0} \mu(d)\nu(d)$

(c) $\sum_{d \mid n,\, d>0} \mu(d)\sigma(d)$

(d) $\sum_{d \mid n,\, d>0} \mu(d)\lambda(d)$

68. Let f be an arithmetic function and, for $n \in \mathbf{Z}$ with $n > 0$, let

$$F(n) = \sum_{d \mid n,\, d>0} f(d)$$

Prove that if F is multiplicative, then f is multiplicative. (Note that this is the converse of Theorem 3.1.)

69. (The following exercise presents the interpretation of the Möbius Inversion Formula from the viewpoint of convolutions. See also Student Project 7.) ***Definition:*** Let $n \in \mathbf{Z}$ with $n > 0$ and let f and g be arithmetic functions. The convolution of f and g, denoted $f*g$, is

$$(f*g)(n) = \sum_{d \mid n,\, d>0} f(d)g\left(\frac{n}{d}\right)$$

(a) Let f and g be arithmetic functions. Prove that $f*g = g*f$. (So the convolution of arithmetic functions is commutative.)

(b) Let f, g, and h be arithmetic functions. Prove that $(f*g)*h = f*(g*h)$. (So the convolution of arithmetic functions is associative.)

(c) For $n \in \mathbf{Z}$ with $n > 0$, let

$$\delta(n) = \begin{cases} 1, & \text{if } n = 1 \\ 0, & \text{if } n > 1 \end{cases}$$

Let f be an arithmetic function. Prove that $f*\delta = \delta*f = f$. (So δ is the identity element under convolution of arithmetic functions.)

(d) For $n \in \mathbf{Z}$ with $n > 0$, let $\mathbf{1}(n) = 1$ so that $\mathbf{1}$ is the arithmetic function mapping every positive integer to 1. Prove that $\mu*\mathbf{1} = \mathbf{1}*\mu = \delta$. (So the Möbius μ-function and $\mathbf{1}$ are inverses to each other under convolution of arithmetic functions.)

(e) Prove the Möbius Inversion Formula from the viewpoint of convolutions. (*Hint*: From the viewpoint of convolutions, show that Theorem 3.15 can be restated as $f = g*\mathbf{1}$ if and only if $g = \mu*f = f*\mu$.)

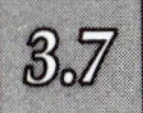

Concluding Remarks

We have encountered many arithmetic functions in this chapter, both in the exposition and in the exercises. We have also encountered several open problems involving arithmetic functions; consult, for example, Exercises 24, 43, 57, and 59 as well as Conjectures 1 and 2. Open problems involving arithmetic functions abound and are usually quite difficult to prove. Some progress is being made, however. For example, an open problem until quite recently was conjectured by Franz Mertens (1840–1927) in 1897 and was thus known as Mertens' Conjecture:

If n is a positive integer, then

$$\left|\sum_{i=1}^{n} \mu(i)\right| < \sqrt{n}$$

Prior to 1984, a computer had verified Mertens' Conjecture for all integers less than 10^9. Then, in 1984, Mertens' Conjecture was disproved by Andrew Odlyzko and Herman te Riele, who proved that the conjecture must be false for some n. Unfortunately, the method of proof was indirect and no specific n was produced; the exact value of such an n is unknown today. Note how the solution of open problems can generate new open problems! An excellent compilation of unsolved problems in number theory, many of which involve the main arithmetic functions ϕ, v, and σ of this chapter, may be found in Guy (1981).

Student Projects

1. (Programming project for calculator or computer)
Given a positive integer n, compute $\phi(n)$, $v(n)$, and $\sigma(n)$.

2. The answer to the following question is unknown:
Are there infinitely many pairs of positive integers m and n such that $\phi(m) = \sigma(n)$?
   (a) Prove that $\phi(m) = \sigma(n)$ for $m = 780$ amd $n = 105$.
   (b) Prove that the truth of the Twin Prime Conjecture (Conjecture 1 of Chapter 1) would affirmatively answer the open question above.
   (c) Prove that the truth of Conjecture 3 of Chapter 1 (the infinitude of the Mersenne primes) would affirmatively answer the open question above.
3. Characterize all positive integers n for which $\phi(n) + \sigma(n) = n\nu(n)$.
4. ***Definition:*** Let n be a positive integer. The *abundancy index* of n, denoted $I(n)$, is

$$I(n) = \frac{\sigma(n)}{n}$$

   (a) Let n be a positive integer. Prove that n is a perfect number if and only if $I(n) = 2$.
   (b) Let n be a positive integer. Prove that n is a deficient number if and only if $I(n) < 2$. (See Exercise 56 for a definition of *deficient.*)
   (c) Let n be a positive integer. Prove that n is an abundant number if and only if $I(n) > 2$. (See Exercise 56 for a definition of *abundant.*)
   (d) Let $S = \{I(n): n \in \mathbf{Z}, n > 1\}$. Prove that the greatest lower bound of S is 1. (The *greatest lower bound* of a set S of real numbers is a real number α such that $\alpha \leq s$ for all $s \in S$ and such that α is the greatest number with this property.)
   (e) Let $S = \{I(n): n \in \mathbf{Z}, n > 0\}$. Prove that the least upper bound of S does not exist. (The *least upper bound* of a set S of real numbers is a real number β such that $s \leq \beta$ for all $s \in S$ and such that β is the least number with this property.)

[A discussion of these and related problems may be found in Richard Laatsch, "Measuring the Abundancy of Integers," *Mathematics Magazine, 59* (1986), 84–92.]

5. Let n be a positive integer and let $f(n)$ be the product of all positive divisors of n. [For example, $f(3) = 3$ and $f(4) = 8$.] Does $f(m) = f(n)$ imply $m = n$? (This problem appears as Elementary Problem E2946 in the May 1982 edition of the *American Mathematical Monthly.* A solution appears in the magazine's December 1984 edition.)
6. The following conjecture is false:

   For every even positive integer n, there is an odd positive integer k such that $\phi(k) = \phi(n)$.

   Prove that $n = 33817088$ is the minimal counterexample to the conjecture. (This problem is discussed in the May 1991 edition of the *American Mathematical Monthly.*)
7. Read the following article: S. K. Berberian, "Number-Theoretic Functions via Convolution Rings," *Mathematics Magazine, 65* (1992), 75–90.

4

Quadratic Residues

The study of the solvability of linear congruences in one variable was undertaken in Chapter 2. You will recall that such congruences take the general form $ax \equiv b \bmod m$. This chapter is devoted to the study of the solvability of quadratic congruences in one variable, the general form of which is $ax^2 + bx \equiv c \bmod m$. Such congruences were studied extensively by Gauss.

Most discussions in this chapter will focus on quadratic congruences that take the special form $x^2 \equiv a \bmod p$ where p is an odd prime number. A symbol introduced in 1798 by Adrien-Marie Legendre will conveniently deal with solutions and nonsolutions of this congruence. Further analyses of these quadratic congruences and their solvability will culminate in one of the most beautiful theorems in elementary number theory, the quadratic reciprocity law, which was first proved by Gauss in *Disquisitiones Arithmeticae.*

4.1

Quadratic Residues

If $a, b, c \in \mathbf{Z}$, the general quadratic congruence in the variable x takes the form $ax^2 + bx \equiv c \bmod m$. It can be shown (see Exercise 9) that finding a solution of this general quadratic congruence is equivalent in certain cases to finding a solution of a simpler quadratic congruence of the form $x^2 \equiv a \bmod m$. We begin by looking more closely at the solutions of these latter congruences.

Definition 1: Let $a, m \in \mathbf{Z}$ with $m > 0$ and $(a, m) = 1$. Then a is said to be a *quadratic residue modulo m* if the quadratic congruence $x^2 \equiv a \bmod m$ is solvable in $\mathbf{Z}$; otherwise, a is said to be a *quadratic nonresidue modulo m.*

We will be interested primarily in quadratic residues and nonresidues modulo odd prime numbers. The solvability of $x^2 \equiv a \bmod m$ for general m is treated in Exercises 5 and 11. We now give an example of the computation of quadratic residues and nonresidues.

Example 1:

Find all incongruent quadratic residues and nonresidues modulo 11.

We are searching for all a-values for which the congruence $x^2 \equiv a \bmod 11$ is solvable. Modulo 11, the only possible values for x are $0, 1, 2, \ldots, 10$; so, we simply insert each x value into the congruence to find all possible quadratic residues modulo 11. Those a-values that do not appear as quadratic residues must then be quadratic nonresidues. We have

$$0^2 \equiv 0 \bmod 11$$

Unfortunately, zero is neither a quadratic residue nor a quadratic nonresidue modulo 11 since $(0, 11) \neq 1$ (see Definition 1). [In fact, the condition $(a, m) = 1$ in Definition 1 excludes zero from being a quadratic residue unless the modulus m happens to be 1 (a rather uninteresting case).] Continuing:

$1^2 \equiv 1 \bmod 11$ implies that 1 is a quadratic residue modulo 11;

$2^2 \equiv 4 \bmod 11$ implies that 4 is a quadratic residue modulo 11;

$3^2 \equiv 9 \bmod 11$ implies that 9 is a quadratic residue modulo 11;

$4^2 \equiv 16 \equiv 5 \bmod 11$ implies that 5 is a quadratic residue modulo 11;

$5^2 \equiv 25 \equiv 3 \bmod 11$ implies that 3 is a quadratic residue modulo 11;

$6^2 \equiv 36 \equiv 3 \bmod 11$ implies that 3 is a quadratic residue modulo 11 (but we already knew this!);

$7^2 \equiv 49 \equiv 5 \bmod 11$ implies that 5 is a quadratic residue modulo 11 (already known);

$8^2 \equiv 64 \equiv 9 \bmod 11$ implies that 9 is a quadratic residue modulo 11 (already known);

$9^2 \equiv 81 \equiv 4 \bmod 11$ implies that 4 is a quadratic residue modulo 11 (already known);

$10^2 \equiv 100 \equiv 1 \bmod 11$ implies that 1 is a quadratic residue modulo 11 (but—guess what—we already knew this!).

So the incongruent quadratic residues modulo 11 are 1, 3, 4, 5, and 9; the incongruent quadratic nonresidues modulo 11 must then be 2, 6, 7, 8, and 10.

In Example 1, the quadratic congruence $x^2 \equiv a \bmod 11$ (where $11 \nmid a$) had either no solutions or exactly two incongruent solutions modulo 11. (For example, the quadratic congruence $x^2 \equiv 1 \bmod 11$ has the two incongruent solutions 1 and 10.) This fact is true in general for quadratic congruences of the form $x^2 \equiv a \bmod p$ where p is an odd prime number and $p \nmid a$ as we now show.

Proposition 4.1: Let p be an odd prime number and let $a \in \mathbf{Z}$ with $p \nmid a$. Then the quadratic congruence $x^2 \equiv a \bmod p$ has either no solutions or exactly two incongruent solutions modulo p.

Proof: Assume that $x^2 \equiv a \bmod p$ has a solution, say x_0. Clearly, $-x_0$ is also a solution. Also, x_0 and $-x_0$ are incongruent modulo p. (If $x_0 \equiv -x_0 \bmod p$, then $p \mid 2x_0$ and, by Lemma 1.14, we have $p \mid 2$ or $p \mid x_0$. Since $p \nmid 2$, we have $p \mid x_0$ or $x_0 \equiv 0 \bmod p$. But x_0 is a solution to $x^2 \equiv a \bmod p$; so $a \equiv 0 \bmod p$, and we have $p \mid a$, a contradiction.) So $x^2 \equiv a \bmod p$ has at least two incongruent solutions modulo p if it has a solution. It remains to show that $x^2 \equiv a \bmod p$ has at most two incongruent solutions modulo p. Let x_0 and x_1 be solutions of $x^2 \equiv a \bmod p$. Then $x_0^2 \equiv x_1^2 \equiv a \bmod p$, and we have $p \mid x_0^2 - x_1^2$ or, equivalently, $p \mid (x_0 - x_1) \times (x_0 + x_1)$. By Lemma 1.14, we have that $p \mid x_0 - x_1$ or $p \mid x_0 + x_1$. So $x_0 \equiv x_1 \bmod p$ or $x_0 \equiv -x_1 \bmod p$ and, consequently, $x^2 \equiv a \bmod p$ has at most two incongruent solutions modulo p as desired. ■

We note here that the proof above establishes the fact that if x_0 is a solution of the congruence $x^2 \equiv a \bmod p$, then $p - x_0$ (which is congruent to $-x_0$ modulo p) is also a solution of this congruence and $p - x_0$ is not congruent to x_0 modulo p. This was clearly the case in Example 1, where 1 and $11 - 1$ (=10) were the incongruent solutions of $x^2 \equiv 1 \bmod 11$, 2 and $11 - 2$ (=9) were the incongruent solutions of $x^2 \equiv 4 \bmod 11$, and so on. So we have the following porism.

Porism 4.2: Let p be an odd prime number and let $a \in \mathbf{Z}$ with $p \nmid a$. If the congruence $x^2 \equiv a \bmod p$ is solvable, say with $x = x_0$, then the two incongruent solutions modulo p of this congruence are given precisely by x_0 and $p - x_0$. ■

In Example 1, the number of incongruent quadratic residues modulo 11 was equal to the number of incongruent quadratic nonresidues modulo 11 (five in each case). This fact is true in general for odd prime moduli as we record in the following proposition.

Proposition 4.3: Let p be an odd prime number. Then there are exactly $\frac{p-1}{2}$ incongruent quadratic residues modulo p and exactly $\frac{p-1}{2}$ incongruent quadratic nonresidues modulo p.

(The statement of Proposition 4.3 with $p = 11$ agrees with the results of Example 1.)

Proof: (of Proposition 4.3) Consider the $p - 1$ quadratic congruences given by

$$\begin{aligned} x^2 &\equiv 1 \bmod p \\ x^2 &\equiv 2 \bmod p \\ &\vdots \\ x^2 &\equiv (p-1) \bmod p \end{aligned}$$

Since each quadratic congruence has either zero or two incongruent solutions modulo p by Proposition 4.1 and no one solution solves more than one of the congruences (obviously!), we have that exactly half of the $p - 1$ congruences are solvable (each having two incongruent solutions modulo p). So there are exactly $\frac{p-1}{2}$ incongruent quadratic residues modulo p; it follows that there are exactly $\frac{p-1}{2}$ incongruent quadratic nonresidues modulo p. ■

At this point, the determination of quadratic residues and nonresidues has the potential for being quite cumbersome. As an example, consider the question "Is 2 a quadratic residue modulo 11?" To answer this question, our current strategy would be to begin finding all quadratic residues modulo 11 as in Example 1. After producing the five quadratic residues modulo 11 (none of which is 2), we would have the answer to our question: "No, 2 is not a quadratic residue modulo 11." Of course, if we had been lucky enough to encounter 2 as a quadratic residue along the way, we could have stopped computing (with an affirmative answer to the question), but the moral here is that answering such questions may involve considerable work, especially with larger prime moduli. (Is 7 a quadratic residue modulo 53?) One goal in the succeeding sections of this chapter will be to develop theory to simplify such computations.

Exercise Set 4.1

1. Find all incongruent quadratic residues and nonresidues modulo each integer below.
   (a) 13
   (b) 17
   (c) 19
   (d) 23
2. How many incongruent quadratic residues exist modulo 4177? (*Note*: 4177 *is* a prime number.)
3. Find all incongruent solutions of the quadratic congruence $x^2 \equiv 1 \bmod 8$. Is it not true that quadratic congruences have either no solutions or exactly two incongruent solutions? Explain.
4. ***Definition:*** Let $a, p \in \mathbf{Z}$ with p a prime number and $p \equiv 1 \bmod 4$. Then a is said to be a *biquadratic residue modulo p* if a is a quadratic residue modulo p and the congruence $x^4 \equiv a \bmod p$ is solvable.
   Find all incongruent biquadratic residues modulo 13.
5. Find all incongruent solutions of each quadratic congruence below.
   (a) $x^2 \equiv 23 \bmod 77$ (*Hint*: Consider the congruences $x^2 \equiv 23 \bmod 7$ and $x^2 \equiv 23 \bmod 11$ and use the Chinese Remainder Theorem.)
   (b) $x^2 \equiv 11 \bmod 39$
   (c) $x^2 \equiv 46 \bmod 105$
6. Let p be an odd prime number. Prove that the $\frac{p-1}{2}$ quadratic residues modulo p are congruent to
$$1^2, 2^2, 3^2, \ldots, \left(\frac{p-1}{2}\right)^2$$
   modulo p.
7. *(a)* Let p be a prime number with $p > 3$. Prove that the sum of the quadratic residues modulo p is divisible by p.
   (b) Let p be a prime number with $p > 5$. Prove that the sum of the squares of the quadratic nonresidues modulo p is divisible by p.

8. Let p be an odd prime number. Prove that the product of the quadratic residues modulo p is congruent to 1 modulo p if and only if $p \equiv 3 \bmod 4$.
9. Let $a, b, c \in \mathbf{Z}$ and let p be an odd prime number with $p \nmid a$. The following exercise shows that finding a solution of the general quadratic congruence $ax^2 + bx \equiv c \bmod p$ is equivalent to finding a solution of a simpler quadratic congruence of the form $x^2 \equiv a \bmod p$.
   (a) Prove that the congruence $ax^2 + bx \equiv c \bmod p$ is equivalent to the congruence $4a(ax^2 + bx - c) \equiv 0 \bmod p$.
   (b) Prove that the congruence $4a(ax^2 + bx - c) \equiv 0 \bmod p$ is equivalent to the congruence $(2ax + b)^2 \equiv b^2 + 4ac \bmod p$. (Note that this latter congruence is of the form $x^2 \equiv a \bmod p$.)
   (c) Put $y = 2ax + b$ and $d = b^2 + 4ac$. Let x_0 be a solution of the congruence $ax^2 + bx \equiv c \bmod p$. Prove that $y_0 = 2ax_0 + b$ is a solution of $y^2 \equiv d \bmod p$.
   (d) With y and d as in part (c) above, let y_0 be a solution of the congruence $y^2 \equiv d \bmod p$. Prove that any solution x_0 of the congruence $2ax \equiv y_0 - b \bmod p$ is a solution of $ax^2 + bx \equiv c \bmod p$.
   (e) With d as in part (c) above, prove that the quadratic congruence $ax^2 + bx \equiv c \bmod p$ has no solution if d is a quadratic nonresidue modulo p, one incongruent solution modulo p if $p \mid d$, and two incongruent solutions modulo p if d is a quadratic residue modulo p.
10. Use Exercise 9 above to solve each quadratic congruence below.
   (a) $x^2 + x \equiv 3 \bmod 13$
   (b) $3x^2 + 2x \equiv 4 \bmod 17$
   (c) $x^2 + 3x \equiv 1 \bmod 19$
   (d) $2x^2 + x \equiv 5 \bmod 23$
11. The following exercise investigates the solvability of the quadratic congruence $x^2 \equiv a \bmod m$ for general m.
   (a) Let p be an odd prime number and let $a \in \mathbf{Z}$ with $p \nmid a$. If k is a positive integer, prove that the congruence $x^2 \equiv a \bmod p^k$ is solvable if and only if the congruence $x^2 \equiv a \bmod p$ is solvable. In the case of solvability, prove that $x^2 \equiv a \bmod p^k$ has exactly two incongruent solutions modulo p^k.
   (b) Let a be an odd integer. Prove that the congruence $x^2 \equiv a \bmod 2$ is always solvable with exactly one incongruent solution modulo 2.
   (c) Let a be an odd integer. Prove that the congruence $x^2 \equiv a \bmod 4$ is solvable if and only if $a \equiv 1 \bmod 4$. In the case of solvability, prove that $x^2 \equiv a \bmod 4$ has exactly two incongruent solutions modulo 4.
   (d) Let a be an odd integer. If k is a positive integer with $k \geq 3$, prove that the congruence $x^2 \equiv a \bmod 2^k$ is solvable if and only if $a \equiv 1 \bmod 8$. In the case of solvability, prove that $x^2 \equiv a \bmod 2^k$ has exactly four incongruent solutions modulo 2^k.
   (e) Let $m \in \mathbf{Z}$ with $m > 1$. If $2^k p_1^{k_1} p_2^{k_2} \cdots p_n^{k_n}$ is the prime factorization of m (where $p_1, p_2, \ldots, p_n$ are distinct odd prime numbers), find a necessary and sufficient condition for the congruence $x^2 \equiv a \bmod m$ to be solvable if $(a, m) = 1$. In the case of solvability, how many incongruent solutions modulo m exist?

Biography

Adrien-Marie Legendre (1752–1833)

Adrien-Marie Legendre, a French mathematician, is known principally for a work entitled *Éléments de géométrie,* which was a simplification and rearrangement of Euclid's *Elements.* This work was very popular and became the prototype for modern geometry textbooks. In addition, he was the author of the first textbook in number theory, a two-volume work entitled *Essai sur la théorie des nombres.* Legendre's name is connected with the solutions of certain second-order differential equations (the so-called Legendre functions) as well as the Legendre symbol below.

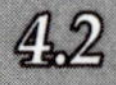

4.2 The Legendre Symbol

We now introduce a convenient symbol for dealing with quadratic residues and nonresidues. The symbol was introduced in 1798 by Adrien-Marie Legendre in his *Essai sur la théorie des nombres* and, appropriately enough, is named for him.

Definition 2: Let p be an odd prime number and let $a \in \mathbf{Z}$ with $p \nmid a$. The *Legendre symbol,* denoted $\left(\frac{a}{p}\right)$, is

$$\left(\frac{a}{p}\right) = \begin{cases} 1, & \text{if } a \text{ is a quadratic residue modulo } p \\ -1, & \text{if } a \text{ is a quadratic nonresidue modulo } p \end{cases}$$

The notation for the Legendre symbol will cause no confusion with the usual notation for rational numbers since the meaning of the symbol $\left(\frac{\cdot}{\cdot}\right)$ will either be stated explicitly or be clear from the context in which it is used. (Note again the potential for ambiguous notation!) We illustrate the use of the Legendre symbol with an example.

Example 2:

Since 1, 3, 4, 5, and 9 are quadratic residues modulo 11 by Example 1 above, we have that

$$\left(\frac{1}{11}\right) = \left(\frac{3}{11}\right) = \left(\frac{4}{11}\right) = \left(\frac{5}{11}\right) = \left(\frac{9}{11}\right) = 1$$

Similarly, since 2, 6, 7, 8, and 10 are quadratic nonresidues modulo 11 by Example 1 above, we have that

$$\left(\frac{2}{11}\right) = \left(\frac{6}{11}\right) = \left(\frac{7}{11}\right) = \left(\frac{8}{11}\right) = \left(\frac{10}{11}\right) = -1$$

By definition, the Legendre symbol $\left(\frac{a}{p}\right)$ indicates whether a is a quadratic residue modulo p. In other words, the Legendre symbol $\left(\frac{a}{p}\right)$ records whether the quadratic congruence $x^2 \equiv a \bmod p$ is solvable. It is important to remember this connection between the Legendre symbol and its associated quadratic congruence; the next example emphasizes this connection.

Example 3:

Find $\left(\frac{3}{7}\right)$.

The Legendre symbol $\left(\frac{3}{7}\right)$ asks the question: Is the congruence $x^2 \equiv 3 \bmod 7$ solvable? We have

$$1^2 \equiv 1 \bmod 7 \text{ implies that 1 is a quadratic residue modulo 7}$$
$$2^2 \equiv 4 \bmod 7 \text{ implies that 4 is a quadratic residue modulo 7}$$
$$3^2 \equiv 9 \equiv 2 \bmod 7 \text{ implies that 2 is a quadratic residue modulo 7}$$

By Proposition 4.3, there are exactly $\frac{p-1}{2} = \frac{7-1}{2} = 3$ incongruent quadratic residues modulo 7; inasmuch as 1, 2, and 4 are quadratic residues modulo 7, it must be the case that 3, 5, and 6 are quadratic nonresidues modulo 7, from which the congruence $x^2 \equiv 3 \bmod 7$ is not solvable and $\left(\frac{3}{7}\right) = -1$.

While a Legendre symbol gives information about whether its associated congruence is solvable, note that the symbol does *not* in any way provide the solutions in the case of solvability. The fact that $\left(\frac{2}{7}\right) = 1$ implies only that the congruence $x^2 \equiv 2 \bmod 7$ is solvable; the actual solutions of this congruence must be found by other means.

Our only method at present for the computation of Legendre symbols requires a possible consideration of $\frac{p-1}{2}$ congruences (unless, of course, we are fortunate enough to encounter the desired quadratic residue along the way). We now work toward a simpler procedure for the computation of Legendre symbols. We first give a criterion from Euler that is not so important for direct computation of the Legendre symbol but is very useful in proving nice computational properties of the Legendre symbol.

Theorem 4.4: (Euler's Criterion) Let p be an odd prime number and let $a \in \mathbf{Z}$ with $p \nmid a$. Then

$$\left(\frac{a}{p}\right) \equiv a^{(p-1)/2} \bmod p$$

Proof: Assume that $(\frac{a}{p}) = 1$. Then $x^2 \equiv a \bmod p$ has a solution, say x_0. We have

$$\begin{aligned} a^{(p-1)/2} &\equiv (x_0^2)^{(p-1)/2} \bmod p \\ &\equiv x_0^{p-1} \bmod p \\ &\equiv 1 \bmod p \quad \text{(by Fermat's Little Theorem; note that } p \nmid x_0) \\ &\equiv \left(\frac{a}{p}\right) \bmod p \end{aligned}$$

So the desired result holds if $(\frac{a}{p}) = 1$. Now assume that $(\frac{a}{p}) = -1$. For each $i \in \mathbf{Z}$ with $1 \leq i \leq p - 1$, there is a unique $j \in \mathbf{Z}$ with $1 \leq j \leq p - 1$ such that $ij \equiv a \bmod p$ by Theorem 2.6. Furthermore, $i \neq j$. [If $i = j$, then $i^2 \equiv a \bmod p$ implies that a is a quadratic residue modulo p and we obtain $(\frac{a}{p}) = 1$, a contradiction.] So we can pair the integers $1, 2, \ldots, p - 1$ into $\frac{p-1}{2}$ pairs, each with product a modulo p. Then

$$(p - 1)! \equiv a^{(p-1)/2} \bmod p$$

implies that

$$-1 \equiv a^{(p-1)/2} \bmod p \quad \text{(why?)}$$

from which

$$\left(\frac{a}{p}\right) \equiv a^{(p-1)/2} \bmod p$$

So the desired result holds if $(\frac{a}{p}) = -1$. ■

Example 4:

Find $(\frac{3}{7})$ by using Euler's Criterion.
We have

$$\begin{aligned} \left(\frac{3}{7}\right) &\equiv 3^{(7-1)/2} \bmod 7 \\ &\equiv 27 \bmod 7 \\ &\equiv -1 \bmod 7 \end{aligned}$$

Hence $(\frac{3}{7}) = -1$. (Compare the answer above with that of Example 3!)

We remark again that Euler's Criterion is more of theoretical importance than practical importance since the direct computation of the quantity $a^{(p-1)/2}$ modulo p may be prohibitive.

We now prove three useful computational properties of the Legendre symbol, the third of which uses Euler's Criterion. We remark here that the first two statements below follow trivially from the definition of the Legendre symbol. The reader may wish to attempt proofs of these statements before reading the easy justifications given.

Proposition 4.5: Let p be an odd prime number and let $a, b \in \mathbf{Z}$ with $p \nmid a$ and $p \nmid b$.

(a) $\left(\frac{a^2}{p}\right) = 1$

(b) If $a \equiv b \bmod p$, then $\left(\frac{a}{p}\right) = \left(\frac{b}{p}\right)$.

(c) $\left(\frac{ab}{p}\right) = \left(\frac{a}{p}\right)\left(\frac{b}{p}\right)$

Proof:

(a) The quadratic congruence $x^2 \equiv a^2 \bmod p$ clearly has a solution, namely a.

(b) Since $a \equiv b \bmod p$, the quadratic congruence to be solved is $x^2 \equiv a \equiv b \bmod p$. Clearly, $x^2 \equiv a \bmod p$ is solvable if and only if $x^2 \equiv b \bmod p$ is solvable.

(c) We have

$$\begin{aligned}\left(\frac{ab}{p}\right) &\equiv (ab)^{(p-1)/2} \bmod p \quad \text{(by Euler's Criterion)}\\ &\equiv a^{(p-1)/2} b^{(p-1)/2} \bmod p\\ &\equiv \left(\frac{a}{p}\right)\left(\frac{b}{p}\right) \bmod p\end{aligned}$$

Since the only possible values of a Legendre symbol are ± 1, we have

$$\left(\frac{ab}{p}\right) = \left(\frac{a}{p}\right)\left(\frac{b}{p}\right) \blacksquare$$

We now illustrate the use of the three properties of Proposition 4.5 above with two examples.

Example 5:

Find $\left(\frac{-11}{7}\right)$.

We have

$$\begin{aligned}\left(\frac{-11}{7}\right) &= \left(\frac{-1}{7}\right)\left(\frac{11}{7}\right) && \text{[by (c) of Proposition 4.5]}\\ &= \left(\frac{-1}{7}\right)\left(\frac{4}{7}\right) && \text{[by (b) of Proposition 4.5, since } 11 \equiv 4 \bmod 7]\\ &= \left(\frac{-1}{7}\right) && \text{[by (a) of Proposition 4.5, since } \left(\tfrac{4}{7}\right) = 1]\\ &= \left(\frac{6}{7}\right) && \text{[by (b) of Proposition 4.5, since } -1 \equiv 6 \bmod 7]\\ &= -1 && \text{(by Example 3)}\end{aligned}$$

Example 6:

Find $\left(\frac{7}{53}\right)$.

Proposition 4.5 is not helpful here. [Convince yourself of this! You may also wish to note that finding $\left(\frac{7}{53}\right)$ is equivalent to answering the parenthetical question in the last paragraph of Section 4.1.] Given our knowledge to date, this computation is cumbersome; our only current method would be to determine whether 7 is a quadratic residue modulo 53 via the procedures of Examples 1 and 3. The computation can, however, be made much easier by a beautiful theorem of number theory called the *law of quadratic reciprocity,* which we will study in Section 4.3. We postpone the computation until then.

Continuing our deeper study of Legendre symbols, let p be an odd prime number and let $a \in \mathbf{Z}$ with $p \nmid a$. Then $a = \pm 2^{a_0} p_1^{a_1} p_2^{a_2} \cdots p_r^{a_r}$ with $p_1, p_2, \ldots, p_r$ odd prime numbers distinct from p and $a_0, a_1, a_2, \ldots, a_r$ nonnegative integers. (What "fundamental" theorem allows us to say this?) By (c) of Proposition 4.5, we have

$$\left(\frac{a}{p}\right) = \left(\frac{\pm 1}{p}\right)\left(\frac{2}{p}\right)^{a_0}\left(\frac{p_1}{p}\right)^{a_1}\left(\frac{p_2}{p}\right)^{a_2} \cdots \left(\frac{p_r}{p}\right)^{a_r}$$

So to compute the Legendre symbol $\left(\frac{a}{p}\right)$, it suffices to compute $\left(\frac{\pm 1}{p}\right)$, $\left(\frac{2}{p}\right)$, and $\left(\frac{q}{p}\right)$ for q an odd prime number distinct from p since all of the Legendre symbols on the right-hand side of the equation above take one of these three forms. By (a) of Proposition 4.5, we have $\left(\frac{1}{p}\right) = 1$. We conclude this section with the computations of $\left(\frac{-1}{p}\right)$ and $\left(\frac{2}{p}\right)$. The symbol $\left(\frac{q}{p}\right)$ will be dealt with in Section 4.3.

The computation of $\left(\frac{-1}{p}\right)$ is given immediately by the following theorem.

Theorem 4.6: Let p be an odd prime number. Then

$$\left(\frac{-1}{p}\right) = (-1)^{(p-1)/2} = \begin{cases} 1, & \text{if } p \equiv 1 \bmod 4 \\ -1, & \text{if } p \equiv 3 \bmod 4 \end{cases}$$

Proof: The first equality follows from Euler's Criterion with $a = -1$. For the second equality, note that $p \equiv 1 \bmod 4$ implies that $p = 4k + 1$ for some $k \in \mathbf{Z}$, from which

$$(-1)^{(p-1)/2} = (-1)^{((4k+1)-1)/2} = (-1)^{2k} = 1$$

The proof for $p \equiv 3 \bmod 4$ is similar. ■

The reader should now provide the proof for the case $p \equiv 3 \bmod 4$ in Theorem 4.6. Note that Theorem 4.6 reduces the computation of the Legendre symbol $\left(\frac{-1}{p}\right)$ to the simpler consideration of whether p is congruent to 1 or 3 modulo 4. We illustrate with an example.

Example 7:

(a) $\left(\frac{-1}{7}\right) = -1$ by Theorem 4.6, since $7 \equiv 3 \bmod 4$. Consequently, the quadratic congruence $x^2 \equiv -1 \bmod 7$ is not solvable.

(b) $\left(\frac{-1}{53}\right) = 1$ by Theorem 4.6, since $53 \equiv 1 \bmod 4$. Consequently, the quadratic congruence $x^2 \equiv -1 \bmod 53$ is solvable.

We now work on the Legendre symbol $(\frac{2}{p})$. We first need a lemma.

Lemma 4.7: (Gauss's Lemma) Let p be an odd prime number and let $a \in \mathbf{Z}$ with $p \nmid a$. Let n be the number of least positive residues of the integers $a, 2a, 3a, \ldots, (\frac{p-1}{2})a$ that are greater than $\frac{p}{2}$. Then

$$\left(\frac{a}{p}\right) = (-1)^n$$

Gauss's Lemma above is similar to Euler's Criterion (Theorem 4.4) in two respects: (a) Gauss's Lemma provides a method for direct evaluation of the Legendre symbol and (b) this method has more significance as a theoretical tool than as a computational tool. Before proving Gauss's Lemma, we give a concrete example of such a computation.

Example 8:

Use Gauss's Lemma to find $(\frac{6}{11})$.
By Gauss's Lemma, $(\frac{6}{11})$ is equal to $(-1)^n$ where n is the number of least positive residues of the integers $6, 2 \cdot 6, 3 \cdot 6, 4 \cdot 6$, and $5 \cdot 6$ that are greater than $\frac{11}{2} = 5.5$. We have

$$\begin{aligned} 6 &\equiv 6 \bmod 11 \\ 2 \cdot 6 &\equiv 12 \equiv 1 \bmod 11 \\ 3 \cdot 6 &\equiv 18 \equiv 7 \bmod 11 \\ 4 \cdot 6 &\equiv 24 \equiv 2 \bmod 11 \\ 5 \cdot 6 &\equiv 30 \equiv 8 \bmod 11 \end{aligned}$$

So three of the aforementioned least positive residues are greater than 5.5, from which

$$\left(\frac{6}{11}\right) = (-1)^3 = -1$$

We now prove Gauss's Lemma.

Proof: (of Lemma 4.7) Let $r_1, r_2, \ldots, r_n$ be the least nonnegative residues of the integers $a, 2a, 3a, \ldots, (\frac{p-1}{2})a$ that are greater than $\frac{p}{2}$ and let $s_1, s_2, \ldots, s_m$ be the least nonnegative residues of these integers that are less than $\frac{p}{2}$. (Why is no least positive residue equal to $\frac{p}{2}$?) Note that no r_i or s_j is equal to zero, since p does not divide any of $a, 2a, 3a, \ldots, (\frac{p-1}{2})a$. Consider the $\frac{p-1}{2}$ integers given by

$$p - r_1, p - r_2, \ldots, p - r_n, s_1, s_2, \ldots, s_m$$

We wish to show that these integers are the integers from 1 to $\frac{p-1}{2}$ inclusive in some order. Since each integer is less than or equal to $\frac{p-1}{2}$ (why?), it suffices to show that no two of these integers are congruent modulo p. No two of the first n integers are congruent modulo p. [If $p - r_i \equiv p - r_j \bmod p$ for some $i \neq j$, then $r_i \equiv r_j \bmod p$; so $k_i a \equiv k_j a \bmod p$ for some $k_i, k_j \in \mathbf{Z}$ with $k_i \neq k_j$, $1 \leq k_i \leq \frac{p-1}{2}$, and $1 \leq k_j \leq \frac{p-1}{2}$. Since $p \nmid a$, the inverse of a modulo p, say

a', exists by Corollary 2.8. Hence $k_i aa' \equiv k_j aa' \bmod p$, from which $k_i \equiv k_j \bmod p$, which is impossible.] Similarly, no two of the second m integers are congruent modulo p. Finally, no one of the first n integers can be congruent modulo p to one of the second m integers. [If $p - r_i \equiv s_j$ for some i and j, then $-r_i \equiv s_j \bmod p$ and so $-k_i a \equiv k_j a \bmod p$ for some $k_i, k_j \in \mathbf{Z}$ with $k_i \neq k_j$, $1 \le k_i \le \frac{p-1}{2}$, and $1 \le k_j \le \frac{p-1}{2}$. Considerations as above yield $-k_i \equiv k_j \bmod p$, which is impossible.] So the $\frac{p-1}{2}$ integers given by

$$p - r_1, p - r_2, \ldots, p - r_n, s_1, s_2, \ldots, s_m$$

are the integers from 1 to $\frac{p-1}{2}$ inclusive in some order. Then

$$(p - r_1)(p - r_2) \cdots (p - r_n) s_1 s_2 \cdots s_m \equiv \left(\frac{p-1}{2}\right)! \bmod p$$

implies that

$$(-1)^n r_1 r_2 \cdots r_n s_1 s_2 \cdots s_m \equiv \left(\frac{p-1}{2}\right)! \bmod p$$

By definition of the r_i and the s_j, we have that

$$(-1)^n a(2a)(3a) \cdots \left(\frac{p-1}{2}\right)a \equiv \left(\frac{p-1}{2}\right)! \bmod p$$

or, equivalently,

$$(-1)^n a^{(p-1)/2}\left(\frac{p-1}{2}\right)! \equiv \left(\frac{p-1}{2}\right)! \bmod p$$

So

$$(-1)^n a^{(p-1)/2} \equiv 1 \bmod p \quad \text{(why?)}$$

which implies that

$$a^{(p-1)/2} \equiv (-1)^n \bmod p$$

By Euler's Criterion, we have that $\left(\frac{a}{p}\right) \equiv (-1)^n \bmod p$. Since $\left(\frac{a}{p}\right) = \pm 1$, we must have $\left(\frac{a}{p}\right) = (-1)^n$ as desired. ■

We strongly suggest that you verify the steps in the proof of Gauss's Lemma above by using Example 8 as a concrete guide. (The proof of the lemma will seem much less mysterious, if it indeed seems mysterious now.)

We now use Gauss's Lemma to compute $\left(\frac{2}{p}\right)$.

Theorem 4.8: Let p be an odd prime number. Then

$$\left(\frac{2}{p}\right) = (-1)^{(p^2-1)/8} = \begin{cases} 1, & \text{if } p \equiv 1, 7 \bmod 8 \\ -1, & \text{if } p \equiv 3, 5 \bmod 8 \end{cases}$$

Proof: By Gauss's Lemma, we have that $\left(\frac{2}{p}\right) = (-1)^n$ where n is the number of least positive residues of the integers 2, (2)2, (3)2, ..., $\left(\frac{p-1}{2}\right)2$ that are greater than $\frac{p}{2}$. Let $k \in \mathbf{Z}$ with $1 \le k \le \frac{p-1}{2}$. Then $(k)2 < \frac{p}{2}$ if and only if $k < \frac{p}{4}$; so $\left[\frac{p}{4}\right]$ of the integers 2, (2)2, (3)2, ..., $\left(\frac{p-1}{2}\right)2$ are less than $\frac{p}{2}$. (Here, $[\cdot]$ denotes the greatest integer function.) So $\frac{p-1}{2} - \left[\frac{p}{4}\right]$ of these integers are greater than $\frac{p}{2}$, from which

$$\left(\frac{2}{p}\right) = (-1)^{(p-1)/2-[p/4]}$$

by Gauss's Lemma. For the first equality, it suffices to show that

$$\frac{p-1}{2} - \left[\frac{p}{4}\right] \equiv \frac{p^2-1}{8} \bmod 2$$

If $p \equiv 1 \bmod 8$, then $p = 8k + 1$ for some $k \in \mathbf{Z}$, from which

$$\frac{p-1}{2} - \left[\frac{p}{4}\right] = \frac{(8k+1)-1}{2} - \left[\frac{8k+1}{4}\right] = 4k - 2k = 2k \equiv 0 \bmod 2$$

and

$$\frac{p^2-1}{8} = \frac{(8k+1)^2-1}{8} = 8k^2 + 2k \equiv 0 \bmod 2$$

so that the desired congruence holds if $p \equiv 1 \bmod 8$. The cases where $p \equiv 3, 5, 7 \bmod 8$ are similar (and make excellent exercises!). So $\frac{p-1}{2} - \left[\frac{p}{4}\right]$ agrees modulo 2 with $\frac{p^2-1}{8}$ in all cases, proving the first equality. The cases above yield

$$\frac{p^2-1}{8} \equiv \begin{cases} 0 \bmod 2, & \text{if } p \equiv 1, 7 \bmod 8 \\ 1 \bmod 2, & \text{if } p \equiv 3, 5 \bmod 8 \end{cases}$$

which implies

$$(-1)^{(p^2-1)/8} = \begin{cases} 1, & \text{if } p \equiv 1, 7 \bmod 8 \\ -1, & \text{if } p \equiv 3, 5 \bmod 8 \end{cases}$$

completing the proof. ■

Note that Theorem 4.8 reduces the computation of the Legendre symbol $\left(\frac{2}{p}\right)$ to the simpler consideration of whether p is congruent to 1, 3, 5, or 7 modulo 8. We illustrate with an example.

Example 9:

(a) $\left(\frac{2}{7}\right) = 1$ by Theorem 4.8, since $7 \equiv 7 \bmod 8$. Consequently, the quadratic congruence $x^2 \equiv 2 \bmod 7$ is solvable.

(b) $\left(\frac{2}{53}\right) = -1$ by Theorem 4.8, since $53 \equiv 5 \bmod 8$. Consequently, the quadratic congruence $x^2 \equiv 2 \bmod 53$ is not solvable.

Exercise Set 4.2

12. Use Euler's Criterion (Theorem 4.4) to evaluate the following Legendre symbols.

(a) $\left(\frac{11}{23}\right)$

(b) $\left(\frac{-6}{11}\right)$

13. Use Gauss's Lemma (Lemma 4.7) to evaluate the following Legendre symbols.

(a) $\left(\frac{12}{23}\right)$
(b) $\left(\frac{-5}{11}\right)$

14. Evaluate the following Legendre symbols.
(a) $\left(\frac{-1}{59}\right)$
(b) $\left(\frac{-1}{41}\right)$
(c) $\left(\frac{2}{43}\right)$
(d) $\left(\frac{2}{71}\right)$
(e) $\left(\frac{-2}{61}\right)$
(f) $\left(\frac{-2}{67}\right)$

15. Let p be an odd prime number and let a be a quadratic residue modulo p. Prove that $-a$ is also a quadratic residue modulo p if and only if $p \equiv 1 \bmod 4$.

16. (a) Does there exist a positive integer n such that $n^2 + 1$ is evenly divisible by 7? Prove your assertion.
(b) Does there exist a positive integer n such that $n^2 + 1$ is evenly divisible by 13? Prove your assertion.
(c) Find a necessary and sufficient condition on an odd prime number p for there to exist a positive integer n such that $n^2 + 1$ is evenly divisible by p.

17. Prove or disprove the following statements.
(a) Let p be an odd prime number and let a and b be quadratic nonresidues modulo p. Then the congruence $x^2 \equiv ab \bmod p$ is solvable.
(b) Let p and q be distinct odd prime numbers and let b be a quadratic nonresidue of each of p and q. Then the congruence $x^2 \equiv b \bmod pq$ is solvable.

18. Let p be an odd prime number and let $a, b \in \mathbf{Z}$ with $p \nmid a$ and $p \nmid b$. Prove that among the congruences $x^2 \equiv a \bmod p$, $x^2 \equiv b \bmod p$, and $x^2 \equiv ab \bmod p$, either all three are solvable or exactly one is solvable.

19. Let p be an odd prime number. Prove that if a and b are both quadratic residues modulo p, then the quadratic congruence $ax^2 \equiv b \bmod p$ is solvable. What happens if a and b are both quadratic nonresidues modulo p? What happens if one of a and b is a quadratic residue modulo p and the other is a quadratic nonresidue modulo p?

20. Let a be a positive integer and let p be an odd prime number with $p \nmid a$. Prove that

$$\left(\frac{a}{p}\right) + \left(\frac{2a}{p}\right) + \left(\frac{3a}{p}\right) + \cdots + \left(\frac{(p-1)a}{p}\right) = 0$$

21. Let p be an odd prime number. Prove that

$$\left(\frac{1 \cdot 2}{p}\right) + \left(\frac{2 \cdot 3}{p}\right) + \left(\frac{3 \cdot 4}{p}\right) + \cdots + \left(\frac{(p-2)(p-1)}{p}\right) = -1$$

22. Let p be a prime number with $p \equiv 1 \bmod 4$. Prove that

$$\sum_{a=1}^{(p-1)/2} \left(\frac{a}{p}\right) = 0$$

23. Let p be an odd prime number and let n be a quadratic nonresidue modulo

p. Prove that

$$\sum_{d \mid n,\, d>0} d^{(p-1)/2} \equiv 0 \bmod p$$

24. *(a)* Let p be a prime number with $p \geq 7$. Prove that at least one of 2, 5, and 10 is a quadratic residue modulo p.

(b) Could exactly two of 2, 5, and 10 be quadratic residues modulo p in part (a) above? Why or why not?

(c) Let p be a prime number with $p \geq 7$. Prove that there are at least two consecutive quadratic residues modulo p. (*Hint*: Use part (a) above.)

25. Let p be a prime number with $p \equiv 3 \bmod 4$. Prove that $(\frac{p-1}{2})! \equiv (-1)^n \bmod p$ where n is the number of positive integers less than $\frac{p}{2}$ that are quadratic nonresidues of p.

26. Prove that there are infinitely many prime numbers expressible in the form $4n + 1$ where n is a positive integer. (*Hint*: Assume, by way of contradiction, that there are only finitely many such prime numbers, say $p_1, p_2, \ldots, p_r$. Consider $4p_1^2p_2^2 \cdots p_r^2 + 1$ and use Theorem 4.6.)

27. Let p and q be prime numbers with $p \equiv 3 \bmod 4$ and $q = 2p + 1$. Prove that $2^p - 1$ is a Mersenne prime if and only if $p = 3$. (*Hint*: The "if" direction is easy. For the "only if" direction, assume, by way of contradiction, that $p \neq 3$ and conclude that in such a case, q divides $2^p - 1$.)

4.3

The Law of Quadratic Reciprocity

We now study one of the most beautiful theorems in elementary number theory: the *law of quadratic reciprocity*. This law was conjectured by Euler in 1783; the first complete proof of the law was given by Gauss in his *Disquisitiones Arithmeticae* in 1801. Today, there are many more than 100 proofs of this theorem; in fact, Gauss himself published five different proofs and found several more. (Alas, we will give but one.) The law of quadratic reciprocity greatly simplifies the evaluation of the Legendre symbol.

We should remark here that the quadratic reciprocity law is the beginning of a long line of reciprocity theorems. For example, the fact that the cubic congruence $y^3 \equiv 2 \bmod 7$ is not solvable (please verify this!) means that 2 is a cubic nonresidue modulo 7; the more general consideration of cubic residues and nonresidues modulo prime numbers would eventually lead us into a *cubic reciprocity law*, just as consideration of quadratic residues and nonresidues will lead us shortly into a quadratic reciprocity law. (In case you are wondering, the development of a cubic reciprocity law is much more difficult than the development of the quadratic reciprocity law. See Student Project 5.) This sequence of reciprocity theorems would eventually culminate in the Artin reciprocity law (perhaps the most beautiful result in all of number theory),

Biography

Emil Artin (1898–1962)

Born in Austria, Emil Artin was one of the leading mathematicians of the twentieth century, greatly influencing his contemporaries through outstanding teaching and research. Artin has made significant contributions to the fields of both algebra and number theory. Appropriately, many of Artin's contributions belong to the area of mathematics known as *algebraic number theory,* which applies algebraic principles to the solution of number-theoretic problems. The stunning *Artin reciprocity law,* a generalization of the quadratic reciprocity law, falls in this domain. Proved by Artin in 1927, this law partially solves one of the 23 significant problems posed by David Hilbert at the International Mathematical Congress in Paris in 1900. In algebra, much of the modern development of Galois theory results from Artin's work. In addition, his name is connected with a class of rings (the so-called Artinian rings).

which is named after the mathematician Emil Artin. The mere statement of the Artin reciprocity law is beyond the scope of this book [the interested reader may, however, consult Garbanati (1981)].

Recall from Section 4.2 that to evaluate the general Legendre symbol $\left(\frac{a}{p}\right)$, it suffices to evaluate the Legendre symbol $\left(\frac{1}{p}\right)$ [which is 1 by (a) of Proposition 4.5], the Legendre symbol $\left(\frac{-1}{p}\right)$ (which we have done — see Theorem 4.6), the Legendre symbol $\left(\frac{2}{p}\right)$ (which we have done — see Theorem 4.8), and the Legendre symbol $\left(\frac{q}{p}\right)$ for q an odd prime number distinct from p (which we have *not* done). The law of quadratic reciprocity relates the Legendre symbols $\left(\frac{q}{p}\right)$ and $\left(\frac{p}{q}\right)$, which makes possible the simplification of Legendre symbols involving two odd prime numbers.

Theorem 4.9: (Law of Quadratic Reciprocity) Let p and q be distinct odd prime numbers. Then

$$\left(\frac{p}{q}\right)\left(\frac{q}{p}\right) = (-1)^{((p-1)/2)((q-1)/2)}$$

$$= \begin{cases} 1, & \text{if } p \equiv 1 \bmod 4 \text{ or } q \equiv 1 \bmod 4 \text{ (or both)} \\ -1, & \text{if } p \equiv q \equiv 3 \bmod 4 \end{cases}$$

If $p \equiv 1 \bmod 4$ or $q \equiv 1 \bmod 4$ (or both) in Theorem 4.9, then $(\frac{p}{q})(\frac{q}{p}) = 1$, so that $(\frac{p}{q}) = (\frac{q}{p})$. If $p \equiv q \equiv 3 \bmod 4$ in Theorem 4.9, then $(\frac{p}{q})(\frac{q}{p}) = -1$, so that $(\frac{p}{q}) = -(\frac{q}{p})$. These facts can be used to simplify the computation of Legendre symbols, as the following examples illustrate.

Example 10:

Find $(\frac{7}{53})$. (Do you remember this problem? If not, see Example 6.)
We have

$$\left(\frac{7}{53}\right) = \left(\frac{53}{7}\right) \quad \text{(by Theorem 4.9, since } 53 \equiv 1 \bmod 4\text{)}$$

$$= \left(\frac{4}{7}\right) \quad \text{[by (b) of Proposition 4.5]}$$

$$= 1 \quad \text{[by (a) of Proposition 4.5]}$$

(So the quadratic congruence $x^2 \equiv 7 \bmod 53$ is solvable.) Are you impressed with the power of quadratic reciprocity?

Example 11:

Find $(\frac{-158}{101})$.
We have

$$\left(\frac{-158}{101}\right) = \left(\frac{-1}{101}\right)\left(\frac{158}{101}\right) \quad \text{[by (b) of Proposition 4.5]}$$

$$= \left(\frac{158}{101}\right) \quad \text{(by Theorem 4.6, since } 101 \equiv 1 \bmod 4\text{)}$$

$$= \left(\frac{57}{101}\right) \quad \text{[by (b) of Proposition 4.5]}$$

$$= \left(\frac{3}{101}\right)\left(\frac{19}{101}\right) \quad \text{[by (c) of Proposition 4.5]}$$

$$= \left(\frac{101}{3}\right)\left(\frac{101}{19}\right) \quad \text{(by Theorem 4.9, since } 101 \equiv 1 \bmod 4\text{)}$$

$$= \left(\frac{2}{3}\right)\left(\frac{6}{19}\right) \quad \text{[by (b) of Proposition 4.5]}$$

$$= (-1)\left(\frac{6}{19}\right) \quad \text{(by Theorem 4.8, since } 3 \equiv 3 \bmod 8)$$

$$= -\left(\frac{2}{19}\right)\left(\frac{3}{19}\right) \quad \text{[by (c) of Proposition 4.5]}$$

$$= -(-1)\left(\frac{3}{19}\right) \quad \text{(by Theorem 4.8, since } 19 \equiv 3 \bmod 8)$$

$$= \left(\frac{3}{19}\right)$$

$$= -\left(\frac{19}{3}\right) \quad \text{(by Theorem 4.9, since } 3 \equiv 3 \bmod 4 \text{ and } 19 \equiv 3 \bmod 4)$$

$$= -\left(\frac{1}{3}\right) \quad \text{[by (b) of Proposition 4.5]}$$

$$= -1$$

(So the quadratic congruence $x^2 \equiv 158 \bmod 101$ is not solvable.) An important remark is in order here: It would *not* be correct in the transition from the third line to the fourth line above to assert that $\left(\frac{57}{101}\right) = \left(\frac{101}{57}\right)$ by Theorem 4.9, even though $101 \equiv 1 \bmod 4$. The reason for this is that 57 is not a prime number. Legendre symbols may be "flipped" only if both components are odd prime numbers! A similar observation applies to $\left(\frac{6}{19}\right)$ in the transition from the seventh line to the eighth line above. See Exercise 29 for another common error that is sometimes made when one uses the quadratic reciprocity law.

The proof of the law of quadratic reciprocity that we give here [which, incidentally, comes from Ferdinand Eisenstein (1823–1852), a Gauss student] requires a preliminary, rather technical lemma.

Lemma 4.10: Let p be an odd prime number and let $a \in \mathbf{Z}$ with $p \nmid a$ and a odd. If

$$N = \sum_{j=1}^{(p-1)/2} \left[\frac{ja}{p}\right]$$

then

$$\left(\frac{a}{p}\right) = (-1)^N$$

Here $[\cdot]$ denotes the greatest integer function.

Lemma 4.10 gives us yet another method for computing Legendre symbols. As with Euler's Criterion and Gauss's Lemma, our interest in Lemma 4.10 is theoretical rather than computational. Before proving this technical lemma, we give a concrete example of a typical computation.

Example 12:

Use Lemma 4.10 to find $(\frac{7}{11})$.
We have

$$N = \sum_{j=1}^{5} \left[\frac{j \cdot 7}{11}\right] = \left[\frac{7}{11}\right] + \left[\frac{14}{11}\right] + \left[\frac{21}{11}\right] + \left[\frac{28}{11}\right] + \left[\frac{35}{11}\right]$$
$$= 0 + 1 + 1 + 2 + 3$$
$$= 7$$

So $(\frac{7}{11}) = (-1)^N = (-1)^7 = -1$.

We now prove Lemma 4.10. As you read the proof, note the connections with Gauss's Lemma (Lemma 4.7).

Proof: (of Lemma 4.10) As in the proof of Gauss's Lemma, let $r_1, r_2, \ldots, r_n$ be the least nonnegative residues of the integers $a, 2a, 3a, \ldots, (\frac{p-1}{2})a$ that are greater than $\frac{p}{2}$ and let $s_1, s_2, \ldots, s_m$ be the least nonnegative residues of these integers that are less than $\frac{p}{2}$. Then, for $j = 1, 2, \ldots, \frac{p-1}{2}$, we have

$$ja = p\left[\frac{ja}{p}\right] + (\text{remainder depending on } j)$$

where each of $r_1, r_2, \ldots, r_n, s_1, s_2, \ldots, s_m$ appears exactly once as a remainder. By adding the $\frac{p-1}{2}$ equations of the type above, we obtain

$$\sum_{j=1}^{(p-1)/2} ja = \sum_{j=1}^{(p-1)/2} p\left[\frac{ja}{p}\right] + \sum_{j=1}^{n} r_j + \sum_{j=1}^{m} s_j. \qquad \textbf{(1)}$$

As in the proof of Gauss's Lemma, the integers $p - r_1, p - r_2, \ldots, p - r_n, s_1, s_2, \ldots, s_m$ are precisely the integers from 1 to $\frac{p-1}{2}$ inclusive in some order; so

$$\sum_{j=1}^{(p-1)/2} j = \sum_{j=1}^{n} (p - r_j) + \sum_{j=1}^{m} s_j = pn - \sum_{j=1}^{n} r_j + \sum_{j=1}^{m} s_j \qquad \textbf{(2)}$$

Subtracting (2) from (1), we obtain

$$\sum_{j=1}^{(p-1)/2} ja - \sum_{j=1}^{(p-1)/2} j = \sum_{j=1}^{(p-1)/2} p\left[\frac{ja}{p}\right] - pn + 2\sum_{j=1}^{n} r_j$$

or, equivalently,

$$(a - 1)\sum_{j=1}^{(p-1)/2} j = \sum_{j=1}^{(p-1)/2} p\left[\frac{ja}{p}\right] - pn + 2\sum_{j=1}^{n} r_j$$

Now, reducing both sides of the equation above modulo 2, we have

$$0 \equiv \sum_{j=1}^{(p-1)/2} \left[\frac{ja}{p}\right] - n \bmod 2$$

or, equivalently,

$$n \equiv \sum_{j=1}^{(p-1)/2} \left[\frac{ja}{p}\right] \bmod 2$$

So $n \equiv N \bmod 2$; now $(\frac{a}{p}) = (-1)^n$ implies $(\frac{a}{p}) = (-1)^N$, as desired. ■

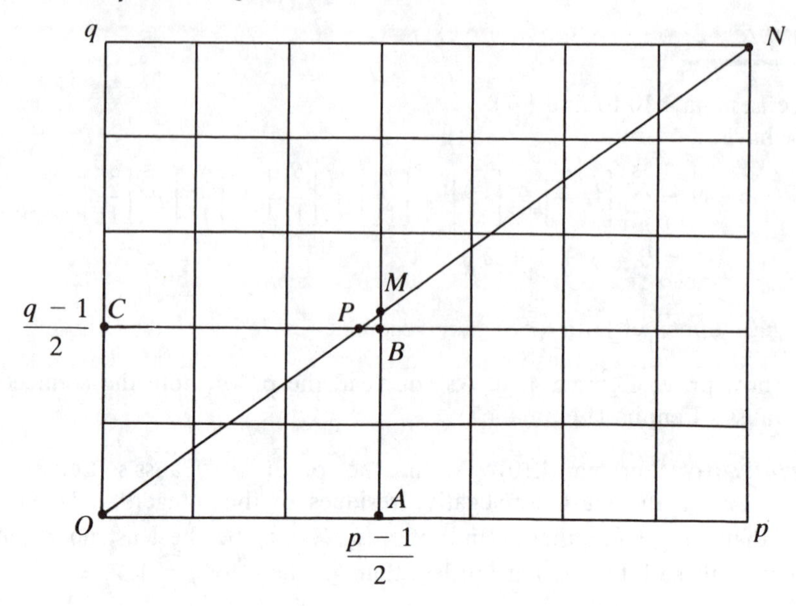

Figure 4.1

We may now prove the law of quadratic reciprocity (Theorem 4.9). As mentioned at the beginning of this section, there are many different proofs from which to choose; we opt for a geometric proof known as the "lattice point proof."

Proof: (of Theorem 4.9) The proof of the second equality is left to the reader. For the first equality, assume without loss of generality that $p > q$ and consider Figure 4.1.

In Figure 4.1, $p = 7$ and $q = 5$; the following proof is written generally and is valid for any prime numbers p and q with $p > q$. We will count, in two different ways, the number of lattice points in rectangle $OABC$, including those points on AB and BC but not those points on OA and OC. In the first place, this number is clearly $(\frac{p-1}{2})(\frac{q-1}{2})$. In the second place, we have the following facts:

(a) ON has slope $\frac{q}{p}$; since p and q are distinct prime numbers, there are no lattice points on ON except at the endpoints;

(b) the y-coordinate of M is

$$\left(\frac{q}{p}\right)\left(\frac{p-1}{2}\right) = \frac{q}{2} - \frac{1}{2}\left(\frac{q}{p}\right)$$

[here the notation $(\frac{q}{p})$ means the usual rational number $\frac{q}{p}$—*not* the Legendre symbol $(\frac{q}{p})$]; and

(c) the y-coordinate of M in (b) lies between the two consecutive integers $\frac{q-1}{2}$ and $\frac{q+1}{2}$, since

$$\frac{q-1}{2} = \frac{q}{2} - \frac{1}{2}$$

$$< \frac{q}{2} - \frac{1}{2}\left(\frac{q}{p}\right) \quad \text{(since } q/p < 1\text{)}$$

$$< \frac{q}{2}$$

$$< \frac{q+1}{2}$$

So there are no lattice points in triangle *MPB* except possibly on *PB*.

Thus, the number of lattice points in rectangle *OABC*, including those points on *AB* and *BC* but not those points on *OA* and *OC*, is the number of lattice points in triangle *OPC*, not including those points on *OC*, plus the number of lattice points in triangle *OAM*, not including those points on *OC*. The number of lattice points in triangle *OPC*, not including those points on *OC*, is

$$N_1 = \sum_{j=1}^{(q-1)/2} \left[\frac{jp}{q}\right]$$

The number of lattice points in triangle *OAM*, not including those points on *OA*, is

$$N_2 = \sum_{j=1}^{(p-1)/2} \left[\frac{jq}{p}\right]$$

So

$$N_1 + N_2 = \sum_{j=1}^{(q-1)/2} \left[\frac{jp}{q}\right] + \sum_{j=1}^{(p-1)/2} \left[\frac{jq}{p}\right] = \left(\frac{p-1}{2}\right)\left(\frac{q-1}{2}\right)$$

Then

$$\left(\frac{p}{q}\right)\left(\frac{q}{p}\right) = (-1)^{N_1}(-1)^{N_2} \quad \text{(by Lemma 4.10)}$$

$$= (-1)^{N_1+N_2}$$

$$= (-1)^{((p-1)/2)((q-1)/2)} \quad \blacksquare$$

As a summary, we wish to stress a subtle but extremely important application of the law of quadratic reciprocity. Consider the problem of finding all quadratic residues modulo an odd prime number p, say 11 (see Example 1). In terms of the Legendre symbol, we want all integers a for which $\left(\frac{a}{11}\right) = 1$. This problem is inherently a *finite* problem since, at the worst, we may compute $\left(\frac{a}{11}\right)$ for all positive integers a less than 11 and observe the results. Now consider the problem of finding all odd prime numbers p for which 11 is a quadratic residue. In terms of the Legendre symbol, we want all odd prime numbers p for which $\left(\frac{11}{p}\right) = 1$. This problem is seemingly an *infinite* problem since there are infinitely many prime numbers. Notice, however, that the law of quadratic reciprocity relates the Legendre symbol $\left(\frac{11}{p}\right)$ to the Legendre symbol $\left(\frac{p}{11}\right)$; we have $\left(\frac{11}{p}\right) = \left(\frac{p}{11}\right)$ if $p \equiv 1 \bmod 4$ and $\left(\frac{11}{p}\right) = -\left(\frac{p}{11}\right)$ if $p \equiv 3 \bmod 4$. Furthermore, the computations of the symbols $\left(\frac{p}{11}\right)$ and $-\left(\frac{p}{11}\right)$ are "finite" problems! In this (very real) sense, the law of quadratic reciprocity reduces an apparently infinite problem to a finite problem! (By the way, it is precisely this fact that is being motivated in Exercises 35 and 36; you owe it to yourself to give these problems very careful consideration.)

Biography

Carl Gustav Jacob Jacobi (1804–1851)

C. G. J. Jacobi was an eminent German mathematician and an outstanding teacher of mathematics, probably the greatest mathematics teacher of his generation. Jacobi believed in engaging in independent mathematical activity as early as possible. He believed that total mastery of prior mathematical achievements was unnecessary; indeed, true mathematical talent was best cultivated through independent research. Jacobi guided and influenced a great many gifted students.

Jacobi's research accomplishments include results concerning elliptic functions as well as contributions to the theory of determinants. Jacobi's name is connected with a certain functional determinant (called the *Jacobian*) and the Jacobi symbol below.

In the context of quadratic congruences, symbols more general than the Legendre symbol exist. Although the quadratic reciprocity law establishes a relationship between Legendre symbols, identical relationships exist between these more general symbols. One such symbol, the Jacobi symbol, is investigated in Exercises 37 and 38 as well as in Student Project 6.

Exercise Set 4.3

28. Evaluate the following Legendre symbols.

(a) $\left(\frac{-79}{101}\right)$
(b) $\left(\frac{87}{131}\right)$
(c) $\left(\frac{91}{127}\right)$
(d) $\left(\frac{-107}{211}\right)$

(e) $\left(\frac{2817}{4177}\right)$ (*Note*: 2817 is *not* a prime number.)

(f) $\left(\frac{2819}{4177}\right)$ (*Note*: 2819 *is* a prime number.)

29. Let p be an odd prime number. Rather than use Theorem 4.8 to compute $\left(\frac{2}{p}\right)$, a common mistake sometimes made by students is to "flip" the symbol $\left(\frac{2}{p}\right)$ to produce either $\left(\frac{p}{2}\right)$ or $-\left(\frac{p}{2}\right)$, both of which are undefined since Legendre symbols must have odd prime "denominators." Provide a good reason why $\left(\frac{p}{2}\right)$ should be defined to be 1 for all odd prime numbers p.

30. Let p and q be odd prime numbers with $p = 4q + 1$. Prove that $\left(\frac{q}{p}\right) = 1$.

31. *(a)* Let p be an odd prime number with $p \neq 2819$. Prove that

$$\left(\frac{2819}{p}\right) = (-1)^{(p-1)/2}\left(\frac{p}{2819}\right)$$

(b) State and prove a result similar to that in part (a) above for the Legendre symbol $\left(\frac{4177}{p}\right)$.

32. Let p and q be distinct prime numbers with $p \equiv q \equiv 3 \bmod 4$. Prove that if the congruence $x^2 \equiv p \bmod q$ is not solvable, then the congruence $x^2 \equiv q \bmod p$ has exactly two incongruent solutions modulo p.

33. Does there exist a positive integer n such that $n^2 - 71$ is evenly divisible by 107? Prove your assertion.

34. Let p and q be odd prime numbers with $p = q + 4a$ for some $a \in \mathbf{Z}$. Prove that

$$\left(\frac{a}{p}\right) = \left(\frac{a}{q}\right)$$

35. Let p be an odd prime number. Prove the following statements.

(a) $\left(\frac{-2}{p}\right) = 1$ if and only if $p \equiv 1, 3 \bmod 8$.

(b) $\left(\frac{3}{p}\right) = 1$ if and only if $p \equiv \pm 1 \bmod 12$.

(c) $\left(\frac{-3}{p}\right) = 1$ if and only if $p \equiv 1 \bmod 6$.

36. Find congruences characterizing all prime numbers p for which the following integers are quadratic residues modulo p (as has been done with $-2, 3$, and -3 in Exercise 35).

(a) 5

(b) -5

(c) 7

(d) -7

37. ***Definition:*** Let n be an odd positive integer with $n > 1$ and let $p_1^{b_1} p_2^{b_2} \cdots p_r^{b_r}$ be the prime factorization of n. Let $a \in \mathbf{Z}$ with $(a, n) = 1$. The *Jacobi symbol,* denoted $\left(\frac{a}{n}\right)$, is

$$\left(\frac{a}{n}\right) = \left(\frac{a}{p_1}\right)^{b_1}\left(\frac{a}{p_2}\right)^{b_2} \cdots \left(\frac{a}{p_r}\right)^{b_r}$$

where $\left(\frac{a}{p_1}\right), \left(\frac{a}{p_2}\right), \ldots, \left(\frac{a}{p_r}\right)$ are Legendre symbols.

(a) Evaluate the following Jacobi symbols: $\left(\frac{-79}{105}\right), \left(\frac{87}{133}\right), \left(\frac{91}{129}\right)$.

(b) Prove that if a is a quadratic residue modulo n, then $\left(\frac{a}{n}\right) = 1$, but not conversely.

38. Let n and n' be odd positive integers with $n, n' > 1$, and let $a, a' \in \mathbf{Z}$ with $(aa', nn') = 1$. Prove the following properties of the Jacobi symbol.
   (a) If $a \equiv a' \bmod n$, then $\left(\frac{a}{n}\right) = \left(\frac{a'}{n}\right)$.
   (b) $\left(\frac{aa'}{n}\right) = \left(\frac{a}{n}\right)\left(\frac{a'}{n}\right)$
   (c) $\left(\frac{a}{nn'}\right) = \left(\frac{a}{n}\right)\left(\frac{a}{n'}\right)$
   (d) $\left(\frac{a^2}{n}\right) = \left(\frac{a}{n^2}\right) = 1$
   (e) $\left(\frac{-1}{n}\right) = (-1)^{(n-1)/2}$ [*Hint*: If r and s are odd positive integers, then

$$\frac{r-1}{2} + \frac{s-1}{2} \equiv \frac{rs-1}{2} \bmod 2$$

   Why?]
   (f) $\left(\frac{2}{n}\right) = (-1)^{(p^2-1)/8}$ [*Hint*: If r and s are odd positive integers, then

$$\frac{r^2-1}{8} + \frac{s^2-1}{8} \equiv \frac{r^2s^2-1}{8} \bmod 2$$

   Why?]
   (g) If a is an odd positive integer, then $\left(\frac{a}{n}\right)\left(\frac{n}{a}\right) = (-1)^{((a-1)/2)((b-1)/2)}$.

Concluding Remarks

A 1963 paper by M. Gerstenhaber claimed to contain the 152nd proof of the law of quadratic reciprocity. We stress that not all proofs of the law of quadratic reciprocity are absolutely different; some differ only in small details from others. It is, however, curious that mathematicians seem to be obsessed with proving this law. Gauss had an underlying motive in his search for new proofs: He hoped to find a proof of quadratic reciprocity that would generalize to establish biquadratic reciprocity. He would never succeed in his search. Readers interested in proofs of the quadratic reciprocity law other than the "lattice-point proof" given in Section 4.3 can find four such proofs in Ireland and Rosen (1972). The 152nd proof mentioned above can be found in Gerstenhaber (1963).

Student Projects

1. (Programming project for calculator or computer)
   Given an odd prime number p and an integer a with $p \nmid a$, compute the Legendre symbol $\left(\frac{a}{p}\right)$ by using the law of quadratic reciprocity.
2. Let p be a prime number with $p \geq 7$. Exercise 24(c) shows that there are at least two consecutive quadratic residues modulo p. In other words, there is at least one pair of quadratic residues modulo p differing by 1.
   (a) Prove that there is at least one pair of quadratic residues differing by 2.
   (b) Prove that there is at least one pair of quadratic residues differing by 3.

(c) Prove that there is at least one pair of quadratic residues differing by 4.

3. Prove or disprove the following conjecture.
   Conjecture: If N is the maximum number of consecutive quadratic nonresidues modulo an odd prime number p, then $N < \sqrt{p}$.
4. Prove or disprove the following conjecture.
   Conjecture: If n is the least positive quadratic nonresidue modulo an odd prime number p, then $n < 1 + \sqrt{p}$.
5. Take a look at the development of the law of cubic reciprocity in Chapter 9 of Ireland and Rosen (1972).
6. Consider the Jacobi symbol $(\frac{a}{n})$ defined in Exercise 37. Provide a reason why this symbol was *not* defined to be

$$\left(\frac{a}{n}\right) = \begin{cases} 1, & \text{if } a \text{ is a quadratic residue modulo } n \\ -1, & \text{if } a \text{ is a quadratic nonresidue modulo } n \end{cases}$$

   in direct parallel with the definition of the Legendre symbol. [See page 146 of Niven, Zuckerman and Montgomery (1991) for a discussion of this problem.]
7. As mentioned at the end of Section 4.3, symbols more general than the Legendre symbol exist. One such symbol is the Jacobi symbol defined in Exercise 37. Another generalization of the Legendre symbol $(\frac{a}{p})$ is given by the *Hilbert residue symbol* $(\frac{a,b}{p})$ discussed on pages 22 and 23 of H. Cohn, *Introduction to the Construction of Class Fields* (Cambridge: Cambridge University Press, 1985). Illustrate the computation of the Hilbert residue symbol with several examples. In particular, show how the Hilbert residue symbol generalizes the Legendre symbol.

5

Primitive Roots

This chapter provides a synthesis of ideas presented in the first four chapters of this book; as such, the chapter provides an excellent opportunity for the reader to test his or her knowledge of Chapters 1 through 4. The primary goal here is to prove an important theorem in elementary number theory known as the *Primitive Root Theorem,* a precise characterization of positive integers possessing a certain property. (You will be invited to predict the statement of this theorem at various points during a rather extensive development.) The chapter concludes with a discussion of index arithmetic, an application of primitive roots to the solution of several congruences more general than those presented in Chapter 4. Further applications of material in this chapter can be found in Section 8.3.

One additional remark is in order here. Use of the binomial theorem is crucial in two lemmata leading to the proof of the Primitive Root Theorem. Accordingly, you may now wish to peruse a quick treatment of this important theorem in Appendix D.

5.1

The Order of an Integer; Primitive Roots

Let m be a positive integer and let a be an integer relatively prime to m. By Euler's Theorem, we have that $a^{\phi(m)} \equiv 1 \bmod m$. It may happen, however, that some smaller power than $\phi(m)$, say n, will result in $a^n \equiv 1 \bmod m$. (Consider, for example, $a = 2$, $m = 7$, and $n = 3$; we will return to this example in a moment.) This observation motivates the following definition.

Definition 1: Let $a, m \in \mathbf{Z}$ with $m > 0$ and $(a, m) = 1$. The *order of a modulo m,* denoted $\text{ord}_m a$, is the least positive integer n for which $a^n \equiv 1 \bmod m$.

Example 1:

(a) The order of 2 modulo 7 can be obtained by considering the positive powers of 2 in turn until one with a residue of 1 (modulo 7) is encountered. Accordingly, $2^1 \equiv 2 \bmod 7$, $2^2 \equiv 4 \bmod 7$, and $2^3 \equiv 1 \bmod 7$; so the order of 2 modulo 7 is 3, which is denoted $\text{ord}_7 2 = 3$. Note that Euler's Theorem guarantees that $\text{ord}_7 2 \leq \phi(7) = 6$ since $2^{\phi(7)} = 2^6 \equiv 1 \bmod 7$.

(b) The order of 3 modulo 7 can be obtained by considering the positive powers of 3 in turn until one with a residue of 1 (modulo 7) is encountered. Accordingly, $3^1 \equiv 3 \bmod 7$, $3^2 \equiv 2 \bmod 7$, $3^3 \equiv 6 \bmod 7$, $3^4 \equiv 4 \bmod 7$, $3^5 \equiv 5 \bmod 7$, and $3^6 \equiv 1 \bmod 7$; so the order of 3 modulo 7 is 6, which is denoted $\text{ord}_7 3 = 6$. Note that Euler's Theorem guarantees that $\text{ord}_7 3 \leq \phi(7) = 6$ since $3^{\phi(7)} = 3^6 \equiv 1 \bmod 7$.

We have seen that Euler's Theorem guarantees that $\text{ord}_m a \leq \phi(m)$. In fact, more is true; $\text{ord}_m a$ always *divides* $\phi(m)$ as given by the proposition below.

Proposition 5.1: Let $a, m \in \mathbf{Z}$ with $m > 0$ and $(a, m) = 1$. Then $a^n \equiv 1 \bmod m$ for some positive integer n if and only if $\text{ord}_m a \mid n$. In particular, $\text{ord}_m a \mid \phi(m)$.

Proof: ($\Rightarrow$) Assume that $a^n \equiv 1 \bmod m$ for some positive integer n. By the division algorithm, there exist $q, r \in \mathbf{Z}$ such that

$$n = (\text{ord}_m a)q + r, \qquad 0 \leq r < \text{ord}_m a$$

Then

$$a^n = a^{(\text{ord}_m a)q + r} = (a^{\text{ord}_m a})^q a^r \equiv a^r \bmod m$$

Since $a^n \equiv 1 \bmod m$, we have $a^r \equiv 1 \bmod m$. Now $0 \leq r < \text{ord}_m a$ and the definition of $\text{ord}_m a$ as the *least* positive power of a congruent to 1 modulo m together imply that $r = 0$, from which $n = (\text{ord}_m a)q$ and $\text{ord}_m a \mid n$ as desired. ($\Leftarrow$) Assume that $\text{ord}_m a \mid n$. Then there exists $b \in \mathbf{Z}$ with $b > 0$ such that $n = (\text{ord}_m a)b$. Now

$$a^n = a^{(\text{ord}_m a)b} = (a^{\text{ord}_m a})^b \equiv 1 \bmod m$$

as desired. ∎

Proposition 5.1 further restricts the possibilities for $\text{ord}_m a$ making its computation easier as motivated by the following example.

Example 2:

(a) By Proposition 5.1, we have $\text{ord}_7 2 \mid \phi(7) = 6$. So $\text{ord}_7 2$ is 1, 2, 3, or 6. [Example 1(a) shows that $\text{ord}_7 2 = 3$.]

(b) By Proposition 5.1, we have $\text{ord}_{13} 2 \mid \phi(13) = 12$. So $\text{ord}_{13} 2$ is 1, 2, 3, 4, 6, or 12. We have $2^1 \equiv 2 \bmod 13$, $2^2 \equiv 4 \bmod 13$, $2^3 \equiv 8 \bmod 13$, $2^4 \equiv 3 \bmod 13$, and $2^6 \equiv 12 \bmod 13$. Since we know that $2^{12} \equiv 1 \bmod 13$ by Euler's Theorem, we have that $\text{ord}_{13} 2 = 12$ (without having to compute 2^5, 2^7, 2^8, 2^9, 2^{10}, and 2^{11} modulo 13).

Another useful fact regarding the order of an integer a modulo m is given by the following proposition, which will be especially useful in Section 5.4.

Proposition 5.2: Let $a, m \in \mathbf{Z}$ with $m > 0$ and $(a, m) = 1$. If i and j are nonnegative integers, then $a^i \equiv a^j \bmod m$ if and only if $i \equiv j \bmod \mathrm{ord}_m a$.

Proof: Without loss of generality, we may assume that $i \geq j$.
($\Rightarrow$) Assume that $a^i \equiv a^j \bmod m$ or, equivalently, $a^j a^{i-j} \equiv a^j \bmod m$ (since $i \geq j$). Since $(a^j, m) = 1$, the inverse of a^j, say $(a^j)'$, exists by Corollary 2.8. Multiplying both sides of the congruence above by $(a^j)'$, we obtain

$$(a^j)'a^j a^{i-j} \equiv (a^j)'a^j \bmod m$$

from which $a^{i-j} \equiv 1 \bmod m$. By Proposition 5.1, we have that $\mathrm{ord}_m a \mid i - j$, from which $i \equiv j \bmod \mathrm{ord}_m a$ as desired.
($\Leftarrow$) Assume that $i \equiv j \bmod \mathrm{ord}_m a$. Then $\mathrm{ord}_m a \mid i - j$ and so there exists $n \in \mathbf{Z}$ with $n > 0$ such that $i - j = (\mathrm{ord}_m a)n$. Now

$$a^i = a^{(\mathrm{ord}_m a)n + j} = a^{(\mathrm{ord}_m a)n} a^j = (a^{\mathrm{ord}_m a})^n a^j \equiv a^j \bmod m$$

as desired. ■

We illustrate Proposition 5.2 with an example.

Example 3:

Let $a = 2$ and $m = 7$. Then $\mathrm{ord}_7 2 = 3$ by Example 1(a). Proposition 5.2 implies that if i and j are nonnegative integers, then $2^i \equiv 2^j \bmod 7$ if and only if $i \equiv j \bmod 3$. In other words, the congruence of 2^i and 2^j modulo 7 is determined precisely by the congruence of the exponents i and j modulo 3. For example, since $2000 \equiv 2 \bmod 3$, we could ascertain that $2^{2000} \equiv 2^2 \equiv 4 \bmod 7$.

As stated before, the largest possible order of an integer a modulo m is $\phi(m)$ by Euler's Theorem. We distinguish those integers having this maximum order with the following terminology.

Definition 2: Let $r, m \in \mathbf{Z}$ with $m > 0$ and $(r, m) = 1$. Then r is said to be a *primitive root modulo m* if $\mathrm{ord}_m r = \phi(m)$.

Example 4:

(a) 3 is a primitive root modulo 7 since $\mathrm{ord}_7 3 = 6 = \phi(7)$ by Example 1(b).
(b) 2 is a primitive root modulo 13 since $\mathrm{ord}_{13} 2 = 12 = \phi(13)$ by Example 2(b).
(c) 2 is not a primitive root modulo 7 since $\mathrm{ord}_7 2 = 3 \neq \phi(7)$ by Example 1(a).

Many questions about primitive roots now arise. Given any positive integer m, does there necessarily exist a primitive root modulo m? If not, what positive integers possess primitive roots? Can we obtain a characterization of those positive integers possessing primitive roots? If a positive integer possesses a primitive root, how many such primitive roots exist? Each question will be answered in forthcoming sections, although not necessarily in the order asked. Before reading further, experiment with primitive roots and formulate your own conjectures! (The following paragraph begins the extensive development that will culminate in an almost trivial proof of the central result of this chapter: the Primitive Root Theorem. This proof does not occur until the end of Section 5.3, so we have much to learn.)

The first question of the paragraph above is answered immediately by the following example.

Example 5:

Prove that there are no primitive roots modulo 8.

The reduced residues modulo 8 are 1, 3, 5, and 7. An easy verification shows that $\text{ord}_8 1 = 1$, $\text{ord}_8 3 = 2$, $\text{ord}_8 5 = 2$, and $\text{ord}_8 7 = 2$. Since none of the above orders are equal to $\phi(8) = 4$, we have that there are no primitive roots modulo 8.

So not all integers possess primitive roots. It *is* possible to determine precisely which integers possess primitive roots; the Primitive Root Theorem provides a complete list of such integers and answers the second and third questions of the paragraph above. (Have you formulated your own conjecture?)

A primitive root modulo m, if it exists, can be thought of as generating the reduced residues modulo m in the following sense.

Proposition 5.3: Let r be a primitive root modulo m. Then

$$\{r, r^2, r^3, \ldots, r^{\phi(m)}\}$$

is a set of reduced residues modulo m.

Proof: Since r is a primitive root modulo m, we have that $(r, m) = 1$, and so each element of the set is relatively prime to m. Since there are $\phi(m)$ such elements, it remains to show that no two elements of the set are congruent modulo m. Assume that $r^i \equiv r^j \bmod m$ where i and j are integers between 1 and $\phi(m)$ inclusive. Then Proposition 5.2 implies that $i \equiv j \bmod \phi(m)$, from which $i = j$ by the conditions on i and j. So no two elements of the set are congruent modulo m, and the proof is complete. ■

We give an explicit example that illustrates reduced residues being generated by a primitive root.

Example 6:

(a) 3 is a primitive root modulo 7 by Example 4(a). So, by Proposition 5.3, we have that $\{3, 3^2, 3^3, 3^4, 3^5, 3^6\}$ is a set of reduced residues modulo 7. Indeed, the reader may verify that the set $\{3, 3^2, 3^3, 3^4, 3^5, 3^6\}$, when viewed modulo 7, is $\{3, 2, 6, 4, 5, 1\}$, which is clearly a set of reduced residues modulo 7.

(b) 2 is a primitive root modulo 13 by Example 4(b). So, by Proposition 5.3, we have that $\{2, 2^2, 2^3, \ldots, 2^{12}\}$ is a set of reduced residues modulo 13. The reader may verify that this set, when viewed modulo 13, is $\{1, 2, 3, \ldots, 12\}$ in some permuted order, which is clearly a set of reduced residues modulo 13.

If a primitive root modulo m exists, it is usually not unique modulo m. In fact, we can determine exactly how many such primitive roots modulo m exist in terms of m. We conclude this section with a sequence of results that establishes this number. We begin with a proposition that expresses the order of an integral power of an integer in terms of the order of the integer.

Proposition 5.4: Let $a, m \in \mathbf{Z}$ with $m > 0$ and $(a, m) = 1$. If i is a positive integer, then

$$\text{ord}_m(a^i) = \frac{\text{ord}_m a}{(\text{ord}_m a, i)}$$

Proof: Put $d = (\text{ord}_m a, i)$. Then there exist $b, c \in \mathbf{Z}$ such that $\text{ord}_m a = db$, $i = dc$, and $(b, c) = 1$. Now

$$(a^i)^b = (a^{dc})^{(\text{ord}_m a)/d} = (a^c)^{\text{ord}_m a} = (a^{\text{ord}_m a})^c \equiv 1 \bmod m$$

By Proposition 5.1, we have that $\text{ord}_m(a^i) \mid b$. Also,

$$a^{i(\text{ord}_m(a^i))} = (a^i)^{\text{ord}_m(a^i)} \equiv 1 \bmod m$$

By Proposition 5.1, we have that $\text{ord}_m a \mid i(\text{ord}_m(a^i))$. So $db \mid dc(\text{ord}_m(a^i))$, and we have that $b \mid c(\text{ord}_m(a^i))$. Inasmuch as $(b, c) = 1$, we obtain $b \mid \text{ord}_m(a^i)$. So

$$\text{ord}_m(a^i) = b = \frac{\text{ord}_m a}{d} = \frac{\text{ord}_m a}{(\text{ord}_m a, i)}$$

as desired. ■

The following corollary of Proposition 5.4 is immediate.

Corollary 5.5: Let $a, m \in \mathbf{Z}$ with $m > 0$ and $(a, m) = 1$. If i is a positive integer, then $\text{ord}_m(a^i) = \text{ord}_m a$ if and only if $(\text{ord}_m a, i) = 1$.

Proof: The proof is immediate upon application of Proposition 5.4. ■

We may now use Corollary 5.5 above to obtain an expression for the number of incongruent primitive roots modulo m provided such primitive roots exist.

Corollary 5.6: If a primitive root modulo m exists, then there are exactly $\phi(\phi(m))$ incongruent primitive roots modulo m.

Proof: Let r be a primitive root modulo m. Then $\text{ord}_m r = \phi(m)$. By Proposition 5.3, any other primitive root must be congruent to exactly one element in the set $\{r, r^2, r^3, \ldots, r^{\phi(m)}\}$. If i is a positive integer such that $1 \le i \le \phi(m)$, then $\text{ord}_m(r^i) = \text{ord}_m r = \phi(m)$ if and only if $(\phi(m), i) = 1$ by Corollary 5.5; the number of such positive integers is $\phi(\phi(m))$ as desired. ■

Example 7:

(a) There are exactly $\phi(\phi(7)) = \phi(6) = 2$ incongruent primitive roots modulo 7 by Corollary 5.6. One of these roots is 3, as has been shown previously. By Corollary 5.5, $\text{ord}_7(3^i) = \text{ord}_7 3$ if and only if $(\text{ord}_7 3, i) = 1$. Inasmuch as $\text{ord}_7 3 = 6$, we take $i = 5$; since $3^5 \equiv 5 \bmod 7$, we have that 5 is the other incongruent primitive root modulo 7. (To test whether you fully understand what has transpired above, answer the following questions: Could we have taken $i = 7$ instead of $i = 5$? What about $i = 11$?)

(b) There are exactly $\phi(\phi(13)) = \phi(12) = 4$ incongruent primitive roots modulo 13 by Corollary 5.6. One of these roots is 2, as has been shown previously; you are invited to find the other primitive roots modulo 13 by using a procedure similar to that used in (a) above.

(c) Since $\phi(\phi(8)) = \phi(4) = 2$, do there exist two primitive roots modulo 8 by Corollary 5.6? The answer is a resounding *no.* Since there are no primitive roots modulo 8 by Example 5, Corollary 5.6 does not apply here!

Note also that Corollary 5.6 provides the answer to the fourth question in the paragraph after Example 4.

Exercise Set 5.1

1. ***(a)*** Find the order of 2 modulo 11.
   (b) Find the order of 3 modulo 13.
   (c) Find the order of 8 modulo 15.
   (d) Find the order of 9 modulo 17.
2. Prove that there are no primitive roots modulo each integer below.
   (a) 12
   (b) 15
   (c) 16
   (d) 20
3. Assuming that each integer below has a primitive root, determine the number of incongruent primitive roots. Find all such primitive roots.
   (a) 14
   (b) 17
   (c) 18
   (d) 19
4. Let $a, m \in \mathbf{Z}$ with $m > 0$. If a' is the inverse of a modulo m, prove that $\text{ord}_m a = \text{ord}_m a'$. Deduce that if r is a primitive root modulo m, then r' is a primitive root modulo m.
5. ***(a)*** Let m be a positive integer and let a and b be integers relatively prime to m with $(\text{ord}_m a, \text{ord}_m b) = 1$. Prove that $\text{ord}_m(ab) = (\text{ord}_m a)(\text{ord}_m b)$.

(b) Show that the hypothesis $(\text{ord}_m a, \text{ord}_m b) = 1$ cannot be eliminated from part (a). What can be said about $\text{ord}_m(ab)$ if $(\text{ord}_m a, \text{ord}_m b) \neq 1$?

6. Let m be a positive integer and let $d \mid \phi(m)$ with $d > 0$. Prove or disprove that there exists $a \in \mathbf{Z}$ with $\text{ord}_m a = d$.
7. Let m be a positive integer and let $a \in \mathbf{Z}$ with $(a, m) = 1$.
   (a) Prove that if $\text{ord}_m a = xy$ (with x and y positive integers), then $\text{ord}_m(a^x) = y$.
   (b) Prove that if $\text{ord}_m a = m - 1$, then m is a prime number.
8. Let a and n be positive integers with $a > 1$. Prove that $n \mid \phi(a^n - 1)$. (*Hint*: Consider $\text{ord}_{(a^n-1)}a$.)
9. Let p be an odd prime number and let r be an integer with $p \nmid r$. Prove that r is a primitive root modulo p if and only if $r^{(p-1)/q} \not\equiv 1 \bmod p$ for all prime divisors q of $p - 1$.

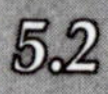

Primitive Roots for Prime Numbers

We have seen that not all positive integers possess primitive roots. In our attempt to characterize precisely those integers for which primitive roots exist, we prove in this section that primitive roots exist for all prime numbers. We will need the following result on polynomial congruences due to Lagrange.

Theorem 5.7: (Lagrange) Let p be a prime number and let

$$f(x) = a_n x^n + a_{n-1}x^{n-1} + \cdots + a_1 x + a_0$$

be a polynomial of degree $n \geq 1$ with integral coefficients such that $p \nmid a_n$. Then the congruence

$$f(x) \equiv 0 \bmod p$$

has at most n incongruent solutions modulo p.

Proof: We use induction on n. If $n = 1$, we have $f(x) = a_1 x + a_0$ with $p \nmid a_1$. Since $(p, a_1) = 1$, the linear congruence $a_1 x \equiv -a_0 \bmod p$ has exactly one incongruent solution modulo p by Theorem 2.6. So $f(x) \equiv 0 \bmod p$ has exactly one incongruent solution modulo p, and the theorem holds for $n = 1$. Assume that $k \geq 1$ and that the theorem holds for polynomials of degree $n = k$. We must show that the theorem holds for polynomials of degree $n = k + 1$. Accordingly, let

$$f(x) = a_{k+1}x^{k+1} + a_k x^k + \cdots + a_1 x + a_0$$

be a polynomial of degree $k + 1$ with integral coefficients such that $p \nmid a_{k+1}$. If the congruence $f(x) \equiv 0 \bmod p$ has no solutions, then the theorem holds and the proof is complete. Assume that the congruence $f(x) \equiv 0 \bmod p$ has at least one solution, say a. Dividing $f(x)$ by $x - a$, we obtain

$$f(x) = (x - a)q(x) + r$$

where $q(x)$ is a polynomial of degree k with integer coefficients and leading coefficient not divisible by p and r is an integer. Now

$$f(a) = (a - a)q(a) + r = r$$

Since $f(a) \equiv 0 \bmod p$, we have $r \equiv 0 \bmod p$ and so

$$f(x) \equiv (x - a)q(x) \bmod p$$

Let b be any solution of $f(x) \equiv 0 \bmod p$. Then

$$f(b) \equiv (b - a)q(b) \bmod p$$

If $b \not\equiv a \bmod p$, then $b - a \not\equiv 0 \bmod p$ and so $q(b) \equiv 0 \bmod p$. By the induction hypothesis, $q(x)$ has at most k incongruent solutions modulo p, from which $f(x)$ has at most $k + 1$ incongruent solutions modulo p and the proof is complete. ■

We now use Theorem 5.7 to prove the proposition below.

Proposition 5.8: Let p be a prime number and let $d \in \mathbf{Z}$ with $d > 0$ and $d \mid p - 1$. Then the congruence

$$x^d - 1 \equiv 0 \bmod p$$

has exactly d incongruent solutions modulo p.

Proof: Since $d \mid p - 1$, there exists $e \in \mathbf{Z}$ such that $p - 1 = de$. Then

$$\begin{aligned} x^{p-1} - 1 &= x^{de} - 1 \\ &= (x^d - 1)(x^{d(e-1)} + x^{d(e-2)} + \cdots + x^d + 1) \end{aligned}$$

The congruence $x^{p-1} - 1 \equiv 0 \bmod p$ has exactly $p - 1$ incongruent solutions modulo p (namely, $1, 2, \ldots, p - 1$) by Fermat's Little Theorem, and any such incongruent solution must be a solution of either the congruence $x^d - 1 \equiv 0 \bmod p$ or the congruence $x^{d(e-1)} + x^{d(e-2)} + \cdots + x^d + 1 \equiv 0 \bmod p$ (why?). Now the congruence

$$x^{d(e-1)} + x^{d(e-2)} + \cdots + x^d + 1 \equiv 0 \bmod p$$

has at most $d(e - 1) = p - 1 - d$ incongruent solutions modulo p by Theorem 5.7; so the congruence

$$x^d - 1 \equiv 0 \bmod p$$

has at least $(p - 1) - (p - 1 - d) = d$ incongruent solutions modulo p. But Theorem 5.7 implies that the congruence $x^d - 1 \equiv 0 \bmod p$ has at most d incongruent solutions modulo p, from which the congruence $x^d - 1 \equiv 0 \bmod p$ has exactly d incongruent solutions modulo p as desired. ■

That primitive roots exist for all prime numbers p will now follow as a trivial corollary of the theorem below, which was first proved by Legendre in 1785. In addition, the number of incongruent primitive roots modulo p will be obtained.

Theorem 5.9: Let p be a prime number and let $d \in \mathbf{Z}$ with $d > 0$ and $d \mid p - 1$. Then there are exactly $\phi(d)$ incongruent integers having order d modulo p.

Proof: Given a positive divisor d of $p - 1$, let $f(d)$ be the number of integers between 1 and $p - 1$ inclusive having order d. We must show that $f(d) = \phi(d)$ for all such d. Inasmuch as each integer between 1 and $p - 1$ inclusive has order dividing $p - 1$ by Proposition 5.1, we have

$$\sum_{d \mid p-1,\, d>0} f(d) = p - 1$$

By Theorem 3.5, we have

$$\sum_{d \mid p-1,\, d>0} \phi(d) = p - 1$$

So

$$\sum_{d \mid p-1,\, d>0} f(d) = \sum_{d \mid p-1,\, d>0} \phi(d)$$

In view of the equation above, to show $f(d) = \phi(d)$ for all positive divisors d of $p - 1$, it suffices to show $f(d) \leq \phi(d)$ for all such d. Given such a d, we have $f(d) = 0$ or $f(d) > 0$. If $f(d) = 0$, then $f(d) < \phi(d)$ is obvious. Assume that $f(d) > 0$. Then there is an integer between 1 and $p - 1$ inclusive, say a, such that $\mathrm{ord}_p a = d$. The d integers $a, a^2, a^3, \ldots, a^d$ are incongruent modulo p by Proposition 5.3; since each such integer is a solution of the congruence $x^d - 1 \equiv 0 \bmod p$ (why?), Proposition 5.8 implies that the incongruent solutions modulo p of the congruence $x^d \equiv 1 \bmod p$ are given precisely by $a, a^2, a^3, \ldots, a^d$. Exactly $\phi(d)$ of these d integers have order d modulo p by Corollary 5.5. So $f(d) = \phi(d)$ [which certainly implies $f(d) \leq \phi(d)$], and the proof is complete. ■

Corollary 5.10: Let p be a prime number. Then there are exactly $\phi(p - 1)$ incongruent primitive roots modulo p. In particular, primitive roots exist for all prime numbers.

Proof: A primitive root modulo p is an integer of order $\phi(p) = p - 1$. There are exactly $\phi(p - 1)$ such incongruent integers modulo p by Theorem 5.9. ■

One remark is in order here. The proof of Theorem 5.9 above, while asserting the existence of integers having certain orders, does not actually construct these integers. Such a construction may involve considerable effort (see Example 9 below). We now provide two examples that illustrate some of the important concepts of this chapter so far.

Example 8:

(a) Let $p = 7$. Theorem 5.9 implies that there exist integers of orders 1, 2, 3, and 6, these being the positive integers evenly dividing 6 (which is $p - 1$ here). In particular, Theorem 5.9 guarantees that there will be $\phi(1) = 1$ integer of order 1, $\phi(2) = 1$ integer of order 2, $\phi(3) = 2$ integers of order 3, and, as reiterated in Corollary 5.10, $\phi(6) = 2$ integers of order 6. The reader may verify that 1 is the integer of order 1, 6 is the integer of order 2, 2 and 4 are the integers of order 3, and, as already seen in Example 7(a), 3 and 5 are the integers of order 6.

(b) Let $p = 13$. Theorem 5.9 implies that there exist integers of orders 1, 2, 3, 4, 6, and 12, these being the positive integers evenly dividing 12 (which is $p - 1$ here). In particular, Theorem 5.9 guarantees that there will be 1 integer of order 1, 1 integer of order 2, 2 integers of order 3, 2 integers of order 4, 2 integers of order 6, and 4 integers of order 12. (Why?) The reader is invited to find all integers of each order.

Example 9:

Find all incongruent integers having order 6 modulo 19.

Note that by Theorem 5.9 there will be $\phi(6) = 2$ such integers. These two integers could be determined, of course, through an exhaustive search of the orders of the integers from 1 to 18 inclusive, but we opt for a different approach. By Exercise 3(d), we have that 2 is a primitive root modulo 19. Now Proposition 5.3 tells us that the first 18 positive integral powers of this primitive root will generate all reduced residues modulo 19. Furthermore, Proposition 5.4 allows us to compute the orders of these integral powers of 2; we have

$$\text{ord}_{19}(2^i) = \frac{\text{ord}_{19}2}{(\text{ord}_{19}2, i)}$$

We want $\text{ord}_{19}(2^i) = 6$; since $\text{ord}_{19}2 = 18$, we have that $\text{ord}_{19}(2^i) = 6$ if and only if $(18, i) = 3$. So $i = 3$ or $i = 15$. We have

$$2^3 \equiv 8 \bmod 19$$

and

$$2^{15} \equiv (2^4)^3 2^3 \equiv (-3)^3 2^3 \equiv (-27)(8) \equiv (-8)(8) \equiv -64 \equiv 12 \bmod 19$$

so that 8 and 12 are the desired incongruent integers having order 6 modulo 19.

The technique of Example 9 above relies on the knowledge of a primitive root modulo 19. Finding such a primitive root is a matter of trial and error; for the reader's convenience, the least positive primitive root modulo p for each prime number p less than 100 is given in Table 5.1.

It can be proven that the least positive primitive root modulo p increases without bound as p increases without bound. (This happens very slowly; for example, the largest least positive primitive root modulo p for prime numbers p less than 1000 occurs when $p = 409$, and the least positive primitive root in

Table 5.1

p	Least positive primitive root modulo p	p	Least positive primitive root modulo p
2	1	43	3
3	2	47	5
5	2	53	2
7	3	59	2
11	2	61	2
13	2	67	2
17	3	71	7
19	2	73	5
23	5	79	3
29	2	83	2
31	3	89	3
37	2	97	5
41	6		

this case is only 21.) The frequency of occurrence of 2 as the least positive primitive root modulo p in the table suggests that perhaps 2 may be the least positive primitive root for infinitely many prime numbers p. The answer to this question is unknown, so we state the following conjecture.

Conjecture 1: There are infinitely many prime numbers p for which 2 is the least positive primitive root modulo p.

It is a fact that 2 is the least positive primitive root for 67 of the 168 prime numbers less than 1000. In 1927, Artin conjectured that any integer, other than -1 or a perfect square, is a primitive root for infinitely many prime numbers p.

Conjecture 2: (Artin) If r is a nonsquare integer other than -1, then there are infinitely many prime numbers p for which r is a primitive root modulo p.

This problem is still open today, although D. R. Heath-Brown proved in 1986 that there are at most two integers for which Artin's Conjecture fails (see also Student Project 7).

Exercise Set 5.2

10. Determine the number of incongruent primitive roots modulo each prime number below.

(a) 41
(b) 43
(c) 47
(d) 53

11. *(a)* Find all incongruent integers having order 6 modulo 31.

(b) Find all incongruent integers having order 4 modulo 37.
(c) Find all incongruent integers having order 14 modulo 43.
(d) Find all incongruent integers having order 10 modulo 61.

12. Let p be an odd prime number.
(a) Prove that any primitive root r modulo p is a quadratic nonresidue modulo p. Deduce that $r^{(p-1)/2} \equiv -1 \bmod p$.
(b) Prove that there are exactly $\frac{p-1}{2} - \phi(p-1)$ incongruent quadratic nonresidues modulo p that are not primitive roots modulo p.

13. (a) Let p be a prime number with $p \equiv 1 \bmod 4$ and let r be a primitive root modulo p. Prove that $-r$ is also a primitive root modulo p.
(b) Let p be a prime number with $p \equiv 3 \bmod 4$ and let r be a primitive root modulo p. Prove that $\text{ord}_p(-r) = \frac{p-1}{2}$.

14. Let p be an odd prime number and let r be a primitive root modulo p.
(a) If r' is a primitive root modulo p, prove that rr' is not a primitive root modulo p.
(b) If r' is an integer such that $rr' \equiv 1 \bmod p$, prove that r' is a primitive root modulo p.
(c) If $p > 3$, prove that primitive roots modulo p occur in pairs r, r' where $rr' \equiv 1 \bmod p$.

15. Let p be a prime number. Prove that the product of all incongruent primitive roots modulo p is congruent to $(-1)^{\phi(p-1)}$ modulo p. Deduce that if $p > 3$, then the product of all incongruent primitive roots modulo p is congruent to 1 modulo p.

16. Let p be an odd prime number and let r be a primitive root modulo p.
(a) Prove that the quadratic residues modulo p are congruent to the even powers of r modulo p and the quadratic nonresidues modulo p are congruent to the odd powers of r modulo p.
(b) Prove that the product of the quadratic residues modulo p is congruent to $r^{(p^2-1)/4}$ modulo p while the product of the quadratic nonresidues modulo p is congruent to $r^{(p-1)^2/4}$.

17. (a) Let n be a positive integer and let $p = 2^n + 1$ be a prime number. Prove that every quadratic nonresidue modulo p is a primitive root modulo p.
(b) Prove that a prime number p is a Fermat prime if and only if every quadratic nonresidue modulo p is a primitive root modulo p.

18. Let p and q be odd prime numbers with $q = 4p + 1$.
(a) If a is any quadratic nonresidue modulo q with $\text{ord}_q a \neq 4$, prove that a is a primitive root modulo q.
(b) Prove that 2 is a primitive root modulo q.
(c) Use part (b) to find five prime numbers for which 2 is a primitive root.

19. Let p and q be odd prime numbers with $q = 2p + 1$.
(a) Prove that -4 is a primitive root modulo q.
(b) If $p \equiv 1 \bmod 4$, prove that 2 is a primitive root modulo q.
(c) If $p \equiv 3 \bmod 4$, prove that -2 is a primitive root modulo q.

20. [This exercise presents a shorter and more elegant way to prove $f(d) = \phi(d)$ in Theorem 5.9 using the Möbius Inversion Formula (Theorem 3.15). All notation in this exercise is identical to that in Theorem 5.9.] Let d be a positive divisor of $p - 1$.

(a) Prove that

$$\sum_{c \mid d,\, d>0} f(c) = d$$

(b) Use part (a) and the Möbius Inversion Formula to prove that

$$f(d) = \sum_{c \mid d,\, d>0} \mu(c)\frac{d}{c}$$

(c) Prove that $f(d) = \phi(d)$. (*Hint*: See Example 11 of Chapter 3.)

21. (a) Find the number of incongruent roots modulo 6 of the polynomial $f(x) = x^2 - x$.
(b) Why does the behavior exhibited in part (a) not violate Lagrange's Theorem (Theorem 5.7)?

22. (a) Let p be a prime number and let $f(x)$ be a polynomial of degree $n \geq 1$ with integral coefficients. Prove that if $f(x)$ has more than n incongruent solutions modulo p, then p divides every coefficient of $f(x)$.
(b) Let p be a prime number. Prove that every coefficient of the polynomial

$$f(x) = (x - 1)(x - 2)\cdots(x - (p - 1)) - (x^{p-1} - 1)$$

is divisible by p.
(c) Using part (b), give an alternate proof of Wilson's Theorem (Theorem 2.11).

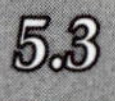

5.3

The Primitive Root Theorem

This section culminates in the Primitive Root Theorem, a precise characterization of those integers for which primitive roots exist. We hope that you have examined the evidence to date quite carefully and made your own conjectures. The Primitive Root Theorem and its proof were first published by Gauss in 1801. We begin with two propositions that limit those integers that we must consider.

Proposition 5.11: There are no primitive roots modulo 2^n where $n \in \mathbf{Z}$ and $n \geq 3$. (See Example 5 for the case $n = 3$.)

Proof: Note that a primitive root modulo 2^n must be odd and have order $\phi(2^n) = 2^{n-1}$. Let a be any odd integer. To prove that no primitive root modulo 2^n exists, it suffices to show that

$$a^{2^{n-2}} \equiv 1 \bmod 2^n$$

We use induction on n. If $n = 3$, we must show that $a^2 \equiv 1 \bmod 8$, which is

Exercise 13(b) of Chapter 2. Assume that $k \geq 3$ and that the desired result holds for $n = k$ so that $a^{2^{k-2}} \equiv 1 \bmod 2^k$. We must show that $a^{2^{k-1}} \equiv 1 \bmod 2^{k+1}$ so that the desired result holds for $n = k + 1$. By the induction hypothesis, we have $2^k \mid a^{2^{k-2}} - 1$, so there exists $b \in \mathbf{Z}$ such that $a^{2^{k-2}} - 1 = 2^k b$. Then $a^{2^{k-2}} = 2^k b + 1$; squaring both sides of this equation, we obtain

$$\begin{aligned} a^{2^{k-1}} &= (2^k b)^2 + 2(2^k b) + 1 \\ &= 2^{k+1}(2^{k-1}b^2 + b) + 1 \\ &\equiv 1 \bmod 2^{k+1} \end{aligned}$$

as desired, and the proof is complete. ■

Proposition 5.12: There are no primitive roots modulo mn where $m, n \in \mathbf{Z}$, $m, n > 2$, and $(m, n) = 1$.

Proof: We will show that the order of any integer modulo mn is less than or equal to $\frac{\phi(mn)}{2}$; since a primitive root modulo mn must have order $\phi(mn)$, this suffices to prove the proposition. Let $a \in \mathbf{Z}$ with $(a, mn) = 1$. Then $(a, m) = 1$ and $(a, n) = 1$. Put $d = (\phi(m), \phi(n))$ and $k = [\phi(m), \phi(n)]$. Since $\phi(m)$ and $\phi(n)$ are both even [Exercise 13(a) of Chapter 3], we have $d \geq 2$. So

$$k = \frac{\phi(m)\phi(n)}{d} = \frac{\phi(mn)}{d} \leq \frac{\phi(mn)}{2}$$

By Euler's Theorem, we have $a^{\phi(m)} \equiv 1 \bmod m$. Taking the $\frac{\phi(n)}{d}$th power of both sides of this congruence, we obtain

$$a^k = a^{\phi(m)\phi(n)/d} = (a^{\phi(m)})^{\phi(n)/d} \equiv 1^{\phi(n)/d} \equiv 1 \bmod m$$

Similarly,

$$a^k \equiv 1 \bmod n$$

Since $(m, n) = 1$, we have

$$a^k \equiv 1 \bmod mn$$

In other words, the order of any integer modulo mn does not exceed k, which, in turn, does not exceed $\frac{\phi(mn)}{2}$. The proof is complete. ■

The cruxes of Propositions 5.11 and 5.12 may be summed up quite nicely in the following corollary.

Corollary 5.13: Let $n \in \mathbf{Z}$ with $n > 0$. If n is divisible by two distinct odd prime numbers or n is divisible by an odd prime number and 4, then there are no primitive roots modulo n. Consequently, if a primitive root modulo n exists, then n must be equal to 1, 2, 4, p^m, or $2p^m$ where p is an odd prime number and m is a positive integer.

Proof: The first statement to be proven follows immediately from Proposition 5.12. The second statement to be proven follows immediately from the first statement and Proposition 5.11. ■

Is your conjecture still valid? (The next paragraph provides a statement of the Primitive Root Theorem if you have not guessed it already.)

The Primitive Root Theorem states that the integers 1, 2, 4, p^m, and $2p^m$ as in Corollary 5.13 are *precisely* those positive integers for which primitive roots exist; so we must prove the truth of the converse of the last statement of Corollary 5.13, namely, that a primitive root exists for each integer 1, 2, 4, p^m, and $2p^m$. (The hard part is proving the existence of a primitive root modulo p^m.) The following series of results culminates in the proof of this important fact. The binomial theorem is used twice in the following discussion; if necessary, see Appendix D for a discussion of the binomial theorem.

Lemma 5.14: Let p be an odd prime number. Then there exists a primitive root r modulo p such that

$$r^{p-1} \not\equiv 1 \bmod p^2$$

Proof: Let r be any primitive root modulo p. If $r^{p-1} \not\equiv 1 \bmod p^2$, then the proof is complete. Assume that $r^{p-1} \equiv 1 \bmod p^2$. Put $r' = r + p$. Since $r' \equiv r \bmod p$, we have that r' is a primitive root modulo p. We will show that $(r')^{p-1} \not\equiv 1 \bmod p^2$. Assume, by way of contradiction, that $(r')^{p-1} \equiv 1 \bmod p^2$. We have

$$\begin{aligned}
(r')^{p-1} &= (r + p)^{p-1} \\
&= r^{p-1} + (p-1)r^{p-2}p + \binom{p-1}{2}r^{p-3}p^2 + \cdots \\
&\quad + \binom{p-1}{p-2}rp^{p-2} + p^{p-1} \\
&\equiv r^{p-1} + (p-1)r^{p-2}p \bmod p^2 \\
&\equiv r^{p-1} + p^2r^{p-2} - pr^{p-2} \bmod p^2 \\
&\equiv r^{p-1} - pr^{p-2} \bmod p^2 \\
&\equiv 1 - pr^{p-2} \bmod p^2 \quad (\text{since } r^{p-1} \equiv 1 \bmod p^2)
\end{aligned}$$

Since $(r')^{p-1} \equiv 1 \bmod p^2$, we have

$$1 - pr^{p-2} \equiv 1 \bmod p^2$$

from which $pr^{p-2} \equiv 0 \bmod p^2$. So $p^2 \mid pr^{p-2}$, which implies that $p \mid r^{p-2}$, giving $p \mid r$, which is a contradiction since r is a primitive root modulo p. So $(r')^{p-1} \not\equiv 1 \bmod p^2$, and the proof is complete. ■

Corollary 5.15: Let p be an odd prime number and let r be a primitive root modulo p. Then either r or $r + p$ is a primitive root modulo p^2.

Proof: Since r is a primitive root modulo p, we have $\text{ord}_p r = \phi(p) = p - 1$. Let $n = \text{ord}_{(p^2)} r$. Then

$$r^n \equiv 1 \bmod p^2$$

and, by Proposition 5.1,

$$n \mid \phi(p^2) = p(p-1) \tag{1}$$

Also, $r^n \equiv 1 \bmod p^2$ implies $r^n \equiv 1 \bmod p$, from which

$$\phi(p) = p - 1 \mid n \tag{2}$$

by Proposition 5.1. Now (1) and (2) imply that either $n = p - 1$ or $n = p(p-1)$. If $n = p(p-1) = \phi(p^2)$, then r is a primitive root modulo p^2 and the proof is complete. If $n = p - 1$, put $r' = r + p$. Since $r' \equiv r \bmod p$, we have that r' is a primitive root modulo p and, by considerations similar to those above, $\text{ord}_{(p^2)} r' = p - 1$ or $\text{ord}_{(p^2)} r' = p(p-1)$. The proof of Lemma 5.14 shows that $\text{ord}_{(p^2)} r' \neq p - 1$. So $\text{ord}_{(p^2)} r' = p(p-1) = \phi(p^2)$, and the proof is complete. ■

Lemma 5.16: Let p be an odd prime number and let r be a primitive root modulo p such that $r^{p-1} \not\equiv 1 \bmod p^2$. If $m \in \mathbf{Z}$ and $m \geq 2$, then

$$r^{p^{m-2}(p-1)} \not\equiv 1 \bmod p^m$$

Proof: We use induction on m. If $m = 2$, the desired statement is obvious by hypothesis. Assume that $k \geq 2$ and that the desired statement holds for $m = k$ so that $r^{p^{k-2}(p-1)} \not\equiv 1 \bmod p^k$. We must show that $r^{p^{k-1}(p-1)} \not\equiv 1 \bmod p^{k+1}$ so that the desired statement holds for $m = k + 1$. Since $(r, p) = 1$, we have $(r, p^{k-1}) = 1$ and, by Euler's Theorem,

$$r^{p^{k-2}(p-1)} = r^{\phi(p^{k-1})} \equiv 1 \bmod p^{k-1}$$

So $p^{k-1} \mid r^{p^{k-2}(p-1)} - 1$; consequently, there exists $a \in \mathbf{Z}$ such that $r^{p^{k-2}(p-1)} - 1 = ap^{k-1}$. Furthermore, $p \nmid a$ for otherwise $r^{p^{k-2}(p-1)} \equiv \bmod p^k$ contrary to the induction hypothesis. Now $r^{p^{k-2}(p-1)} = 1 + ap^{k-1}$, and, raising each side of this equation to the pth power, we obtain

$$\begin{aligned} r^{p^{k-1}(p-1)} &= (1 + ap^{k-1})^p \\ &= 1 + p(ap^{k-1}) + \binom{p}{2}(ap^{k-1})^2 + \cdots \\ &\quad + \binom{p}{p-1}(ap^{k-1})^{p-1} + (ap^{k-1})^p \\ &\equiv 1 + ap^k \bmod p^{k+1} \\ &\not\equiv 1 \bmod p^{k+1} \quad (\text{since } p \nmid a) \end{aligned}$$

as desired, and the proof is complete. ■

Proposition 5.17: Let p be an odd prime number and let m be a positive integer. Then there exists a primitive root modulo p^m.

Proof: The proof for $m = 1$ is a consequence of Corollary 5.10. By Lemma 5.14 and Lemma 5.16, there exists a primitive root r modulo p such that

$$r^{p^{m-2}(p-1)} \not\equiv 1 \bmod p^m \tag{3}$$

for all $m \in \mathbf{Z}$ with $m \geq 2$. We show that r is a primitive root modulo p^m. Let $n = \text{ord}_{(p^m)}r$. Then

$$r^n \equiv 1 \bmod p^m$$

and, by Proposition 5.1,

$$n \mid \phi(p^m) = p^{m-1}(p-1) \tag{4}$$

Also, $r^n \equiv 1 \bmod p^m$ implies $r^n \equiv 1 \bmod p$, from which

$$\phi(p) = p - 1 \mid n \tag{5}$$

by Proposition 5.1. Now (4) and (5) imply that $n = p^k(p-1)$ for some integer k between 0 and $m-1$ inclusive. If $0 \leq k \leq m-2$, we have

$$r^{p^{m-2}(p-1)} = (r^{p^k(p-1)})^{p^{m-2-k}} = (r^n)^{p^{m-2-k}} \equiv (1)^{p^{m-2-k}} \equiv 1 \bmod p^m$$

contradicting (3). So $k = m - 1$ and

$$\text{ord}_{(p^m)}r = n = p^{m-1}(p-1) = \phi(p^m)$$

from which r is a primitive root modulo p^m as desired. ■

Corollary 5.18: Let p be an odd prime number and let m be a positive integer. Then there exists a primitive root modulo $2p^m$.

Proof: Let r be a primitive root modulo p^m. Without loss of generality, we may assume that r is odd. (If r is even, then $r + p^m$ is odd and a primitive root modulo p^m.) So $(r, 2p^m) = 1$. Let $n = \text{ord}_{(2p^m)}r$. Then

$$r^n \equiv 1 \bmod 2p^m$$

and, by Proposition 5.1,

$$n \mid \phi(2p^m) = \phi(2)\phi(p^m) = \phi(p^m) \tag{6}$$

Also, $r^n \equiv 1 \bmod 2p^m$ implies $r^n \equiv 1 \bmod p^m$, from which

$$\phi(p^m) \mid n \tag{7}$$

by Proposition 5.1. Now (6) and (7) imply

$$\text{ord}_{(2p^m)}r = n = \phi(p^m) = \phi(2p^m)$$

from which r is a primitive root modulo $2p^m$ as desired. ■

We are now able to conclude this section with the formal statement of the Primitive Root Theorem, the central result of this chapter. We warn you again that the proof will be extremely brief due to the vast amount of prior work undertaken.

Theorem 5.19: (Primitive Root Theorem) Let n be a positive integer. Then a primitive root modulo n exists if and only if n is equal to 1, 2, 4, p^m, or $2p^m$ where p is an odd prime number and m is a positive integer.

Proof: ($\Rightarrow$) This is the last statement of Corollary 5.13.
($\Leftarrow$) It is easily checked that 1 is a primitive root modulo 1, 1 is a primitive root

modulo 2, and 3 is a primitive root modulo 4. Proposition 5.17 and Corollary 5.18 prove the existence of primitive roots modulo p^m and $2p^m$. ■

The least positive primitive root modulo n for each of those integers n with $1 \le n \le 1000$ for which such a primitive root exists (in accordance with Theorem 5.19 above) is given in Table 3 of Appendix E.

Exercise Set 5.3

23. Determine the number of incongruent primitive roots modulo each integer below.
(a) 60
(b) 61
(c) 62
(d) 63
(e) 64
(f) 65

24. Find a primitive root, for all positive integral m, modulo each integer below.
(a) 7^m (*Hint*: Using Corollary 5.15, find a common primitive root r modulo 7 and 7^2. The proof of Proposition 5.17 then guarantees that r is a primitive root modulo 7^m for all positive integral m.)
(b) 11^m
(c) 13^m
(d) 17^m

25. Find a primitive root, for all positive integral m, modulo each integer below.
(a) $2 \cdot 7^m$ (*Hint*: Use the result of Exercise 24(a) along with the proof of Corollary 5.18.)
(b) $2 \cdot 11^m$
(c) $2 \cdot 13^m$
(d) $2 \cdot 17^m$

26. Let p be an odd prime number and let m be a positive integer.
(a) Prove that the number of incongruent primitive roots modulo $2p^m$ is the same as the number of incongruent primitive roots modulo p^m.
(b) Prove that any primitive root modulo p^m is also a primitive root modulo p.

27. Let $m \in \mathbf{Z}$ with $m > 2$. If a primitive root modulo m exists, prove that the only incongruent solutions of the congruence $x^2 \equiv 1 \bmod m$ are $x \equiv \pm 1 \bmod m$.

28. (Gauss) Let m be a positive integer.
(a) If a primitive root modulo m exists, prove that the product of all positive integers not exceeding m and relatively prime to m is congruent to -1 modulo m. [The special case of this result for m a prime number is Wilson's Theorem (Theorem 2.11).]
(b) What is the least nonnegative residue modulo m of the product of all positive integers not exceeding m and relatively prime to m if no primitive root modulo m exists? Prove your assertion.

5.4

Index Arithmetic; nth Power Residues

This section gives an application of primitive roots to the solution of certain congruences. We first recall a pertinent fact. Let m be a positive integer possessing a primitive root and let r be a primitive root modulo m. Then the powers of r given by $r, r^2, r^3, \ldots, r^{\phi(m)}$ are congruent, in some order, to a set of reduced residues modulo m by Proposition 5.3. Stated a bit differently, any reduced residue modulo m will be congruent (modulo m) to exactly one of the powers of r given by $r, r^2, r^3, \ldots, r^{\phi(m)}$. This observation allows us to make the following definition.

Definition 3: Let r be a primitive root modulo m. If $a \in \mathbf{Z}$ with $(a, m) = 1$, then the *index of a relative to r*, denoted $\text{ind}_r a$, is the least positive integer n for which $r^n \equiv a \bmod m$.

In view of the observation immediately preceding Definition 3, we note that $\text{ind}_r a$ must exist and that $1 \leq \text{ind}_r a \leq \phi(m)$. We illustrate with an example.

Example 10:

Example 6 shows that 3 is a primitive root modulo 7 and that $3^1 \equiv 3 \bmod 7$, $3^2 \equiv 2 \bmod 7$, $3^3 \equiv 6 \bmod 7$, $3^4 \equiv 4 \bmod 7$, $3^5 \equiv 5 \bmod 7$, and $3^6 \equiv 1 \bmod 7$. So modulo 7, we have $\text{ind}_3 1 = 6$, $\text{ind}_3 2 = 2$, $\text{ind}_3 3 = 1$, $\text{ind}_3 4 = 4$, $\text{ind}_3 5 = 5$, and $\text{ind}_3 6 = 3$.

Note also that if a and b are integers relatively prime to m and $a \equiv b \bmod m$, then $\text{ind}_r a = \text{ind}_r b$.

Indices enjoy properties similar to the properties of logarithms. These properties are summarized in the proposition below. Here the primitive root plays the role of the base of the logarithm and, instead of equalities as in logarithms, we obtain congruences modulo $\phi(m)$.

Proposition 5.20: Let r be a primitive root modulo m and let $a, b \in \mathbf{Z}$ with $(a, m) = 1$ and $(b, m) = 1$.

(a) $\text{ind}_r 1 \equiv 0 \bmod \phi(m)$

(b) $\text{ind}_r r \equiv 1 \bmod \phi(m)$

(c) $\text{ind}_r(ab) \equiv \text{ind}_r a + \text{ind}_r b \bmod \phi(m)$

(d) $\text{ind}_r(a^n) \equiv n\, \text{ind}_r a \bmod \phi(m)$, if n is a positive integer.

Proof: (a) and (b) are obvious. For (c), by definition, we have $r^{\text{ind}_r a} \equiv a \bmod m$ and $r^{\text{ind}_r b} \equiv b \bmod m$. Multiplying the left-hand and right-hand sides of these congruences, we obtain

$$r^{\text{ind}_r a + \text{ind}_r b} \equiv ab \bmod m$$

Since $r^{\mathrm{ind}_r(ab)} \equiv ab \bmod m$, we have

$$r^{\mathrm{ind}_r a + \mathrm{ind}_r b} \equiv r^{\mathrm{ind}_r(ab)} \bmod m$$

Now Proposition 5.2 gives

$$\mathrm{ind}_r a + \mathrm{ind}_r b \equiv \mathrm{ind}_r(ab) \bmod \phi(m)$$

as desired. For (d), by definition we have $r^{\mathrm{ind}_r(a^n)} \equiv a^n \bmod m$. Also, we have

$$r^{n\,\mathrm{ind}_r a} = (r^{\mathrm{ind}_r a})^n \equiv a^n \bmod m$$

So

$$r^{\mathrm{ind}_r(a^n)} \equiv r^{n\,\mathrm{ind}_r a} \bmod m$$

and Proposition 5.2 gives

$$\mathrm{ind}_r(a^n) \equiv n\,\mathrm{ind}_r a \bmod \phi(m)$$

as desired. ■

Example 11:

We illustrate Proposition 5.20(c) [which is read "the index of a product is congruent to the sum of the indices modulo $\phi(m)$"]. Modulo 7, we have $\mathrm{ind}_3 2 = 2$ and $\mathrm{ind}_3 3 = 1$ (see Example 10). So, by Proposition 5.20(c), we have

$$\begin{aligned} \mathrm{ind}_3 6 = \mathrm{ind}_3(2 \cdot 3) &\equiv \mathrm{ind}_3 2 + \mathrm{ind}_3 3 \bmod \phi(7) \\ &\equiv 2 + 1 \bmod 6 \\ &\equiv 3 \bmod 6 \end{aligned}$$

So $\mathrm{ind}_3 6 = 3$. (Compare with Example 10!)

We now show how indices may be used to solve certain congruences more general than those considered to date. Let m be an integer possessing a primitive root, say r, and let $a, b \in \mathbf{Z}$ with $(a, m) = 1$ and $(b, m) = 1$. Consider the congruence in the variable x given by

$$ax^n \equiv b \bmod m \qquad \textbf{(8)}$$

where n is a positive integer. We have $\mathrm{ind}_r(ax^n) = \mathrm{ind}_r b$; since

$$\mathrm{ind}_r(ax^n) \equiv \mathrm{ind}_r a + \mathrm{ind}_r(x^n) \equiv \mathrm{ind}_r a + n\,\mathrm{ind}_r x \bmod \phi(m)$$

by (c) and (d) of Proposition 5.20, respectively, we obtain

$$\mathrm{ind}_r a + n\,\mathrm{ind}_r x \equiv \mathrm{ind}_r b \bmod \phi(m)$$

or, equivalently,

$$n\,\mathrm{ind}_r x \equiv \mathrm{ind}_r b - \mathrm{ind}_r a \bmod \phi(m) \qquad \textbf{(9)}$$

Hence, solving (8) is equivalent to solving (9). But (9) is a linear congruence in the variable $\mathrm{ind}_r x$, which can be solved using the techniques in Section 2.2. Do you see what has happened here? Indices have transformed a congruence involving an arbitrary power of x (which we currently *cannot* solve elegantly) into a linear congruence in the one variable $\text{ind}_r x$ (which we *can* solve elegantly)! We now illustrate the solution procedure with a concrete example.

Example 12:

Use indices to find all incongruent solutions of the congruence $6x^4 \equiv 3 \bmod 7$.

We use indices relative to the primitive root 3 modulo 7 as given in Example 10; indices relative to any primitive root modulo 7 could be used. By virtue of the discussion above, we have that the congruence

$$6x^4 \equiv 3 \bmod 7$$

is equivalent to the congruence

$$4\,\text{ind}_3 x \equiv \text{ind}_3 3 - \text{ind}_3 6 \bmod \phi(7)$$

(We strongly suggest that you actually perform the steps in this derivation.) Since $\text{ind}_3 3 = 1$ and $\text{ind}_3 6 = 3$ (see Example 10) as well as $\phi(7) = 6$, we have

$$4\,\text{ind}_3 x \equiv 4 \bmod 6$$

Treating the congruence above as a linear congruence in the variable $\text{ind}_3 x$ and using the techniques in Section 2.2, we obtain $\text{ind}_3 x \equiv 1 \bmod 6$ or $\text{ind}_3 x \equiv 4 \bmod 6$; so $x \equiv 3^1 \equiv 3 \bmod 7$ or $x \equiv 3^4 \equiv 4 \bmod 7$. Consequently, there are two incongruent solutions of the congruence $6x^4 \equiv 3 \bmod 7$, namely, 3 and 4.

We now consider congruences that are slightly less general than those considered above. At this point, the reader is requested to recall the definition of a quadratic residue modulo a positive integer m. (See Definition 1 of Chapter 4 if necessary.) In view of this definition, the definition below seems natural.

Definition 4: Let $a, m, n \in \mathbf{Z}$ with $m, n > 0$ and $(a, m) = 1$. Then a is said to be an nth *power residue modulo* m if the congruence $x^n \equiv a \bmod m$ is solvable in $\mathbf{Z}$.

Example 13:

(a) 6 is a 3rd power residue modulo 7 since the congruence $x^3 \equiv 6 \bmod 7$ is solvable ($x = 3$ is a solution; are there any others?);

(b) 3 is a 4th power residue modulo 13 since the congruence $x^4 \equiv 3 \bmod 13$ is solvable ($x = 2$ is a solution; are there any others?);

(c) 3 is not a 4th power residue modulo 7 since the congruence $x^4 \equiv 3 \bmod 7$ is not solvable (none of 1, 2, 3, 4, 5, or 6 yields a true congruence when substituted for x).

Is there any way to predict whether a given integer will be an nth power residue modulo m beyond the potentially cumbersome substitution procedure used in Example 13(c)? The following theorem presents a necessary and sufficient condition for an integer to be an nth power residue modulo m *provided that a primitive root modulo m exists.* Furthermore, if the integer is an nth power residue modulo m, the theorem gives the number of incongruent solutions modulo m of the associated congruence.

Theorem 5.21: Let $a, m, n \in \mathbf{Z}$ with $m, n > 0$ and $(a, m) = 1$. If a primitive root modulo m exists, then a is an nth power residue modulo m if and only if

$$a^{\phi(m)/d} \equiv 1 \bmod m$$

where $d = (n, \phi(m))$. Furthermore, if a is an nth power residue modulo m, the congruence $x^n \equiv a \bmod m$ has exactly d incongruent solutions modulo m.

Proof: Let r be a primitive root modulo m. Then the congruence

$$x^n \equiv a \bmod m \tag{10}$$

is equivalent to the congruence

$$n \operatorname{ind}_r x \equiv \operatorname{ind}_r a \bmod \phi(m) \tag{11}$$

(Why?) If $d = (n, \phi(m))$, the linear congruence (11) is solvable (for $\operatorname{ind}_r x$) if and only if $d \mid \operatorname{ind}_r a$ by Theorem 2.6; $d \mid \operatorname{ind}_r a$ if and only if

$$\left(\frac{\phi(m)}{d}\right) \operatorname{ind}_r a \equiv 0 \bmod \phi(m)$$

which is equivalent to the congruence

$$a^{\phi(m)/d} \equiv 1 \bmod m$$

(Why?) This proves the first part of the theorem. [Why are we assured of finding an x that solves (10) if we have found an $\operatorname{ind}_r x$ that solves (11)?] Furthermore, Theorem 2.6 also implies that if $d \mid \operatorname{ind}_r a$, then (11) has exactly d incongruent solutions modulo $\phi(m)$. So if $d \mid \operatorname{ind}_r a$, then (10) has exactly d incongruent solutions modulo m as desired (why?), which completes the proof. ■

We note immediately that Theorem 5.21 has the following corollary.

Corollary 5.22: Let p be an odd prime number and let $a \in \mathbf{Z}$ such that $p \nmid a$. Then a is a quadratic residue modulo p if and only if

$$a^{(p-1)/2} \equiv 1 \bmod p$$

Furthermore, if a is a quadratic residue modulo p, the congruence $x^2 \equiv a \bmod p$ has exactly two incongruent solutions modulo p.

Proof: Let $m = p$ and $n = 2$ in Theorem 5.21. (Convince yourself of this!) ■

It is particularly instructive at this point for the reader to compare Corollary 5.22 above with Euler's Criterion (Theorem 4.4) and Proposition 4.1. Are you impressed with the power of Theorem 5.21 above?

We conclude this section with several examples.

Example 14:

(a) Let $a = 6$, $m = 7$, and $n = 3$ in Theorem 5.21 and note that a primitive root modulo m exists by Theorem 5.19. Then $d = (3, \phi(7)) = 3$ and, by Theorem 5.21, we have that 6 is a 3rd power residue modulo 7 if and only if $6^{6/3} \equiv 1 \bmod 7$. Since the congruence is true, we see that 6 is a 3rd power residue modulo 7. [Compare with Example 13(a).] Furthermore, the congruence $x^3 \equiv 6 \bmod 7$ has exactly three incongruent solutions modulo 7. [Did you find these solutions in Example 13(a)?]

(b) Let $a = 3$, $m = 13$, and $n = 4$ in Theorem 5.21 and note that a primitive root modulo m exists by Theorem 5.19. Then $d = (4, \phi(13)) = 4$ and, by Theorem 5.21, we have that 3 is a 4th power residue modulo 13 if and only if $3^{12/4} \equiv 1 \bmod 13$. Since the congruence is true, we see that 3 is a 4th power residue modulo 13. [Compare with Example 13(b).] Furthermore, the congruence $x^4 \equiv 3 \bmod 13$ has exactly four incongruent solutions modulo 13. [Did you find these solutions in Example 13(b)?]

(c) Let $a = 3$, $m = 7$, and $n = 4$ in Theorem 5.21 and note that a primitive root modulo m exists by Theorem 5.19. Then $d = (4, \phi(7)) = 2$ and, by Theorem 5.21, we have that 3 is a 4th power residue modulo 7 if and only if $3^{6/2} \equiv 1 \bmod 7$. Since the congruence is *not* true, we see that 3 is *not* a 4th power residue modulo 7. [Compare with Example 13(c).]

Example 15:

Find all 15th power residues modulo 9.

Note that a primitive root modulo 9 exists by Theorem 5.19. Then $d = (15, \phi(9)) = 3$ and, by Theorem 5.21, we have that a is a 15th power residue modulo 9 if and only if $a^{6/3} \equiv 1 \bmod 9$ or, equivalently, $a^2 \equiv 1 \bmod 9$. The only integers of 1, 2, 3, 4, 5, 6, 7, and 8 that satisfy this congruence are 1 and 8. So there are two incongruent 15th power residues modulo 9, namely, 1 and 8. The 15th power residues modulo 9 are precisely the 3rd power residues modulo 9; indeed, if $a \in \mathbf{Z}$ with $(a, 9) = 1$, we have $a^6 \equiv 1 \bmod 9$ by Euler's Theorem, from which $a^{15} \equiv a^3 \bmod 9$.

Exercise Set 5.4

29. *(a)* Find all indices relative to 2 modulo 13.
(b) Find all indices relative to 3 modulo 17.

30. Use indices to find all incongruent solutions of each congruence below.
(a) $8x^7 \equiv 5 \bmod 13$
(b) $7x^9 \equiv 6 \bmod 13$
(c) $9^x \equiv 10 \bmod 13$

(d) $2^x \equiv x \bmod 13$
(e) $7x^5 \equiv 2 \bmod 17$
(f) $8x^{12} \equiv 9 \bmod 17$
(g) $6^x \equiv 11 \bmod 17$
(h) $x^x \equiv x \bmod 17$

31. Find all positive integers a for which each congruence is solvable.
(a) $ax^9 \equiv 6 \bmod 13$
(b) $ax^{12} \equiv 9 \bmod 17$

32. Find all positive integers b for which each congruence is solvable.
(a) $7x^9 \equiv b \bmod 13$
(b) $9^x \equiv b \bmod 13$
(c) $8x^{12} \equiv b \bmod 17$
(d) $6^x \equiv b \bmod 17$

33. Use Theorem 5.21 to determine whether each congruence is solvable. If the congruence is solvable, find the number of incongruent solutions.
(a) $x^3 \equiv 11 \bmod 14$
(b) $x^4 \equiv 12 \bmod 17$
(c) $x^5 \equiv 13 \bmod 18$
(d) $x^6 \equiv 7 \bmod 19$

34. Let p be an odd prime number and let r be a primitive root modulo p. Prove that $\text{ind}_r(p-1) = \frac{p-1}{2}$.

35. Let p be a prime number and let r and s be primitive roots modulo p. Let $a \in \mathbf{Z}$ with $p \nmid a$.
(a) Prove that $\text{ind}_s a \equiv (\text{ind}_r a)(\text{ind}_s r) \bmod (p-1)$. (*Note*: This congruence corresponds to the change-of-base formula for logarithms.)
(b) Prove that $\text{ind}_r(p-a) \equiv \text{ind}_r a + \frac{p-1}{2} \bmod (p-1)$. (*Note*: This congruence yields all indices relative to r after half of these indices are computed.)

36. (a) Let p be an odd prime number. Prove that the congruence $x^4 \equiv -1 \bmod p$ is solvable if and only if $p \equiv 1 \bmod 8$.
(b) Prove that there are infinitely many prime numbers expressible in the form $8n+1$ where n is a positive integer. (*Hint*: Parallel Exercise 26 from Chapter 4.)

37. Let $a \in \mathbf{Z}$ and let p be a prime number with $p \geq 5$ and $p \nmid a$.
(a) If $p \equiv 1 \bmod 6$, prove that the congruence $x^3 \equiv a \bmod p$ has either no solutions or exactly three incongruent solutions modulo p.
(b) If $p \equiv 5 \bmod 6$, prove that the congruence $x^3 \equiv a \bmod p$ has a unique solution modulo p.

38. Let p be a prime number. If n is a positive integer with $(n, p-1) = 1$, prove that

$$\{1^n, 2^n, 3^n, \ldots, (p-1)^n\}$$

is a reduced residue system modulo p.

39. Let p be an odd prime number and let r be a primitive root modulo p. Let n be a positive integer and let $d = (n, p-1)$. Prove that the congruence $x^n \equiv a \bmod p$ is solvable if and only if a is congruent to one of $r^d, r^{2d}, r^{3d}, \ldots, r^{((p-1)/d)d}$ modulo p.

40. Let p be an odd prime number and let r be a primitive root modulo p. If a

and b are integers and $d = (\text{ind}_r a, p - 1)$, prove that the congruence $a^x \equiv b \bmod p$ is solvable if and only if $d \mid \text{ind}_r b$. In such a case of solvability, prove that there are exactly d incongruent solutions modulo p.

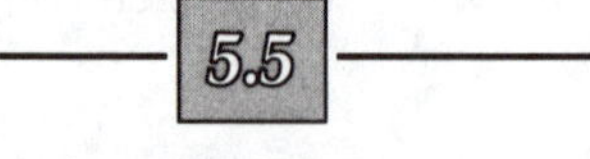

Concluding Remarks

The study of primitive roots is essential in elementary number theory. Besides being important in their own right, primitive roots play important roles in primality testing algorithms (see Section 8.3) and in many other applications (see, for example, Student Project 7 in Chapter 8). We wish, however, to take this time to make an apology. The true beauty of the material in this chapter cannot be appreciated without knowing some group and ring theory from abstract algebra. (A brief treatment of the necessary concepts can be found in Appendix C.) Inasmuch as a prior course in abstract algebra is not assumed as a prerequisite for this book, we have treated primitive roots without regard to their more natural algebraic setting. If you have had the benefit of such a course, we hope that you will translate the material of this chapter into the algebraic setting (if you have not done so already!). If you have not yet taken such a course, you are encouraged to undertake such a translation after a first course in abstract algebra (or after reading Appendix C). (We cannot resist any longer for those readers who know the necessary algebra: Notice that the primitive root modulo m is simply a generator of the group of units $\mathbf{Z}_m^\times$; in this context, the Primitive Root Theorem characterizes those moduli m for which this group of units is cyclic!)

Student Projects

1. (Programming project for calculator or computer)
   Given positive integers a and m with $(a, m) = 1$, determine whether a primitive root modulo m exists. If so, compute the least positive such primitive root, say r, and use this primitive root to compute $\text{ind}_r a$.
2. The following statement, similar in form to Proposition 5.2, is false:
   Let $m \in \mathbf{Z}$ with $m > 0$. If i and j are positive integers such that $i \equiv j \bmod m$, then $a^i \equiv a^j \bmod m$ for all integers a.
   In fact, this statement is true for precisely five values of m. Find as many of these values as you can.
   [A discussion of this problem can be found in J. Dyer-Bennet, "A Theorem on Partitions of the Set of Positive Integers," *American Mathematical Monthly, 47* (1940), 152–154.]
3. ***(a)*** Let m be a positive integer with prime factorization $p_1^{a_1}p_2^{a_2}\cdots p_t^{a_t}$ and let $a \in \mathbf{Z}$ with $(a, m) = 1$. If $n = [\phi(p_1^{a_1}), \phi(p_2^{a_2}), \ldots, \phi(p_t^{a_t})]$, prove that $a^n \equiv 1 \bmod m$.
   (b) ***Definition:*** Let m be a positive integer. A positive integer n such that $a^n \equiv 1 \bmod m$ for all integers a relatively prime to m is said to be a

universal exponent of m. The minimal such universal exponent of m is denoted $\lambda(m)$.

Let m be a positive integer. Prove that $\lambda(m) \leq \phi(m)$. Characterize those positive integers m for which $\lambda(m) = \phi(m)$.

(c) Let m be a positive integer with prime factorization $2^{a_0}p_1^{a_1}p_2^{a_2}\cdots p_t^{a_t}$. Prove that

$$\lambda(m) = [\lambda(2^{a_0}), \phi(p_1^{a_1}), \phi(p_2^{a_2}), \ldots, \phi(p_t^{a_t})]$$

and that there exists $a \in \mathbf{Z}$ such that $\operatorname{ord}_m a = \lambda(m)$.

4. If n is a Carmichael number (see Exercise 65 of Chapter 2), prove that n is a product of distinct prime numbers, say $n = p_1p_2\cdots p_m$, and $p_i - 1 \mid n - 1$ for all i. [*Hint*: Prove that $\lambda(n) \mid n - 1$ where $\lambda(n)$ is defined as in Student Project 3 above.]

5. Prove that a Carmichael number must have at least three prime factors. (*Hint*: Use Student Project 4 above.)

[Much more on Carmichael numbers can be found in Ribenboim (1991).]

6. Let r be a primitive root modulo the odd prime number p. For integral i with $1 \leq i \leq p - 1$, let a_i be the integer in the set $\{0, 1, 2, \ldots, p - 2\}$ such that

$$r^{a_i} \equiv i \bmod p$$

Prove that the integers

$$a_2 - a_1, a_3 - a_2, a_4 - a_3, \ldots, a_{p-1} - a_{p-2}$$

are congruent modulo $p - 1$ to the integers $1, 2, 3, \ldots, p - 2$ in some permuted order.

(The problem above appears as Elementary Problem B-459 in the October 1981 edition of the *Fibonacci Quarterly*. A solution appears in that magazine's November 1982 edition.)

7. Read the following article: M. Ram Murty, "Artin's Conjecture for Primitive Roots," *Mathematical Intelligencer, 10* (1988), 59–67.

6

Diophantine Equations

Diophantine equations are named after the Greek mathematician Diophantus of Alexandria and refer to any equations with one or more variables to be solved in the integers. This is a convenient time to state this as a formal definition.

Definition 0: Any equation with one or more variables to be solved in the integers is said to be a *Diophantine equation.*

Diophantus studied similar equations in his great work *Arithmetica,* the earliest known systematic treatment of algebra. As we will see, a given Diophantine equation may or may not be solvable; our goal in this chapter is to investigate the solvability of certain important Diophantine equations in elementary number theory. Both linear and nonlinear equations will be studied; this latter class will lead to a discussion of several important topics including Pythagorean triples, Fermat's Last Theorem (perhaps the most famous unsolved problem in all of number theory), and representations of integers as sums of squares.

6.1 Linear Diophantine Equations

We begin with a well-known class of Diophantine equations.

Definition 1: Let $a_1, a_2, \dots, a_n,\ b \in \mathbf{Z}$ with $a_1, a_2, \dots, a_n$ nonzero. A Diophantine equation of the form

$$a_1x_1 + a_2x_2 + \cdots + a_nx_n = b$$

is said to be a *linear Diophantine equation* in the *n variables* $x_1, x_2, \dots, x_n$.

Biography

Diophantus of Alexandria (circa A.D. 250)

Very little is known about Diophantus's life beyond the fact that he prospered at Alexandria (but see Exercise 1). His most famous work, *Arithmetica,* consists essentially of the solution of approximately 130 problems, some of which are quite challenging. Most of these problems concern solutions of indeterminate equations in varying degrees and numbers of variables. Although Diophantus was not the first to work with such equations, he may have been the first to use a type of algebraic notation in their solutions. Curiously enough, Diophantus did not restrict his solutions to integers as is conventional today when dealing with "Diophantine" equations; indeed, Diophantus recognized rational solutions.

Linear refers to the fact that all variables appear to the first power and that no products of variables are involved. For example, the Diophantine equation $x + 2y = 3$ is a linear equation in the two variables x and y, while the Diophantine equations $x^2 + 2y = 3$ and $2xy = 3$ are not linear since, in the former case, the variable x appears with exponent 2 and, in the latter case, the product xy appears.

We now give criteria for the solvability of linear Diophantine equations in cases with one or two variables. A remark on the solvability of such equations with more than two variables will conclude this section. The case for linear Diophantine equations with one variable is immediate.

Theorem 6.1: Let $ax = b$ be a linear Diophantine equation in the one variable x. If $a \nmid b$, then the equation has no solutions; *if* $a \mid b$, then the equation has exactly one solution, namely, $x = \frac{b}{a}$.

Proof: The proof is left to the reader. ■

Examples illustrating the use of Theorem 6.1 are also left to the reader.

The case for linear Diophantine equations with two variables is given by the following theorem. The proof of this theorem bears a resemblance to a proof encountered previously in this book. See if you can figure out which prior proof is similar to the proof of the theorem below as well as the reason for the similarity.

Theorem 6.2: Let $ax + by = c$ be a linear Diophantine equation in the two variables x and y and let $d = (a, b)$. If $d \nmid c$, then the equation has no solutions; if $d \mid c$, then the equation has infinitely many solutions. Furthermore, if x_0, y_0 is a particular solution of the equation, then all solutions are given by x, y where $x = x_0 + (\frac{b}{d})n$ and $y = y_0 - (\frac{a}{d})n$ where $n \in \mathbf{Z}$.

Proof: Since $d \mid a$ and $d \mid b$, we have $d \mid c$ by Proposition 1.2. So if $d \nmid c$, then the given linear Diophantine equation has no solutions. Assume that $d \mid c$. By Proposition 1.11, there exist $r, s \in \mathbf{Z}$ such that

$$d = (a, b) = ar + bs$$

Furthermore, $d \mid c$ implies $c = de$ for some $e \in \mathbf{Z}$. Then

$$c = de = (ar + bs)e = a(re) + b(se)$$

So $x = re$ and $y = se$ solve $ax + by = c$. Let x_0, y_0 be any particular solution of $ax + by = c$. Then, if $n \in \mathbf{Z}$, $x = x_0 + (\frac{b}{d})n$, and $y = y_0 - (\frac{a}{d})n$, it is straightforward to show that x, y is a solution of $ax + by = c$. We now show that every solution x, y of $ax + by = c$ takes this form. Let x, y be any solution of $ax + by = c$. Then

$$(ax + by) - (ax_0 + by_0) = c - c = 0$$

So

$$a(x - x_0) = b(y_0 - y)$$

Dividing both sides of this equation by d, we have

$$\left(\frac{a}{d}\right)(x - x_0) = \left(\frac{b}{d}\right)(y_0 - y) \qquad \textbf{(1)}$$

Now $\frac{b}{d} \mid (\frac{a}{d})(x - x_0)$; since $(\frac{a}{d}, \frac{b}{d}) = 1$ by Proposition 1.10, we have $\frac{b}{d} \mid x - x_0$ so that $x - x_0 = (\frac{b}{d})n$ for some $n \in \mathbf{Z}$. So $x = x_0 + (\frac{b}{d})n$ for some $n \in \mathbf{Z}$. Substituting $x = x_0 + (\frac{b}{d})n$ in (1) yields $y = y_0 - (\frac{a}{d})n$ as desired, which completes the proof. ■

The two examples below illustrate the use of Theorem 6.2.

Example 1:

Determine whether the Diophantine equation $24x + 60y = 15$ is solvable or not solvable. If it is solvable, find all solutions.

By Example 11(a) of Chapter 1, we have $(24, 60) = 12$; since $12 \nmid 15$, we have that the given Diophantine equation is not solvable by Theorem 6.2.

Example 2:

Determine whether the Diophantine equation $803x + 154y = 11$ is solvable or not solvable. If it is solvable, find all solutions.
By Example 14 of Chapter 1, we have $(803, 154) = 11$; since $11 \mid 11$, we have that the given Diophantine equation is solvable by Theorem 6.2. By Example 15 of Chapter 1, a particular solution of $803x + 154y = 11$ is given by $x_0 = 5$ and $y_0 = -26$. So by Theorem 6.2 again, all solutions of $803x + 154y = 11$ are given by

$$x = 5 + \left(\frac{154}{11}\right)n = 5 + 14n$$

and

$$y = -26 - \left(\frac{803}{11}\right)n = -26 - 73n$$

where $n \in \mathbf{Z}$. (In Example 15 of Chapter 1 we stated that 11 had infinitely many expressions as an integral linear combination of 803 and 154; this example provides all of these linear combinations.)

By now, the perceptive reader has noticed the resemblance of the proof of Theorem 6.2 to the proof of Theorem 2.6. This resemblance is no coincidence. In the proof of Theorem 2.6, the linear congruence in one variable given by $ax \equiv b \bmod m$ was immediately translated into the equivalent equation $ax - my = b$. But this equation is a linear Diophantine equation in the two variables x and y! So the proof of Theorem 6.2 could have appeared in Chapter 2. We opted instead to collect all the results concerning Diophantine equations in one place; we hope that such a focus facilitates understanding. The connections between Diophantine equations and congruences should be clear, however. In fact, the use of congruence techniques in the solution of Diophantine equations will be investigated in Section 6.2.

Theorem 6.2 can be generalized to linear Diophantine equations in n variables. This generalization is investigated in Exercises 6 and 8.

Exercise Set 6.1

1. Very little is known about Diophantus's life. One of the only known details comes from the following epigram found in the *Greek Anthology*: "Diophantus passed one sixth of his life in childhood, one twelfth in youth, and one seventh more as a bachelor. Five years after his marriage was born a son who died four years before his father, at half his father's age." Use this epigram to determine Diophantus's age upon his death.
2. Find all integral solutions of each linear Diophantine equation.

(*a*) $18x + 28y = 10$
(*b*) $18x + 27y = 15$
(*c*) $28x + 12y = 20$
(*d*) $12x + 9y = 27$
(*e*) $17x + 29y = 10$

3. Apples and oranges at a particular grocery store cost 14¢ and 17¢ each, respectively. A customer spends a total of \$2.90 for apples and oranges. How many pieces of each fruit did the customer buy?
4. On the night of a certain banquet, a caterer offered the choice of two dinners — a steak dinner for \$8 and a lobster dinner for \$13. At the end of the evening, the caterer's receipts totaled \$1571. What is the minimum number of people who could have attended the banquet? What is the maximum number of people who could have attended the banquet?
5. (*a*) A woman wishes to mail a package. The postal clerk has determined the postage for the package to be \$1.83. If the clerk has only 12¢ and 15¢ stamps available, how many of each type of stamp should be used if exact postage is desired?
   (*b*) If the postal clerk of part (a) desires to use the minimum number of stamps possible to mail the package, how many of each type of stamp should be used?
   (*c*) If a man wishes to mail a package for which the postage has been determined by the postal clerk in part (a) to be \$2.00, how many of each type of stamp should be used if exact postage is desired?
6. Find all integral solutions of each linear Diophantine equation.
   (*a*) $9x_1 + 12x_2 + 16x_3 = 13$
   (*Hint*: The Diophantine equation $9x_1 + 12x_2 = 3y$ is solvable for any integer y. Solve the Diophantine equation $3y + 16x_3 = 13$ and use the general value of y so obtained to solve $9x_1 + 12x_2 = 3y$.)
   (*b*) $8x_1 - 4x_2 + 6x_3 = 6$
   (*c*) $3x_1 + 5x_2 + 2x_3 + 6x_4 = 15$
7. At the same grocery store as in Exercise 3, plums cost 9¢ each. If a customer spends a total of \$4.04 for apples, oranges, and plums, how many pieces of each fruit does the customer buy?
8. Let $a_1x_1 + a_2x_2 + \cdots + a_nx_n = b$ be a linear Diophantine equation in the $n\ (\geq 2)$ variables $x_1, x_2, \ldots, x_n$ and let $d = (a_1, a_2, \ldots, a_n)$.
   (*a*) If $d \nmid b$, prove that the equation has no solutions.
   (*b*) If $d \mid b$, prove that the equation has infinitely many solutions. (*Hint*: Use the idea motivated in Exercise 6 and mathematical induction.)
9. Let a, b, and c be positive integers with $(a, b) = 1$ and consider the linear Diophantine equation $ax + by = c$.
   (*a*) Prove that the equation has a solution in positive integers x and y if $c > ab$. (*Hint*: Using the notation of Theorem 6.2, we have that $x > 0$ if and only if $n > -dx_0/b$ and that $y > 0$ if and only if $n < dy_0/a$. So the solutions of $ax + by = c$ in positive integers are given by those integers n such that $-dx_0/b < n < dy_0/a$. Now use the fact that $ax_0 + by_0 = c$ to show that at least one such n exists.)
   (*b*) Prove that the equation has no solution in positive integers x and y if $c = ab$.

(c) Prove that the equation has at most one solution in positive integers x and y if $c < ab$.

10. Let a, b, and c be positive integers with $(a, b) = 1$ and consider the linear Diophantine equation $ax + by = c$.

(a) Prove that the equation has a solution in nonnegative integers x and y if $c > ab - a - b$.

(b) Prove that the equation has no solution in nonnegative integers x and y if $c = ab - a - b$.

(c) Prove that there are exactly $\frac{(a-1)(b-1)}{2}$ values of c with $c < ab - a - b$ for which the equation has a solution in nonnegative integers x and y.

Nonlinear Diophantine Equations; a Congruence Method

Having just discussed *linear* Diophantine equations, you might be expecting the definition below.

Definition 2: A Diophantine equation is said to be *nonlinear* if it is not linear.

We give several examples of important nonlinear Diophantine equations to be studied in subsequent sections.

Example 3:

(a) The Diophantine equation $x^2 + y^2 = z^2$ is nonlinear since the variables are squared. This Diophantine equation will be studied more deeply in Section 6.3.

(b) Let $n \in \mathbf{Z}$ with $n \geq 3$. The Diophantine equation $x^n + y^n = z^n$ is nonlinear since the variables are raised to powers other than 1. This Diophantine equation will be studied more deeply in Section 6.4.

(c) Let $n \in \mathbf{Z}$. The Diophantine equation $x^2 + y^2 = n$ is nonlinear. (Why?) This Diophantine equation will be studied more deeply in Section 6.5.

(d) Let $d, n \in \mathbf{Z}$. The Diophantine equation $x^2 - dy^2 = n$ is nonlinear. (Why?) This Diophantine equation will be studied more deeply in Section 8.4. Note that the Diophantine equation in (c) is a special case of this Diophantine equation (with $d = -1$).

We now describe a method that can sometimes be used to show that a given Diophantine equation has no solutions. [You have perhaps used this method before; see, for example, Exercise 13(c) of Chapter 2.] The method is based on the fact that if a Diophantine equation has a solution, then the corresponding congruence obtained by considering the equation modulo any positive integer m also has a solution. By contraposition then, if a positive integer m can be found so that a Diophantine equation viewed as a congruence modulo m has no solutions, then the original Diophantine equation also has no solutions. We illustrate this congruence method with two examples.

Example 4:

Prove that the Diophantine equation $3x^2 + 2 = y^2$ is not solvable.
Assume, by way of contradiction, that $3x^2 + 2 = y^2$ is solvable. Any solution of this Diophantine equation must also be a solution of the congruence $3x^2 + 2 \equiv y^2 \bmod 3$. Now $3x^2 + 2 \equiv y^2 \bmod 3$ being solvable implies that the congruence $2 \equiv y^2 \bmod 3$ is solvable from which it follows that 2 is a quadratic residue modulo 3, a contradiction. So the Diophantine equation $3x^2 + 2 = y^2$ is not solvable.

An important remark is in order here. Viewing the given Diophantine equation modulo 2 would result in the assumption that $3x^2 + 2 \equiv y^2 \bmod 2$ is solvable and so imply that the congruence $x^2 \equiv y^2 \bmod 2$ is solvable. *This is true* (simply take $x = y$). Unfortunately, this tells you *nothing* about the solvability of the original Diophantine equation! In other words, a successful congruence test in no way implies that the original Diophantine equation is solvable. In general, *the congruence method outlined prior to Example 4 is useful only for proving that a Diophantine equation has no solutions.* Remember this!

Example 5:

Prove that the Diophantine equation $7x^3 + 2 = y^3$ is not solvable.
Assume, by way of contradiction, that $7x^3 + 2 = y^3$ is solvable. Any solution of this Diophantine equation must also be a solution of the congruence $7x^3 + 2 \equiv y^3 \bmod 7$. Now $7x^3 + 2 \equiv y^3 \bmod 7$ being solvable implies that the congruence $2 \equiv y^3 \bmod 7$ is solvable. Modulo 7, there are seven choices for y, namely, 0, 1, 2, 3, 4, 5, and 6. None of these values for y solves the congruence $2 \equiv y^3 \bmod 7$, so this congruence is not solvable, a contradiction. (One could also invoke Theorem 5.21 here.) So the Diophantine equation $7x^3 + 2 = y^3$ is not solvable. (Could we also prove the nonsolvability of $7x^3 + 2 = y^3$ by viewing this equation modulo 2? What about modulo 3?)

Exercise Set 6.2

11. Prove or disprove the following statements.
(a) The Diophantine equation $3x^2 - 7y^2 = 2$ has no integral solutions.
(b) The Diophantine equation $3x^2 - 5y^2 = 7$ has no integral solutions.
(c) The Diophantine equation $4x^2 + 3 = 5y^2$ has no integral solutions.
(d) The Diophantine equation $x^3 - 5 = 7y^3$ has no integral solutions.
(e) The Diophantine equation $2x^3 - 7y^3 = 3$ has no integral solutions.

12. Prove or disprove the following statements.

(a) The Diophantine equation $x^2 + y^2 + 1 = 4z$ has no integral solutions.

(b) The Diophantine equation $x^2 + y^2 + 3 = 4z$ has no integral solutions.

(c) The Diophantine equation $x^2 + 2y^2 + 1 = 8z$ has no integral solutions.

(d) The Diophantine equation $x^2 + 2y^2 + 3 = 8z$ has no integral solutions.

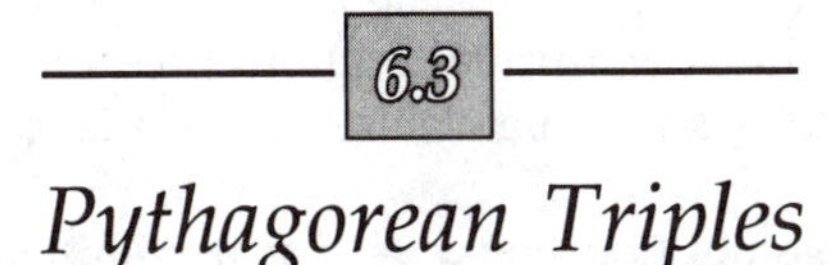

Pythagorean Triples

Perhaps one of the most memorable results from high school geometry and trigonometry is the Pythagorean Theorem, which states that the square of the length of the hypotenuse of a right triangle is equal to the sum of the squares of the lengths of the other two sides. If z denotes the length of the hypotenuse of this right triangle and x and y denote the lengths of the other two sides, respectively, we obtain the relationship $z^2 = x^2 + y^2$. This observation motivates the following definition.

Definition 3: A triple x, y, z of positive integers satisfying the Diophantine equation $x^2 + y^2 = z^2$ is said to be a *Pythagorean triple.*

Can you recall any Pythagorean triples from high school geometry or trigonometry?

Example 6:

(a) The triple 3, 4, 5 is a Pythagorean triple since $3^2 + 4^2 = 5^2$. The 3-4-5 right triangle is a triangle encountered frequently in high school mathematics textbooks.

(b) Any positive integral multiple of a Pythagorean triple is a Pythagorean triple. For example 6, 8, 10 is a Pythagorean triple since it is twice the Pythagorean triple of (a).

(c) The triple 5, 12, 13 is a Pythagorean triple since $5^2 + 12^2 = 13^2$. The 5-12-13 right triangle is another triangle encountered frequently in high school mathematics textbooks.

(d) The triples -3, 4, 5 and 0, 1, 1 are solutions of the Diophantine equation $x^2 + y^2 = z^2$ but are *not* Pythagorean triples, since integers in a Pythagorean triple must all be positive by definition.

Unlike most nonlinear Diophantine equations, all integral solutions of $x^2 + y^2 = z^2$ can be explicitly described; in other words, it is possible to obtain a characterization of these solutions. (In fact, Euclid essentially gave such a description in Book X of his *Elements*, perhaps without realizing it.) We make three preliminary conventions. First, we will only describe solutions of $x^2 + y^2 = z^2$ with $x, y, z > 0$. Equivalently, our description will only give solutions of $x^2 + y^2 = z^2$ that are Pythagorean triples; we will *not* characterize solutions as in Example 6(d). It is clear that all other solutions are obtainable by changing the sign of one or more of the components of a Pythagorean triple x, y, z (since x, y, and z are all squared) and by considering the cases where $x = 0$, $y = 0$, or both.

Second, it is clear that, if x, y, z is a Pythagorean triple and $(x, y, z) = d$, then $\frac{x}{d}, \frac{y}{d}, \frac{z}{d}$ is also a Pythagorean triple; so we will only describe Pythagorean triples x, y, z where $(x, y, z) = 1$. Such Pythagorean triples are said to be *primitive*. All other Pythagorean triple solutions are obtainable from primitive Pythagorean triple solutions in a manner similar to that illustrated in Example 6(b).

Finally, under the assumption that x, y, z is a primitive Pythagorean triple, we show that exactly one of x and y is even. Assume that x and y are both even. Then z is even and $(x, y, z) = 1$ is contradicted. Now assume that x and y are both odd. Then z is even. (Why?) Now Exercise 13 of Chapter 2 yields $x^2 \equiv 1 \bmod 4$, $y^2 \equiv 1 \bmod 4$, and $z^2 \equiv 0 \bmod 4$; since $x^2 + y^2 = z^2$, we have that $x^2 + y^2 \equiv z^2 \bmod 4$ or $1 + 1 \equiv 0 \bmod 4$, a contradiction. (Note the use of congruences here; similar methods will be useful in the exercises). So exactly one of x and y is even; we will describe primitive Pythagorean triples where, without loss of generality, y is even. All other primitive Pythagorean triple solutions x, y, z are obtainable from those with y even by interchanging the roles of x and y.

We may now state the desired characterization of primitive Pythagorean triples x, y, z with y even.

Theorem 6.3: There are infinitely many primitive Pythagorean triples x, y, z with y even. Furthermore, they are given precisely by the equations

$$x = m^2 - n^2$$

$$y = 2mn$$

$$z = m^2 + n^2$$

where $m, n \in \mathbf{Z}$, $m > n > 0$, $(m, n) = 1$, and exactly one of m and n is even.

Before proving Theorem 6.3, we illustrate it with an example.

Example 7:

(a) $m = 2$ and $n = 1$ satisfy the conditions on m and n of Theorem 6.3. Substituting these values for m and n in the equations for x, y, and z in Theorem 6.3, we obtain the primitive Pythagorean triple 3, 4, 5.

(b) $m = 3$ and $n = 2$ satisfy the conditions on m and n of Theorem 6.3. Substituting these values for m and n in the equations for x, y, and z in Theorem 6.3, we obtain the primitive Pythagorean triple 5, 12, 13.

Pick your own values for m and n to satisfy the conditions of Theorem 6.3 above and generate your own primitive Pythagorean triples!

We now prove Theorem 6.3. Be prepared to provide several fundamental justifications for yourself during the proof.

Proof: (of Theorem 6.3) We first show that given a primitive Pythagorean triple with y even, there exist m and n as described. Since y is even, x and z are both odd. (Why?) Moreover, $(x, y) = 1$, $(y, z) = 1$, and $(x, z) = 1$. (Why?) Now

$$y^2 = z^2 - x^2 = (z + x)(z - x)$$

implies that

$$\left(\frac{y}{2}\right)^2 = \left(\frac{z + x}{2}\right)\left(\frac{z - x}{2}\right)$$

Also, $(\frac{z+x}{2}, \frac{z-x}{2}) = 1$. [Let $(\frac{z+x}{2}, \frac{z-x}{2}) = d$. Then $d \mid \frac{z+x}{2}$ and $d \mid \frac{z-x}{2}$. By Proposition 1.2, we have that $d \mid \frac{z+x}{2} + \frac{z-x}{2}$ and $d \mid \frac{z+x}{2} - \frac{z-x}{2}$, or, equivalently, $d \mid z$ and $d \mid x$. Since $(x, z) = 1$, we have that $d = 1$.] So $\frac{z+x}{2}$ and $\frac{z-x}{2}$ are perfect squares. (Why?) Let $m, n \in \mathbf{Z}$ be such that

$$\frac{z + x}{2} = m^2$$

and

$$\frac{z - x}{2} = n^2$$

Then $m > n > 0$, $(m, n) = 1$, $m^2 - n^2 = x$, $2mn = y$, and $m^2 + n^2 = z$. Also, $(m, n) = 1$ implies that not both of m and n are even; if both of m and n are odd, we have that z and x are both even, contradicting the fact that $(x, z) = 1$. So exactly one of m and n is even. This completes the first part of the proof. Second, we show that given m and n as described and $x = m^2 - n^2$, $y = 2mn$, and $z = m^2 + n^2$, we have a primitive Pythagorean triple with y even. Obviously, x, y, z is a Pythagorean triple (please check it!) with y even. It remains to show that $(x, y, z) = 1$. Let $(x, y, z) = d$. Since exactly one of m and n is even, we have that x and z are both odd. Then d is odd and so $d = 1$ or d is divisible by some odd prime number p. Assume that $p \mid d$. Now $p \mid x$ and $p \mid z$; by Proposition 1.2, we have that $p \mid z + x$ and $p \mid z - x$. So $p \mid (m^2 + n^2) + (m^2 - n^2)$ and $p \mid (m^2 + n^2) - (m^2 - n^2)$ or, equivalently, $p \mid 2m^2$ and $p \mid 2n^2$. Since p is odd, we have that $p \mid m^2$ and $p \mid n^2$, from which $p \mid m$ and $p \mid n$ by Lemma 1.14. Then $(m, n) \neq 1$, a contradiction. So $d = 1$ as desired. ■

Note that while the proof of Theorem 6.3 is somewhat intricate, it involves nothing but concepts from Chapter 1 of this book. Ingenuity is required to put the concepts together correctly, but no advanced techniques are needed. In this sense, a mathematician would term the proof above as *elementary*. (You may wish to review the discussion of the "elementary proof" of the Prime Number Theorem in Section 1.2.) Such proofs may involve considerable ingenuity and intricacy but require only elementary number-theoretic concepts (like those in this book). Exercise 16 provides several excellent opportunities for the reader to parallel the proof of Theorem 6.3 above.

Exercise Set 6.3

13. ***(a)*** Find all primitive Pythagorean triples x, y, z with $z \leq 50$.
(b) Find all Pythagorean triples x, y, z with $z \leq 50$.

14. Let x, y, z be a primitive Pythagorean triple with y even.
(a) Prove that exactly one of x and y is divisible by 3.
(b) Prove that exactly one of x and y is divisible by 4.
(c) Prove that exactly one of x, y, and z is divisible by 5.
(d) Prove that $x + y \equiv x - y \equiv 1, 7 \bmod 8$.
(e) Prove that $60 \mid xyz$.

15. Prove that every positive integer greater than 2 is part of at least one Pythagorean triple. [*Hint*: Let $n \in \mathbf{Z}$ with $n > 2$. If n is odd, consider the triple, n, $(\frac{1}{2})(n^2 - 1)$, $(\frac{1}{2})(n^2 + 1)$; if n is even, consider the triple n, $(n^2/4) - 1$, $(n^2/4) + 1$.]

16. Find all solutions in positive integers of each Diophantine equation below.
(a) $x^2 + 2y^2 = z^2$
(*Hint*: Parallel the general technique used in the proof of Theorem 6.3.)
(b) $x^2 + 3y^2 = z^2$
(c) $x^2 + 4y^2 = z^2$
(d) $x^2 + py^2 = z^2$, p a prime number

17. ***(a)*** Prove that there are infinitely many primitive Pythagorean triples x, y, z such that $y = x + 1$. (*Hint*: If x, y, z is such a Pythagorean triple, consider the triple $3x + 2z + 1$, $3x + 2z + 2$, $4x + 3z + 2$.)
(b) Prove that every primitive Pythagorean triple x, y, z with $y = x + 1$ can be generated using the procedure motivated in part (a) above.

18. ***(a)*** Prove that there are infinitely many primitive Pythagorean triples x, y, z such that $z = x + 1$. (*Hint*: If n is a positive integer, consider the triple $2n^2 + 2n$, $2n + 1$, $2n^2 + 2n + 1$.)
(b) Prove that every primitive Pythagorean triple x, y, z with $z = x + 1$ is of the form given in the hint of part (a) above.

19. Find all solutions in relatively prime positive integers x, y, and z of the equation $1/x^2 + 1/y^2 = 1/z^2$.

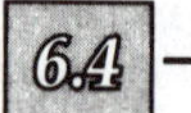

Fermat's Last Theorem

After achieving success with the Diophantine equation $x^2 + y^2 = z^2$, it is somewhat natural to investigate the solvability of the Diophantine equations $x^3 + y^3 = z^3$, $x^4 + y^4 = z^4$, $x^5 + y^5 = z^5$, ... Fermat's Last Theorem (which was proven in 1994 by Andrew Wiles of Princeton University with help from Cambridge mathematician Richard Taylor) comments on the solvability of these latter equations and was, until 1994, probably the most famous unsolved problem in all of number theory. Attempts to prove this conjecture had resulted in new branches of mathematics and had spawned many other open problems. Fermat's Last Theorem itself is rather simply stated.

Fermat's Last Theorem: The Diophantine equation $x^n + y^n = z^n$ has no solutions in nonzero integers x, y, z for $n \in \mathbf{Z}$ with $n \geq 3$.

In other words, it is claimed that none of the Diophantine equations $x^3 + y^3 = z^3$, $x^4 + y^4 = z^4$, $x^5 + y^5 = z^5, \ldots$ have solutions in nonzero integers x, y, and z. Fermat claimed to have a proof of this conjecture. In the margin of his copy of the works of Diophantus next to Problem 8 of Book II (a problem involving the equation $x^2 + y^2 = z^2$), Fermat wrote (in Latin): "On the other hand, it is impossible for a cube to be written as a sum of two cubes or a fourth power to be written as a sum of two fourth powers or, in general, for any number which is a power greater than the second to be written as a sum of two like powers. I have a truly marvelous demonstration of this proposition which this margin is too narrow to contain."

Although many mathematicians had attempted to prove Fermat's Last Theorem, none had succeeded prior to 1994. Andrew Wiles' conclusive (150 page!) proof actually solves a problem from which the truth of Fermat's Last Theorem follows. His work, initially viewed with much skepticism, is one of the greatest mathematical achievements of the twentieth century.

We now show that it suffices to prove Fermat's Last Theorem for the cases of n an odd prime number and $n = 4$. Let $n \in \mathbf{Z}$ with $n \geq 3$. Then either n is divisible by an odd prime number or n is divisible by 4. Write $n = ab$ where $a, b \in \mathbf{Z}$ and b is either an odd prime number or 4. If x, y, z is a solution to $x^n + y^n = z^n$, then x^a, y^a, z^a is a solution to $x^b + y^b = z^b$. (Do you see why?) So, by contraposition, if $x^b + y^b = z^b$ has no solutions, then $x^n + y^n = z^n$ has no solutions. In other words, Fermat's Last Theorem is true for all integral n with $n \geq 3$ if (and only if) it is true for n an odd prime number and $n = 4$.

In the remainder of this section, we give a proof of Fermat's Last Theorem for the case $n = 4$, which is, curiously, the only case for which Fermat himself provided a proof. This proof illustrates a powerful proof technique from Fermat called the *method of descent.* Unfortunately, such a proof technique does not generally handle Fermat's Last Theorem for odd prime exponents n; the known proofs (for those particular n having proofs) vary greatly. The interested reader is referred to Edwards (1977) or Ribenboim (1979) for some of these proofs as well as fascinating accounts of the history of Fermat's Last Theorem.

The idea behind Fermat's method of descent is rather simple: One shows that a given Diophantine equation has no solutions in positive integers by assuming the existence of such a solution and constructing another solution in positive integers having one component strictly smaller than that same component of the original solution. (This construction procedure may be quite cumbersome.) The construction procedure is then applied to this new solution to produce a third solution in positive integers having an even smaller like component and so on. This process cannot be continued ad infinitum, since it is not possible to construct an infinite strictly decreasing sequence of positive integers; so the initial assumption is contradicted, and the given Diophantine equation has no solutions in positive integers.

We now give a proof of Fermat's Last Theorem for $n = 4$.

Theorem 6.4: (Fermat) The Diophantine equation $x^4 + y^4 = z^2$ has no solutions in nonzero integers x, y, z.

You may say "This is not Fermat's Last Theorem for $n = 4$!" You would be correct (since the exponent on z is 2, not 4). But note that if x, y, z is a solution of $x^4 + y^4 = z^4$, then x, y, z^2 is a solution of $x^4 + y^4 = z^2$. So, by contraposition, if $x^4 + y^4 = z^2$ has no solutions, then $x^4 + y^4 = z^4$ has no solutions. Consequently, Theorem 6.4 is stronger than proving Fermat's Last Theorem for $n = 4$. In other words, Fermat's Last Theorem for $n = 4$ follows from Theorem 6.4.

Proof: (of Theorem 6.4) Assume, by way of contradiction, that $x^4 + y^4 = z^2$ has a solution x_1, y_1, z_1 in nonzero integers. Without loss of generality, we may assume that $x_1, y_1, z_1 > 0$ and $(x_1, y_1) = 1$. (Why?) We will show that there is another solution x_2, y_2, z_2 in positive integers such that $(x_2, y_2) = 1$ and $0 < z_2 < z_1$. Now x_1^2, y_1^2, z_1 is a Pythagorean triple with $(x_1^2, y_1^2, z_1) = 1$ and, without loss of generality, y_1^2 even. So y_1 is even and, by Theorem 6.3, there exist $m, n \in \mathbf{Z}$ such that $m > n > 0$, $(m, n) = 1$, exactly one of m and n is even, $x_1^2 = m^2 - n^2$, $y_1^2 = 2mn$, and $z_1 = m^2 + n^2$. Now $x_1^2 = m^2 - n^2$ implies $x_1^2 + n^2 = m^2$ so that x_1, n, m is a Pythagorean triple with $(x_1, n, m) = 1$ and n even. So m is odd and, by Theorem 6.3 again, there exist $a, b \in \mathbf{Z}$ such that $a > b > 0$, $(a, b) = 1$, exactly one of a and b is even, $x_1 = a^2 - b^2$, $n = 2ab$, and $m = a^2 + b^2$. We wish to show that m, a, and b are perfect squares. Since $y_1^2 = 2mn = m(2n)$ and $(m, 2n) = 1$, we have that m and $2n$ are perfect squares. Since $2n$ is a perfect square, there exists $c \in \mathbf{Z}$ such that $2n = 4c^2$ or, equivalently, $n = 2c^2$. Now $n = 2ab$ implies that $c^2 = ab$; since $(a, b) = 1$ we have that a and b are perfect squares. So m, a, and b are perfect squares, and there exist $x_2, y_2, z_2 \in \mathbf{Z}$ such that $m = z_2^2$, $a = x_2^2$, and $b = y_2^2$. Without loss of generality, we may assume that $x_2, y_2, z_2 > 0$. Then $m = a^2 + b^2$ implies $z_2^2 = x_2^4 + y_2^4$ so that x_2, y_2, z_2 is a solution in positive integers of $x^4 + y^4 = z^2$. Also, $(x_2, y_2) = 1$ and

$$0 < z_2 \le z_2^2 = m \le m^2 < m^2 + n^2 = z_1$$

So we have constructed another solution x_2, y_2, z_2 of $x^4 + y^4 = z^2$ as desired. In other words (and this is the crux of Fermat's method of descent), assuming the existence of a solution of $x^4 + y^4 = z^2$ in positive integers, we can construct another solution of $x^4 + y^4 = z^2$ in positive integers with strictly smaller "z value". By repeatedly applying the procedure above ad infinitum, we obtain an infinite sequence of solutions of $x^4 + y^4 = z^2$ with strictly decreasing positive "z values." This is a contradiction since there are only finitely many positive integers between a given positive integer (the initial "z value") and zero. So $x^4 + y^4 = z^2$ has no solutions in nonzero integers x, y, z.

■

Biography

Sophie Germain (1776–1831)

Born in Paris, Sophie Germain was deeply interested in mathematics; unfortunately, women were not allowed to enroll at the École Polytechnique. By obtaining lecture notes from various professors at the school, however, she studied mathematics on her own. Under the pseudonym of Monsieur Leblanc, she corresponded with Lagrange and Gauss, receiving high praise from both. In 1816, the French Academy awarded her a prize for a paper on elasticity. She subsequently proved the result discussed below relating to Fermat's Last Theorem. and introduced the useful concept of mean curvature into the field of differential geometry. Gauss eventually discovered that Monsieur Leblanc was a woman and recommended that the University of Göttingen award her an honorary doctorate. Unfortunately, Germain's death prevented her from receiving this recognition she so rightly deserved. Today a school and a street in Paris bear her name.

Exercise 21 provides an excellent opportunity for the reader to construct a proof by the method of descent using the proof of Theorem 6.4 as a guide.

We conclude this section with a brief discussion of a class of prime numbers related to Fermat's Last Theorem. A prime number p is said to be a *Sophie Germain prime* (in honor of mathematician Sophie Germain) if $2p + 1$ is also a prime number. It has been conjectured that there are infinitely many such prime numbers; the largest known Sophie Germain prime was discovered in 2001: $109433307 \cdot 2^{66452} - 1$. Around 1825, Sophie Germain proved that if p was a Sophie Germain prime, then the Diophantine equation $x^p + y^p = z^p$ had no solutions in nonzero integers x, y, z that were not multiples of p. In other words, for Sophie Germain prime exponents, *part* of Fermat's Last Theorem was true.

Many of the known results concerning Fermat's Last Theorem were similar in nature prior to 1994, giving partial information supporting the truth of Fermat's Last Theorem. A chronological account of some of these results may be found in Edwards (1977).

Exercise Set 6.4

20. Let $x, y, z \in \mathbf{Z}$ and let p be a prime number.
  (a) Prove that if $x^{p-1} + y^{p-1} = z^{p-1}$, then $p \mid xyz$.
  (b) Prove that if $x^p + y^p = z^p$, then $p \mid (x + y - z)$.

21. Prove that the Diophantine equation $x^4 - y^4 = z^2$ has no solutions in nonzero integers x, y, z. (*Hint*: Parallel the proof of Theorem 6.4.)

22. Prove that there is at most one perfect square in any Pythagorean triple.

23. Prove that the Diophantine equation $x^4 - y^4 = 2z^2$ has no solutions in nonzero integers x, y, z. (*Hint*: Note that x and y must be either both odd or both even. By factoring $x^4 - y^4$, argue that there exist positive integers m, n, and k such that $x^2 + y^2 = 2m^2$, $x + y = 2n^2$, and $x - y = 2k^2$. Now, what relationship do m, n, and k satisfy?)

24. Prove that the Diophantine equation $x^4 - 4y^4 = z^2$ has no solutions in nonzero integers x, y, z.

25. Prove that the Diophantine equation $1/x^4 + 1/y^4 = 1/z^2$ has no solutions in nonzero integers x, y, z.

26. Find all solutions in relatively prime positive integers x, y, z of the Diophantine equation $x^4 + y^4 = 2z^2$.

27. Prove that the Diophantine equation $x^2 + y^2 = z^3$ has infinitely many integral solutions. (*Hint*: For $n \in \mathbf{Z}$, consider $x = n^3 - 3n$ and $y = 3n^2 - 1$.)

6.5

Representation of an Integer as a Sum of Squares

We begin by posing this question: Which positive integers are expressible as the sums of two squares of integers?

Example 8:

Among the first 30 positive integers, half are expressible as the sums of two squares of integers and half are not. To wit, 1 $(= 1^2 + 0^2)$, 2 $(= 1^2 + 1^2)$, 4 $(= 2^2 + 0^2)$, 5 $(= 2^2 + 1^2)$, 8 $(= 2^2 + 2^2)$, 9 $(= 3^2 + 0^2)$, 10 $(= 3^2 + 1^2)$, 13 $(= 3^2 + 2^2)$, 16 $(= 4^2 + 0^2)$, 17 $(= 4^2 + 1^2)$, 18 $(= 3^2 + 3^2)$, 20 $(= 4^2 + 2^2)$, 25 $(= 5^2 + 0^2)$, 26 $(= 5^2 + 1^2)$, and 29 $(= 5^2 + 2^2)$ are expressible as the sums of two squares of integers, while 3, 6, 7, 11, 12, 14, 15, 19, 21, 22, 23, 24, 27, 28, and 30 are not. The expression of an integer as a sum of two squares of integers is not necessarily unique if it exists. (One need only consider $25 = 5^2 + 0^2 = 4^2 + 3^2$, for example.) It is instructive at this point for you to make a conjecture as to the answer of our motivating question. For example, would you say that the integer 374^{695} is expressible as the sum of two squares of integers or not?

Exercises 13(c) and 13(d) in Chapter 2 establish a necessary, but not sufficient, condition for a positive integer to be expressible as the sum of two squares of integers, namely that the positive integer must not be congruent to 3 modulo 4. The next several results will culminate in a theorem that *characterizes* those positive integers expressible as the sums of two squares of integers. Equivalently, we will obtain a necessary *and* sufficient condition on n in order for the Diophantine equation $x^2 + y^2 = n$ to be solvable. We hope that your conjecture will be revised if necessary as subsequent results are proved. We begin with an important proposition whose proof relies on a simple algebraic identity.

Proposition 6.5: Let $n_1, n_2 \in \mathbf{Z}$ with $n_1, n_2 > 0$. If n_1 and n_2 are expressible as the sums of two squares of integers, then $n_1 n_2$ is expressible as the sum of two squares of integers.

Proof: Let $x_1, y_1, x_2, y_2 \in \mathbf{Z}$ be such that $n_1 = x_1^2 + y_1^2$ and $n_2 = x_2^2 + y_2^2$. Then

$$n_1 n_2 = (x_1^2 + y_1^2)(x_2^2 + y_2^2) = (x_1 x_2 + y_1 y_2)^2 + (x_1 y_2 - y_1 x_2)^2$$

(expand and see!) as desired. ■

Example 9:

Since $13 = 3^2 + 2^2$ and $17 = 4^2 + 1^2$ are each expressible as the sums of two squares of integers, Proposition 6.5 guarantees that $13 \cdot 17 = 221$ is so expressible. Indeed, the proof of Proposition 6.5 gives

$$\begin{aligned} 221 = 13 \cdot 17 &= (3^2 + 2^2)(4^2 + 1^2) = (3 \cdot 4 + 2 \cdot 1)^2 + (3 \cdot 1 - 2 \cdot 4)^2 \\ &= 14^2 + (-5)^2 = 14^2 + 5^2, \end{aligned}$$

as desired.

The fact that every prime number congruent to 1 modulo 4 is expressible as the sum of two squares of integers will ultimately follow from the lemma below. [Note that *no* prime number congruent to 3 modulo 4 is expressible as the sum of two squares of integers by Exercise 13(c) of Chapter 2.]

Lemma 6.6: If p is a prime number such that $p \equiv 1 \bmod 4$, then there exist $x, y \in \mathbf{Z}$ such that $x^2 + y^2 = kp$ where $k \in \mathbf{Z}$ and $0 < k < p$.

Proof: Since $p \equiv 1 \bmod 4$, we have $\left(\frac{-1}{p}\right) = 1$; so there exists $x \in \mathbf{Z}$ with $0 < x \leq \frac{p-1}{2}$ such that $x^2 \equiv -1 \bmod p$. Then $p \mid x^2 + 1$, and we have that $x^2 + 1 = kp$ for some $k \in \mathbf{Z}$. This proves the existence of x and y satisfying the desired equation (note that $y = 1$ here). Clearly, $k > 0$. Furthermore,

$$kp = x^2 + y^2 < \left(\frac{p}{2}\right)^2 + 1 < p^2$$

implies that $k < p$ as desired. ■

The following proposition proves that 2 as well as all prime numbers congruent to 1 modulo 4 are expressible as the sums of two squares of integers.

Proposition 6.7: If p is a prime number such that $p \not\equiv 3 \bmod 4$, then p is expressible as the sum of two squares of integers.

Proof: The result is clear if $p = 2$. Assume that $p \equiv 1 \bmod 4$. Let m be the least integer such that there exists $x, y \in \mathbf{Z}$ with $x^2 + y^2 = mp$ and $0 < m < p$ in accordance with Lemma 6.6. We show that $m = 1$. Assume, by way of contradiction, that $m > 1$. Let $a, b \in \mathbf{Z}$ be such that

$$a \equiv x \bmod m, \frac{-m}{2} < a \leq \frac{m}{2}$$

and

$$b \equiv y \bmod m, \frac{-m}{2} < b \leq \frac{m}{2}$$

Then

$$a^2 + b^2 \equiv x^2 + y^2 = mp \equiv 0 \bmod m$$

and so there exists $k \in \mathbf{Z}$ with $k > 0$ such that $a^2 + b^2 = km$. (As an exercise, prove that $k \neq 0$). Now

$$(a^2 + b^2)(x^2 + y^2) = (km)(mp) = km^2p$$

By the proof of Proposition 6.5, we have

$$(a^2 + b^2)(x^2 + y^2) = (ax + by)^2 + (ay - bx)^2$$

so

$$(ax + by)^2 + (ay - bx)^2 = km^2p$$

Since $a \equiv x \bmod m$ and $b \equiv y \bmod m$, we have

$$ax + by \equiv x^2 + y^2 \equiv 0 \bmod m$$

and

$$ay - bx \equiv xy - yx \equiv 0 \bmod m$$

from which $\frac{ax+by}{m}, \frac{ay-bx}{m} \in \mathbf{Z}$ and

$$\left(\frac{ax + by}{m}\right)^2 + \left(\frac{ay - bx}{m}\right)^2 = \frac{km^2p}{m^2} = kp \qquad \textbf{(2)}$$

Now $\frac{-m}{2} < a \leq \frac{m}{2}$ and $\frac{-m}{2} < b \leq \frac{m}{2}$ imply that $a^2 \leq \frac{m^2}{4}$ and $b^2 \leq \frac{m^2}{4}$; then

$$km = a^2 + b^2 \leq \frac{m^2}{2}$$

and we have $k \leq \frac{m}{2} < m$. But $0 < k < m$ contradicts the minimality of m for which the equation $x^2 + y^2 = mp$ is solvable since (2) gives such a solution with k instead of m. So $m = 1$, $x^2 + y^2 = p$, and the proof is complete. ■

And now the answer to our motivating question! If your conjecture coincides with the statement of the theorem below, you should feel extremely proud.

Theorem 6.8: Let $n \in \mathbf{Z}$ with $n > 0$. Then n is expressible as the sum of two squares of integers if and only if every prime factor congruent to 3 modulo 4 occurs to an even power in the prime factorization of n.

(Convince yourself that the positive integers *not* expressible as the sums of two squares of integers in Example 8 are precisely those integers containing at least one prime factor congruent to 3 modulo 4 occurring to an *odd* power in the corresponding prime factorizations.)

Proof: ($\Rightarrow$) Assume that p is an odd prime number and that p^{2i+1}, $i \in \mathbf{Z}$, occurs in the prime factorization of n. We will show that $p \equiv 1 \bmod 4$. Since n is expressible as the sum of two squares of integers, there exist $x, y \in \mathbf{Z}$ such that

$$n = x^2 + y^2$$

Let $(x, y) = d$, $a = \frac{x}{d}$, $b = \frac{y}{d}$, and $m = n/d^2$. Then $(a, b) = 1$ by Proposition 1.10 and

$$a^2 + b^2 = m$$

Let p^j, $j \in \mathbf{Z}$, be the largest power of p dividing d. Then $p^{(2i+1)-2j} \mid m$; since $(2i + 1) - 2j \geq 1$, we have $p \mid m$. Now $p \nmid a$. [If $p \mid a$, then $p \mid b$, violating $(a, b) = 1$.] By Theorem 2.6, there exists $z \in \mathbf{Z}$ such that

$$az \equiv b \bmod p$$

Then

$$m = a^2 + b^2 \equiv a^2 + (az)^2 \equiv a^2(1 + z^2) \bmod p$$

Since $p \mid m$, we have

$$a^2(1 + z^2) \equiv 0 \bmod p$$

or, equivalently, $p \mid a^2(1 + z^2)$. Since $p \nmid a$, we have $p \mid 1 + z^2$ or, equivalently, $z^2 \equiv -1 \bmod p$. In other words, -1 is a quadratic residue modulo p from which $p \equiv 1 \bmod 4$. So any odd prime factor occurring to an odd power in the prime factorization of n must be congruent to 1 modulo 4. By contraposition, then, any prime factor congruent to 3 modulo 4 occurs to an even power in the prime factorization of n as desired.

($\Leftarrow$) Assume that every prime factor congruent to 3 modulo 4 occurs to an even power in the prime factorization of n. Then n can be written as $n = m^2 p_1 p_2 \cdots p_r$ where $m \in \mathbf{Z}$ and $p_1, p_2, \ldots, p_r$ are distinct prime numbers equal to 2 or congruent to 1 modulo 4. Now m^2 is clearly expressible as the sum of two squares of integers $(m^2 + 0^2)$ and each p_i is so expressible by Proposition 6.7. So Proposition 6.5 implies that n is so expressible, and the proof is complete. ■

We now give two examples that illustrate the use of the results in this section so far.

Example 10:

Determine whether 374^{695} is expressible as the sum of two squares of integers. (See the last sentence of Example 8.)
The prime factorization of 374 is $2 \cdot 11 \cdot 17$; so the prime factorization of 374^{695} is $2^{695}11^{695}17^{695}$. Inasmuch as the prime factor 11 (which is congruent to 3 modulo 4) occurs to an odd power in the prime factorization of 374^{695}, we have that 374^{695} is not expressible as the sum of two squares of integers by Theorem 6.8.

Example 11:

Express 4410 as the sum of two squares of integers if possible.
The prime factorization of 4410 is $2 \cdot 3^2 5 \cdot 7^2$; since the prime factors 3 and 7 occur to even powers in this prime factorization, we have that 4410 is expressible as the sum of two squares of integers by Theorem 6.8. To determine such an expression, split 4410 into a product of two (or more) smaller factors, each expressible as the sum of two squares of integers. For example, $4410 = (2 \cdot 7^2)(3^2 5) = 98 \cdot 45$; by inspection, we have $98 = 7^2 + 7^2$ and $45 = 6^2 + 3^2$. The technique of Example 9 now yields $4410 = 63^2 + 21^2$ as desired. (One could, of course, use trial and error to obtain the desired expression, but such a method will generally become increasingly cumbersome with larger integers.)

Theorem 6.8 shows that not all positive integers are expressible as the sum of two squares of integers (see also Examples 8 and 10). Furthermore, Theorem 6.8 characterizes precisely those integers that are so expressible. What about analogous results for three squares of integers? A partial answer is provided by the following proposition.

Proposition 6.9: Let $m, n \in \mathbf{Z}$ with $m, n \geq 0$. If $N = 4^m(8n + 7)$, then N is not expressible as the sum of three squares of integers.

Proof: We first prove the desired result for $m = 0$. Assume, by way of contradiction, that $N = 8n + 7$ is expressible as the sum of three squares of integers. Then there exist $x, y, z \in \mathbf{Z}$ such that

$$8n + 7 = x^2 + y^2 + z^2$$

Now $8n + 7 \equiv 7 \bmod 8$, while $x^2 + y^2 + z^2 \not\equiv 7 \bmod 8$ (this latter fact is left as an exercise for the reader), a contradiction. So $N = 8n + 7$ is not expressible as the sum of three squares of integers. Now assume that $m > 0$ and, by way of contradiction, that $N = 4^m(8n + 7)$ is expressible as the sum of three squares of integers. Then, as before, there exist $x, y, z \in \mathbf{Z}$ such that

$$4^m(8n + 7) = x^2 + y^2 + z^2 \qquad \textbf{(3)}$$

Now x, y, and z must be even (yet another fact left as an exercise for the reader!). So there exist $x', y', z' \in \mathbf{Z}$ such that $x = 2x'$, $y = 2y'$, and $z = 2z'$. Substituting in (3) and simplifying, we obtain

$$4^{m-1}(8n + 7) = (x')^2 + (y')^2 + (z')^2$$

By repeating the above procedure $m - 1$ more times, we obtain that $8n + 7$ is expressible as the sum of three squares of integers, a contradiction. So $N = 4^m(8n + 7)$ is not expressible as the sum of three squares of integers, which completes the proof. ■

The converse of Proposition 6.9 is true and was proven by Legendre in 1798. This converse is much more difficult to prove since there is no formula analogous to Proposition 6.5; consequently, we omit such a proof. [The reader interested in such a proof may consult Landau (1909).] Proposition 6.9 and its converse provide a precise characterization of those positive integers expressible as the sum of three squares of integers. Note that the smallest positive integer not expressible as the sum of three squares of integers (and thus not expressible as the sum of two squares of integers) is 7 (take $m = n = 0$ in Proposition 6.9 above). We now provide another example of the use of Proposition 6.9.

Example 12:

Determine whether 1584 is expressible as the sum of three squares of integers.

We wish to see whether 1584 can be expressed in the form $4^m(8n + 7)$ of Proposition 6.9. The highest power of 4 dividing 1584 evenly is 16; the quotient after division of 1584 by 16 is 99. Dividing 99 by 8, we obtain a remainder of 3. So 1584 is *not* expressible in the form $4^m(8n + 7)$ of Proposition 6.9, from which 1584 *is* expressible as the sum of three squares of integers by the (unproven) converse of Proposition 6.9. As an exercise you should find an expression of 1584 as the sum of three squares of integers and show that 1584 is *not* expressible as the sum of two squares of integers.

Even though not all positive integers are expressible as the sum of two squares of integers or three squares of integers, *every* positive integer is expressible as the sum of four squares of integers. Euler made substantial progress on this problem but was unable to find a complete proof. With proper acknowledgement to Euler's contributions, Lagrange published the first complete proof of the four-square theorem in 1770. The following sequence of results, simplified from Lagrange's proof by Euler, culminates in a proof of this theorem. You will see a resemblance between this sequence and the sequence of results that was used to prove Theorem 6.8. Our first result is analogous to Proposition 6.5; notice the algebraic identity used in the proof.

Proposition 6.10: (Euler) Let $n_1, n_2 \in \mathbf{Z}$ with $n_1, n_2 > 0$. If n_1 and n_2 are expressible as the sums of four squares of integers, then $n_1 n_2$ is expressible as the sum of four squares of integers.

Proof: Let $w_1, x_1, y_1, z_1, w_2, x_2, y_2, z_2 \in \mathbf{Z}$ be such that

$$n_1 = w_1^2 + x_1^2 + y_1^2 + z_1^2$$

and

$$n_2 = w_2^2 + x_2^2 + y_2^2 + z_2^2$$

Then

$$\begin{aligned} n_1n_2 &= (w_1^2 + x_1^2 + y_1^2 + z_1^2)(w_2^2 + x_2^2 + y_2^2 + z_2^2) \\ &= (w_1w_2 + x_1x_2 + y_1y_2 + z_1z_2)^2 + (-w_1x_2 + x_1w_2 - y_1z_2 + z_1y_2)^2 \\ &\quad + (-w_1y_2 + y_1w_2 + x_1z_2 - z_1x_2)^2 + (-w_1z_2 + z_1w_2 - x_1y_2 + y_1x_2)^2 \end{aligned}$$

(expand and see!) as desired. ■

Example 13:

Since $78 = 1^2 + 2^2 + 3^2 + 8^2$ and $126 = 4^2 + 5^2 + 6^2 + 7^2$ are each expressible as the sums of four squares of integers, Proposition 6.10 guarantees that $78 \cdot 126 = 9828$ is so expressible. Indeed, the proof of Proposition 6.10 gives

$$\begin{aligned} 9828 = 78 \cdot 126 &= (1^2 + 2^2 + 3^2 + 8^2)(4^2 + 5^2 + 6^2 + 7^2) \\ &= (1 \cdot 4 + 2 \cdot 5 + 3 \cdot 6 + 8 \cdot 7)^2 + (-1 \cdot 5 + 2 \cdot 4 - 3 \cdot 7 + 8 \cdot 6)^2 \\ &\quad + (-1 \cdot 6 + 3 \cdot 4 + 2 \cdot 7 - 8 \cdot 5)^2 + (-1 \cdot 7 + 8 \cdot 4 - 2 \cdot 6 + 3 \cdot 5)^2 \\ &= 88^2 + 30^2 + (-20)^2 + (-28)^2 \\ &= 88^2 + 30^2 + 20^2 + 28^2 \end{aligned}$$

as desired.

The lemma below is analogous to Lemma 6.6. Note in particular how the proof of Lemma 6.6 and Case I of the proof below proceed virtually verbatim.

Lemma 6.11: (Euler) If p is an odd prime number, then there exist $x, y \in \mathbf{Z}$ such that $x^2 + y^2 + 1 = kp$ where $k \in \mathbf{Z}$ and $0 < k < p$.

Proof: We consider two cases.

Case I: $p \equiv 1 \bmod 4$

Then $\left(\frac{-1}{p}\right) = 1$; so there exists $x \in \mathbf{Z}$ with $0 < x \leq \frac{p-1}{2}$ such that $x^2 \equiv -1 \bmod p$. Then $p \mid x^2 + 1$, and we have that $x^2 + 1 = kp$ for some $k \in \mathbf{Z}$. This proves the existence of x and y satisfying the desired equation (note that $y = 0$ here). Clearly, $k > 0$. Furthermore,

$$kp = x^2 + 1 < \left(\frac{p}{2}\right)^2 + 1 < p^2$$

implies that $k < p$ as desired.

Case II: $p \equiv 3 \bmod 4$

Let a be the least positive quadratic nonresidue modulo p. Note that $a \geq 2$. Then

$$\left(\frac{-a}{p}\right) = \left(\frac{-1}{p}\right)\left(\frac{a}{p}\right) = (-1)(-1) = 1$$

and so there exists $x \in \mathbf{Z}$ with $0 < x \leq \frac{p-1}{2}$ such that $x^2 \equiv -a \bmod p$. Now $a - 1$, being positive and less than a, must be a quadratic residue modulo p, and so there exists $y \in \mathbf{Z}$ with $0 < y \leq \frac{p-1}{2}$ such that $y^2 \equiv a - 1 \bmod p$. So

$$x^2 + y^2 + 1 \equiv (-a) + (a - 1) + 1 \equiv 0 \bmod p$$

or, equivalently, $x^2 + y^2 + 1 = kp$ for some $k \in \mathbf{Z}$. This proves the existence of x and y satisfying the desired equation. Clearly, $k > 0$. Furthermore,

$$kp = x^2 + y^2 + 1 < \left(\frac{p}{2}\right)^2 + \left(\frac{p}{2}\right)^2 + 1 < p^2$$

implies that $k < p$ as desired. ■

We may now show that every prime number is expressible as the sum of four squares of integers. Compare the statement and proof of the proposition below with the statement and proof of Proposition 6.7.

Proposition 6.12: All prime numbers are expressible as the sums of four squares of integers.

Proof: The result is clear if $p = 2$ since $2 = 1^2 + 1^2 + 0^2 + 0^2$. Assume that p is an odd prime number. Let m be the least positive integer such that there exists $w, x, y, z \in \mathbf{Z}$ such that $w^2 + x^2 + y^2 + z^2 = mp$ and $0 < m < p$ in accordance with Lemma 6.11. (Note that Lemma 6.11 guarantees such an expression by taking, for example, $w = 0$ and $z = 1$.) We show that $m = 1$. Assume, by way of contradiction, that $m > 1$. We consider two cases.

Case I: m is even

Then there are three possibilities: w, x, y, and z are all even; w, x, y, and z are all odd; or two of w, x, y, and z are even and two are odd. In any event, we may assume, without loss of generality, that $w \equiv x \bmod 2$ and $y \equiv z \bmod 2$. Then $\frac{w+x}{2}$, $\frac{w-x}{2}$, $\frac{y+z}{2}$, and $\frac{y-z}{2}$ are integers and

$$\left(\frac{w+x}{2}\right)^2 + \left(\frac{w-x}{2}\right)^2 + \left(\frac{y+z}{2}\right)^2 + \left(\frac{y-z}{2}\right)^2 = \left(\frac{m}{2}\right)p$$

which contradicts the minimality of m for which the equation $w^2 + x^2 + y^2 + z^2 = mp$ is solvable.

Case II: m is odd

Then $m \geq 3$. Let $a, b, c, d \in \mathbf{Z}$ be such that

$$a \equiv w \bmod m, \frac{-m}{2} < a < \frac{m}{2}$$

$$b \equiv x \bmod m, \frac{-m}{2} < b < \frac{m}{2}$$

$$c \equiv y \bmod m, \frac{-m}{2} < c < \frac{m}{2}$$

and

$$d \equiv z \bmod m, \frac{-m}{2} < d < \frac{m}{2}$$

Then

$$a^2 + b^2 + c^2 + d^2 \equiv w^2 + x^2 + y^2 + z^2 = mp \equiv 0 \bmod m$$

and so there exists $k \in \mathbf{Z}$ with $k > 0$ such that $a^2 + b^2 + c^2 + d^2 = km$. (As an exercise, prove that $k \neq 0$.) Now

$$(a^2 + b^2 + c^2 + d^2)(w^2 + x^2 + y^2 + z^2) = (km)(mp) = km^2p$$

By the proof of Proposition 6.10, we have

$$\begin{aligned}(a^2 + b^2 + c^2 + d^2)(w^2 + x^2 + y^2 + z^2) = {} & (aw + bx + cy + dz)^2 \\ & + (-ax + bw - cz + dy)^2 \\ & + (-ay + cw + bz - dx)^2 \\ & + (-az + dw - by + cx)^2\end{aligned}$$

so

$$\begin{aligned}&(aw + bx + cy + dz)^2 + (-ax + bw - cz + dy)^2 \\ &\qquad + (-ay + cw + bz - dx)^2 + (-az + dw - by + cx)^2 = km^2p\end{aligned}$$

Since $a \equiv w \bmod m$, $b \equiv x \bmod m$, $c \equiv y \bmod m$, and $d \equiv z \bmod m$, we have

$$aw + bx + cy + dz \equiv w^2 + x^2 + y^2 + z^2 \equiv 0 \bmod m$$

$$-ax + bw - cz + dy \equiv -wx + xw - yz + zy \equiv 0 \bmod m$$

$$-ay + cw + bz - dx \equiv -wy + yw + xz - zx \equiv 0 \bmod m$$

and

$$-az + dw - by + cx \equiv -wz + zw - xy + yx \equiv 0 \bmod m$$

Put

$$W = \frac{aw + bx + cy + dz}{m}$$

$$X = \frac{-ax + bw - cz + dy}{m}$$

$$Y = \frac{-ay + cw + bz - dx}{m}$$

and

$$Z = \frac{-az + dw - by + cx}{m}$$

Then $W, X, Y, Z \in \mathbf{Z}$ and

$$W^2 + X^2 + Y^2 + Z^2 = \frac{km^2p}{m^2} = kp \tag{4}$$

Now $\frac{-m}{2} < a < \frac{m}{2}$, $\frac{-m}{2} < b < \frac{m}{2}$, $\frac{-m}{2} < c < \frac{m}{2}$, and $\frac{-m}{2} < d < \frac{m}{2}$ imply that $a^2 < \dfrac{m^2}{4}$, $b^2 < \dfrac{m^2}{4}$, $c^2 < \dfrac{m^2}{4}$, and $d^2 < \dfrac{m^2}{4}$; then

$$km = a^2 + b^2 + c^2 + d^2 < 4\left(\frac{m^2}{4}\right)$$

and we have $k < m$. But $0 < k < m$ contradicts the minimality of m for which the equation $w^2 + x^2 + y^2 + z^2 = mp$ is solvable since (4) gives such a solution with k instead of m. So $m = 1$, $w^2 + x^2 + y^2 + z^2 = p$, and the proof is complete. ■

The four-square theorem may now be stated and proved quite easily.

Theorem 6.13: (Lagrange) All positive integers are expressible as the sums of four squares of integers.

Proof: Let $n \in \mathbf{Z}$ with $n > 0$. If $n = 1$, the desired result is obvious. If $n > 1$, then n is a product of prime numbers by the Fundamental Theorem of Arithmetic. Since each such prime number is expressible as the sum of four squares of integers by Proposition 6.12, then n is so expressible by Proposition 6.10, and the proof is complete. ■

The expression of a given positive integer as the sum of four squares of integers may be accomplished in a manner similar to that illustrated in Example 11 (with the use of Proposition 6.10 instead of Proposition 6.5). Examples are left to the reader.

We conclude this section with the discussion of a classical (circa 1770) problem in Diophantine analysis from English mathematician Edward Waring (therefore known as Waring's Problem). We have seen that every positive integer can be expressed as the sum of four squares of integers, but that some positive integers cannot be expressed as a sum of three squares of integers (and so cannot be expressed as a sum of less than three squares of integers either). In other words, every positive integer is expressible as the sum of at most four squares of integers, and four is the minimal such number. What is the minimum number of cubes required in order for every positive integer to be expressible as a sum of this number of cubes? (Here the cubes are cubes of nonnegative integers.) In fact, does such a minimum number necessarily exist? What about the analogous questions for minimum numbers of fourth powers, minimum numbers of fifth powers, and so on? In his book *Meditationes Algebraicae,* Waring posed the following problem.

Waring's Problem: Let $k \in \mathbf{Z}$ with $k > 0$. Does there exist a minimum integer $g(k)$ such that every positive integer is expressible as the sum of at most $g(k)$ kth powers of nonnegative integers?

For example, $g(1) = 1$, since every positive integer is expressible as the first power of a nonnegative integer (namely, itself). Similarly, $g(2) = 4$, since every positive integer is expressible as the sum of four squares of integers (by Theorem 6.13), and no smaller number of squares will suffice (since three squares are not sufficient by Proposition 6.9). Does $g(3)$ exist? If so, how many cubes of nonnegative integers does it take to guarantee an expression of every positive integer as a sum of this number of cubes? In general, does $g(k)$ exist for arbitrary positive k?

The answer to Waring's Problem is given by the following theorem, which was proved by German mathematician David Hilbert in 1906.

Theorem 6.14: (Hilbert) Let $k \in \mathbf{Z}$ with $k > 0$. Then there exists a minimum integer $g(k)$ such that every positive integer is expressible as the sum of at most $g(k)$ kth powers of nonnegative integers.

So $g(k)$ exists for all positive integers k. Unfortunately, the proof (which we omit) is not a constructive proof: It does not provide a formula for $g(k)$ or a means for determining such a formula. There is currently strong numerical evidence to suggest that $g(k) = [(\frac{3}{2})^k] + 2^k - 2$ where $[\cdot]$ denotes the greatest integer function. For example, it is known that this formula is valid for all but possibly finitely many k and, in particular, that it is valid for all k with $k \leq 471600000$. So $g(3) = 9$; in other words, nine cubes of nonnegative integers are required to be able to represent every positive integer as a sum of these cubes. Similarly, $g(4) = 19$, $g(5) = 37$, and so on. It should be stressed that these results are nonelementary—proofs of these facts require a branch of number theory incorporating complex analysis, which is called *analytic number theory.*

Exercise Set 6.5

28. Determine whether each integer is expressible as the sum of two squares of integers.
(a) 103
(b) 207
(c) 637
(d) 6!
(e) 10!
(f) 1989^{1989}

29. Express each integer as the sum of two squares of integers.
(a) 98
(b) 181
(c) 522
(d) 605
(e) 1105
(f) 16133

30. Determine whether each integer is expressible as the sum of three squares of integers.
(a) 324
(b) 496
(c) 3008
(d) 28672
(e) 10!
(f) a prime number p with $(\frac{2}{p}) = -1$ [$(\frac{\cdot}{\cdot})$ is the Legendre symbol]

31. Find all integers between 100 and 200 that are not expressible as the sum of three squares of integers.

32. Express each integer as the sum of four squares of integers.
(a) 207
(b) 605 [Do you really have to do any work here? See Exercise 29(d) above!]
(c) 1638
(d) 2926

33. Find the minimum number of cubes of nonnegative integers needed to express each integer as a sum of these cubes. (See also Student Project 4.)
(a) 23
(b) 25

(*c*) 27
(*d*) 29

34. Prove that a positive integer is expressible as the difference of two squares of integers if and only if it is not of the form $4n + 2, n \in \mathbf{Z}$.

35. Prove or disprove the following statement: The sum of two integers expressible as the sums of three squares of integers is expressible as the sum of three squares of integers.

36. Prove that any odd positive integer is expressible as the sum of four squares of integers, two of which are consecutive. [*Hint*: Any positive integer of the form $4n + 1$ is expressible as $(2a + 1)^2 + (2b)^2 + (2c)^2$ for some integers a, b, and c. (Why?) Now show that $2n + 1$ is expressible as the sum of four squares of integers, two of which are a and $a + 1$.]

37. Let $n \in \mathbf{Z}$ with $n \geq 170$. Prove that n is expressible as the sum of five squares of positive integers. (*Hint*: Express $n - 169$ as the sum of four squares of nonnegative integers, say $n - 169 = a^2 + b^2 + c^2 + d^2$. If none of these squares are zero, then $n = a^2 + b^2 + c^2 + d^2 + 13^2$. If exactly one of these squares is zero, say d^2, then $n = a^2 + b^2 + c^2 + 5^2 + 12^2$. Continue.)

38. Let $n \in \mathbf{Z}$ with $n > 0$. Prove that $8n$ is expressible as the sum of eight squares of odd integers.

39. Prove that the only prime number that is expressible as the sum of two cubes of nonnegative integers is 2. [*Hint*: $a^3 + b^3 = (a + b) \times ((a - b)^2 + ab)$.]

40. (*a*) Prove that every integer is expressible as the sum of five cubes of integers. [*Hint*: For $n \in \mathbf{Z}$, show that $n^3 \equiv n \bmod 6$ and use the identity $n^3 - 6a = n^3 + a^3 + a^3 - (a + 1)^3 - (a - 1)^3$.]

(*b*) Why does the result of part (a) above not contradict the fact that $g(3) = 9$?

6.6

Concluding Remarks

The solution of equations is important in all branches of mathematics. We have seen that characterizing all solutions of Diophantine equations or proving the unsolvability of Diophantine equations may require considerable effort. (In fact, if you have worked diligently through the problems of this chapter, you have constructed several proofs of considerable length.) The Diophantine equations encountered in this chapter are in no way meant as a comprehensive list of the important Diophantine equations in number theory; for example, the interested reader is referred to Section 8.4 for the discussion of Pell's Equation, another noteworthy Diophantine equation. A more comprehensive treatment of Diophantine equations may be found in Hardy and Wright (1979).

Student Projects

1. (Programming project for calculator or computer)
   Given a positive integer n, express n as the sum of four squares of integers.
2. ***Definition:*** A right triangle with sides of integral length is said to be a *Pythagorean triangle.*
   (a) Prove that the radius of a circle inscribed in a Pythagorean triangle is an integer.
   (b) Prove or disprove the following statement: The area of a Pythagorean triangle is never a perfect square.
   (c) Prove or disprove the following statement: The areas of Pythagorean triangles with distinct hypotenuses are distinct.
3. Attempt a proof of Fermat's Last Theorem for the case $n = 3$ by using Fermat's method of descent. [Such a proof can be found in Chapter 2 of Edwards (1977).]
4. The minimum number of cubes of nonnegative integers needed to express the integer 23 as a sum of these cubes [see Exercise 33(a)] may be obtained by finding the largest possible perfect cube less than or equal to 23, subtracting this cube from 23, finding the largest possible perfect cube less than or equal to the difference, subtracting this cube from the difference, and so on. This procedure yields

$$23 = 2^3 + 2^3 + 1^3 + 1^3 + 1^3 + 1^3 + 1^3 + 1^3 + 1^3$$

   and so the desired minimum number of cubes is nine. However, *the procedure motivated above will not work for all positive integers* (try, for example, the integer 34). Characterize those positive integers whose expression as a minimum number of cubes of nonnegative integers is *not* obtained by subtracting the largest perfect cube less than or equal to the integer first.
5. Read the following article: Penn & Teller, "Penn & Teller's Impossible Number Prediction," *GAMES, 16* (1992), 9–11. Prove that the trick works precisely for the numbers given in the article by considering the solutions of an appropriate Diophantine equation.
6. Read the following article: Blair K. Spearman & Kenneth S. Williams, "Representing Primes by Binary Quadratic Forms," *American Mathematical Monthly, 99* (1992), 423–426.
7. Let a, b, and c be nonzero integers not all of the same sign and such that abc is squarefree. (Here, *squarefree* means "not divisible by any perfect square other than 1.") Prove that the Diophantine equation $ax^2 + by^2 + cz^2 = 0$ has a solution in integers x, y, z, not all zero, if and only if $-bc$, $-ca$, and $-ab$ are quadratic residues modulo $|a|$, $|b|$, and $|c|$, respectively. [This classical result comes from Legendre. A proof of this result may be found in Niven, Zuckerman, and Montgomery (1991).]

7

Continued Fractions

The reader is presumably well aware that there are important systems of numbers besides the integers. Two such systems are those for rational numbers and real numbers. Up to this point, we have scarcely mentioned these latter systems; indeed, our focus has been on the properties of the system of *integers*. It is now time to remedy this situation.

In this chapter, we will deal extensively with rational numbers and real numbers through a concept called *continued fractions*. The first types of continued fractions were introduced by Leonardo Fibonacci in his work *Liber abaci* published in 1202. They provide a unique way of viewing rational numbers and real numbers as sequences of integers; as such, continued fractions may be thought of as a way of "reducing" rational numbers and real numbers to considerations of integers. Continued fractions are useful in many advanced areas of number theory. One application of continued fractions in the solution of an important Diophantine equation is given in Section 8.4.

7.1

Rational and Irrational Numbers

It is presumably well known to the reader that the set of real numbers ($\mathbf{R}$) is the disjoint union of the set of rational numbers ($\mathbf{Q}$) and the set of irrational numbers ($\mathbf{R} - \mathbf{Q}$). We begin by making these latter concepts precise.

Definition 1: Let $\alpha \in \mathbf{R}$. Then α is said to be a *rational number* if $\alpha = \frac{a}{b}$ where $a, b \in \mathbf{Z}$ and $b \neq 0$. If α is not a rational number, then α is said to be an *irrational number.*

Example 1:

(a) The real number 0.5 is a rational number since $0.5 = \frac{5}{10} = \frac{1}{2}$.

(b) The real number 0.666... (here, the 6's repeat ad infinitum) is a

Biography

Leonardo of Pisa (circa 1180–1250)

Leonardo of Pisa is better known as Leonardo Fibonacci (literally translated, "Leonardo, son of Bonacci"). Generally regarded as the greatest mathematician of the Middle Ages, he wrote works on arithmetic, elementary algebra, geometry, trigonometry, and Diophantine equations. His work of 1202, *Liber abaci,* was largely devoted to computations using the Hindu-Arabic system of numeration, which Fibonacci felt to be superior to other systems. *Liber abaci* was instrumental in the spread of the Hindu-Arabic system throughout Europe. In addition to illustrating various ways of reading and writing numbers with the symbols 0 through 9 (including a precursor of present-day continued fractions), the work contained a wealth of problems that occupied mathematicians for centuries. One of these was the famous "rabbit problem," which gave rise to the equally famous Fibonacci sequence $1, 1, 2, 3, 5, 8, 13, \ldots$. (See Example 3 in Appendix A for a definition of the Fibonacci numbers.)

rational number since $0.666\ldots = \frac{2}{3}$. (You probably know this fact, but do you know how to *prove* it? A technique for proving that $0.666\ldots = \frac{2}{3}$ will be developed in this section.)

(c) The real number $\sqrt{2}$ is an irrational number since $\sqrt{2}$ cannot be represented as the quotient of two integers (see Proposition 7.1 below).

(d) The real constants π and e are irrational numbers. [For a proof of the irrationality of π, see Hardy and Wright (1979). The (much simpler) proof of the irrationality of e is motivated in Exercise 8.]

(e) The real numbers $2^{\sqrt{2}}$, e^{π}, and πe are irrational numbers. The proofs of these facts are relatively recent (circa 1929).

(f) It is currently unknown whether the real numbers $\pi^{\sqrt{2}}$, π^e, and 2^e (for example) are rational or irrational.

As should be clear from Example 1, proving the rationality or irrationality of a real number may be no easy matter. As we progress in this chapter, we will develop methods for discerning between rational numbers and irrational numbers.

Before continuing, we should produce a bona fide irrational number and its proof of irrationality. The following proof of the irrationality of $\sqrt{2}$ is so important that any person having had a course in number theory should be able to reproduce it on the spot.

Proposition 7.1: $\sqrt{2} \notin \mathbf{Q}$

Proof: Assume, by way of contradiction, that $\sqrt{2} \in \mathbf{Q}$. Then $\sqrt{2} = \frac{a}{b}$ for some $a, b \in \mathbf{Z}$ with $b \neq 0$ and, without loss of generality, $(a, b) = 1$ (in other words, the fraction $\frac{a}{b}$ is in lowest terms). Squaring both sides of $\sqrt{2} = \frac{a}{b}$, we obtain $2 = a^2/b^2$, which implies that $2b^2 = a^2$, from which $2 \mid a^2$ and $2 \mid a$. So $a = 2c$ for some $c \in \mathbf{Z}$. Then $2b^2 = a^2 = 4c^2$, and we have that $b^2 = 2c^2$, from which $2 \mid b^2$ and $2 \mid b$. But $2 \mid a$ and $2 \mid b$ imply that $(a, b) \neq 1$, a contradiction. So $\sqrt{2} \notin \mathbf{Q}$ as desired. ■

Two remarks are in order here. First, note the usefulness of proof by contradiction to prove the irrationality of a number. This usefulness results from the fact that the assumption of rationality (by way of contradiction) gives us important information with which to work, namely, the equation $\sqrt{2} = \frac{a}{b}$. The approach of Proposition 7.1 will establish the irrationality of other real numbers; the reader is referred to Exercise 4. Second, note the importance of the assumption $(a, b) = 1$ in deriving the contradiction above; the contradiction is realized at the point where $(a, b) \neq 1$. It is possible to prove Proposition 7.1 without the assumption $(a, b) = 1$; the reader is referred to Exercise 5.

The set of rational numbers $\mathbf{Q}$ is closed under the four basic arithmetic operations of addition, subtraction, multiplication, and division as given by the following proposition.

Proposition 7.2: Let $\alpha, \beta \in \mathbf{Q}$. Then $\alpha \pm \beta \in \mathbf{Q}$, $\alpha\beta \in \mathbf{Q}$, and, if $\beta \neq 0$, then $\frac{\alpha}{\beta} \in \mathbf{Q}$.

Proof: Since $\alpha, \beta \in \mathbf{Q}$, we have $\alpha = \frac{a}{b}$ and $\beta = \frac{c}{d}$ for some $a, b, c, d \in \mathbf{Z}$ with $b, d \neq 0$. Then $bd \neq 0$; so

$$\alpha \pm \beta = \frac{a}{b} \pm \frac{c}{d} = \frac{ad \pm bc}{bd} \in \mathbf{Q}$$

and

$$\alpha\beta = \left(\frac{a}{b}\right)\left(\frac{c}{d}\right) = \frac{ac}{bd} \in \mathbf{Q}$$

If $\beta \neq 0$, then $c \neq 0$; so $bc \neq 0$ and

$$\frac{\alpha}{\beta} = \frac{\left(\frac{a}{b}\right)}{\left(\frac{c}{d}\right)} = \frac{ad}{bc} \in \mathbf{Q} \blacksquare$$

The set of irrational numbers $\mathbf{R} - \mathbf{Q}$ does not enjoy the fourfold closure given by Proposition 7.2. One need only note that the product of $\sqrt{2}$ with itself yields a rational number, namely, 2. Can you produce an example showing that $\mathbf{R} - \mathbf{Q}$ does not possess closure under addition?

Must we construct a proof as in Proposition 7.1 every time we wish to prove the irrationality of a real number? The following proposition provides an alternate method of proving such irrationality in certain cases.

Proposition 7.3: Let $\alpha \in \mathbf{R}$ be a root of the polynomial

$$f(x) = x^n + c_{n-1}x^{n-1} + c_{n-2}x^{n-2} + \cdots + c_1x + c_0$$

where $c_0, c_1, \ldots, c_{n-1} \in \mathbf{Z}$ and $c_0 \neq 0$. Then $\alpha \in \mathbf{Z}$ or $\alpha \in \mathbf{R} - \mathbf{Q}$.

Proof: Assume that $\alpha \in \mathbf{Q}$. We must show that $\alpha \in \mathbf{Z}$. Now $\alpha = \frac{a}{b}$ for some $a, b \in \mathbf{Z}$ with $b \neq 0$ and, without loss of generality, $(a, b) = 1$. Then $f(\alpha) = 0$ implies that

$$\left(\frac{a}{b}\right)^n + c_{n-1}\left(\frac{a}{b}\right)^{n-1} + c_{n-2}\left(\frac{a}{b}\right)^{n-2} + \cdots + c_1\left(\frac{a}{b}\right) + c_0 = 0$$

Multiplying both sides of this equation by b^n, we obtain

$$a^n + c_{n-1}a^{n-1}b + c_{n-2}a^{n-2}b^2 + \cdots + c_1ab^{n-1} + c_0b^n = 0$$

Then

$$a^n = b(-c_{n-1}a^{n-1} - c_{n-2}a^{n-2}b - \cdots - c_1ab^{n-2} - c_0b^{n-1})$$

and so $b \mid a^n$. Since $(a, b) = 1$, we have that $b = \pm 1$; now

$$\alpha = \frac{a}{b} = \frac{a}{\pm 1} = \pm a \in \mathbf{Z}$$

as desired. $\blacksquare$

The use of Proposition 7.3 is illustrated in the following example.

Example 2:

(a) $\sqrt{3}$ is a root of the polynomial $f(x) = x^2 - 3$. Since $\sqrt{3} \notin \mathbf{Z}$ (see Exercise 4), we have that $\sqrt{3} \in \mathbf{R} - \mathbf{Q}$ by Proposition 7.3, and so $\sqrt{3}$ is an irrational number. A similar argument may be used to prove that $\sqrt{a}$ is an irrational number if a is any positive integer that is not a perfect square.

(b) $2 + \sqrt{7}$ is a root of the polynomial $f(x) = x^2 - 4x - 3$. (Please check it!) Since $\sqrt{7} \notin \mathbf{Z}$ (see Exercise 4), we have that $2 + \sqrt{7} \notin \mathbf{Z}$, and so $2 + \sqrt{7} \in \mathbf{R} - \mathbf{Q}$ by Proposition 7.3, from which $2 + \sqrt{7}$ is an irrational number.

(c) $\sqrt[3]{5}$ is a root of the polynomial $f(x) = x^3 - 5$. Since $1^3 < 5 < 2^3$, we have that $1 < \sqrt[3]{5} < 2$, and so $\sqrt[3]{5} \notin \mathbf{Z}$. Then $\sqrt[3]{5} \in \mathbf{R} - \mathbf{Q}$ by Proposition 7.3, from which $\sqrt[3]{5}$ is an irrational number. A similar argument may be used to prove that $\sqrt[3]{a}$ is an irrational number if a is any positive integer that is not a perfect cube.

Using Proposition 7.3 to prove the irrationality of a number requires that one find a polynomial (satisfying the given hypotheses) for which this number is a root. This amounts to nothing more than a simple algebra problem. For example, finding a polynomial for $2 + \sqrt{7}$ in Example 2(b) above is achieved by putting $x = 2 + \sqrt{7}$, subtracting 2 from both sides to obtain $x - 2 = \sqrt{7}$, squaring both sides to obtain $x^2 - 4x + 4 = 7$, and subtracting 7 from both sides to obtain the desired result.

The converse of Proposition 7.3 is false. For example, $\pi, e \in \mathbf{R} - \mathbf{Q}$, but neither π nor e is a root of a monic polynomial with integral (or even rational) coefficients. Such irrational numbers are said to be *transcendental numbers.* For proofs of the transcendence of π and e, see Hardy and Wright (1979).

Every real number has a decimal representation. For a rational number $\frac{a}{b}$, such a representation is obtained simply by dividing the denominator b into the numerator a. Decimal representations of irrational numbers are more difficult to obtain. In any event, the forms taken by these decimal representations can be used to determine the rationality or irrationality of the numbers. We first define a special type of decimal representation.

Definition 2: Let $\alpha \in \mathbf{R}$ with $0 \leq \alpha < 1$ and let $\sum_{n=1}^{\infty} a_n/10^n = 0.a_1a_2a_3\cdots$ be a decimal representation of α. If there exist positive integers ρ and N such that $a_n = a_{n+\rho}$ for all $n \geq N$, then α is said to be *eventually periodic*; the sequence $a_Na_{N+1}\cdots a_{N+(\rho-1)}$ with ρ minimal is said to be the *period* of α, and ρ is said to be the *period length.* If N can be taken to be 1, then α is said to be *periodic.* An eventually periodic real number

$$\alpha = 0.a_1a_2\cdots a_{N-1}a_Na_{N+1}\cdots a_{N+(\rho-1)}a_Na_{N+1}\cdots a_{N+(\rho-1)}\cdots$$

is denoted $\alpha = 0.a_1a_2\cdots a_{N-1}\overline{a_Na_{N+1}\cdots a_{N+(\rho-1)}}$.

Definition 2 does nothing more than formalize what is commonly referred to as a "repeating" decimal expansion. Several examples will clarify this concept.

Example 3:

(a) A decimal representation of the real number $\frac{1}{2}$ is 0.5 (or $0.5\overline{0}$). So $\frac{1}{2}$ is eventually periodic, with period 0 and period length 1. Note that a terminating decimal expansion (as in 0.5) may always be considered to be eventually periodic, with period 0 and period length 1 (as in $0.5\overline{0}$).

(b) A decimal representation of the real number $\frac{2}{3}$ is $0.\bar{6}$. (We will obtain a method for proving this shortly.) So $\frac{2}{3}$ is periodic, with period 6 and period length 1.

(c) A decimal representation of the real number $\frac{2}{7}$ is $0.\overline{285714}$. (We will obtain a method for proving this shortly.) So $\frac{2}{7}$ is periodic, with period 285714 and period length 6.

(d) A decimal representation of $\sqrt{2}$ to 20 places (obtainable by using techniques in the theory of infinite series) is 1.414 213 562 373 095 048 80 . . . . The real number given by the part of this representation to the right of the decimal point does not *appear* to be eventually periodic. (However, it could be that we have not computed enough digits in the decimal representation to detect the period.)

(e) A decimal representation of π to 20 places (obtainable by using techniques in the theory of infinite series) is 3.141 592 653 589 793 238 46 . . . . The real number given by the part of this representation to the right of the decimal point does not *appear* to be eventually periodic. (Again, it could be that we have not computed enough digits in the decimal representation to detect the period.)

(f) A decimal representation of e to 20 places (obtainable by using techniques in the theory of infinite series) is 2.718 281 828 459 045 235 36 . . . . The real number given by the part of this representation to the right of the decimal point does not *appear* to be eventually periodic. (Yet again, it could be that we have not computed enough digits in the decimal representation to detect the period.)

One further remark concerning the above proceedings is in order here. Decimal representations of real numbers may not be unique; it is for this reason that the phrase "*a* decimal representation" was used in Definition 2 and Example 3 above rather than the phrase "*the* decimal representation." An example illustrating this nonuniqueness will conclude this section.

In Example 3 above, the rational numbers [in (a) through (c)] were eventually periodic (note that periodic numbers are eventually periodic), while the fractional parts of the irrational numbers [in (d) through (f)] *appeared* not to be eventually periodic. The following proposition shows that this behavior was no coincidence by providing a characterization of rational numbers in terms of the eventual periodicity of their decimal representations.

Proposition 7.4: Let $\alpha \in \mathbf{R}$ with $0 \leq \alpha < 1$. Then $\alpha \in \mathbf{Q}$ if and only if α is eventually periodic.

Proof: ($\Rightarrow$) Assume that $\alpha \in \mathbf{Q}$. Then $\alpha = \frac{a}{b}$ for some $a, b \in \mathbf{Z}$ with $b \neq 0$; since $0 \leq \alpha < 1$, we additionally have $0 \leq a < b$. Now divide b into a by using usual long division; let the resulting decimal representation of α be

$$\sum_{n=1}^{\infty} \frac{q_n}{10^n} = 0.q_1q_2q_3\cdots$$

By the division algorithm, the possible remainders upon division by b are

$0, 1, 2, \ldots, b - 1$. At each stage of the long-division process, b is being divided into one of these remainders times 10 to obtain a quotient. Note that the first such remainder is a itself. Accordingly, let $r_1 = a, r_2, r_3, \ldots$ be the sequence of such remainders corresponding to quotients $q_1, q_2, q_3, \ldots$ (so that $\frac{a}{b} = 0.q_1q_2q_3 \cdots$). Since the number of possible remainders is finite, $r_N = r_M$ for some N and M with $N < M$. If $p = M - N$, then $r_n = r_{n+p}$ for all $n \geq N$, from which $q_n = q_{n+p}$ for all $n \geq N$, and α is eventually periodic.

($\Leftarrow$) Assume that α is eventually periodic. Then there exist positive integers p and N such that

$$\alpha = 0.a_1a_2 \cdots a_{N-1}\overline{a_Na_{N+1} \cdots a_{N+(p-1)}}$$

Now

$$10^{N-1}\alpha = a_1a_2 \cdots a_{N-1}.\overline{a_Na_{N+1} \cdots a_{N+(p-1)}}$$

and

$$10^p10^{N-1}\alpha = a_1a_2 \cdots a_{N-1}a_Na_{N+1} \cdots a_{N+(p-1)}.\overline{a_Na_{N+1} \cdots a_{N+(p-1)}}$$

Furthermore, $10^p10^{N-1}\alpha - 10^{N-1}\alpha$ is an integer since the identical repeating blocks to the right of the decimal points cancel. Since

$$10^p10^{N-1}\alpha - 10^{N-1}\alpha = (10^p - 1)10^{N-1}\alpha$$

we have that $(10^p - 1)10^{N-1}\alpha \in \mathbf{Z}$, say $(10^p - 1)10^{N-1}\alpha = m \in \mathbf{Z}$. Then

$$\alpha = \frac{m}{(10^p - 1)10^{N-1}}$$

and, since $(10^p - 1)10^{N-1}$ is a nonzero integer, we have $\alpha \in \mathbf{Q}$ as desired. ■

By Proposition 7.4, the fractional parts of rational numbers are characterized precisely by being eventually periodic, while the fractional parts of irrational numbers are characterized precisely by not being eventually periodic. In addition, the procedure in the second half of the proof of Proposition 7.4 can be used to express any decimal with eventually periodic fractional part as the quotient of two integers; the requirement in Proposition 7.4 that $0 \leq \alpha < 1$ may be ignored for the sake of this procedure. We illustrate with an example.

Example 4:

Use the procedure in the second half of the proof of Proposition 7.4 to express the decimal $1.2\overline{34}$ as the quotient of two integers.

Let $\alpha = 1.2\overline{34}$. Then $10\alpha = 12.\overline{34}$ and $100(10\alpha) = 1234.\overline{34}$. Now

$$100(10\alpha) - 10\alpha = 1234.\overline{34} - 12.\overline{34} = 1222$$

so that

$$1000\alpha - 10\alpha = 1222.$$

So $990\alpha = 1222$ and $\alpha = \frac{1222}{990} = \frac{611}{495}$. Consequently, $1.2\overline{34} = \frac{611}{495}$.

Incidentally, the procedure of Example 4 may be used to *prove* that $0.\overline{6} = \frac{2}{3}$ or $0.\overline{285714} = \frac{2}{7}$ [see (b) and (c) of Example 3]. The interested reader is invited to perform these computations.

In sum, any rational number $\frac{a}{b}$ expressed as a quotient of two integers can be expressed in decimal form (necessarily with eventually periodic fractional part) by dividing the denominator b into the numerator a; any decimal with eventually periodic fractional part can be expressed as a quotient of two integers by using the procedure in the second half of the proof of Proposition 7.4.

We conclude this section with an example that illustrates the nonuniqueness of decimal representation of real numbers.

Example 5:

Prove that the decimal representation of a real number may not be unique by exhibiting a real number having two distinct decimal representations.

We show that the real number 1 has two distinct decimal representations, $1.\overline{0}$ and $0.\overline{9}$. The first representation is obvious. We now show that $0.\overline{9} = 1$. Let $\alpha = 0.\overline{9}$. Using the procedure in the second half of the proof of Proposition 7.4, we have $10\alpha = 9.\overline{9}$ and

$$10\alpha - \alpha = 9.\overline{9} - 0.\overline{9} = 9$$

from which $9\alpha = 9$ and $\alpha = 1$. So $0.\overline{9} = 1$ as desired, and the real number 1 has two distinct decimal representations.

Considerations similar to that in Example 5 above establish $0.1\overline{0} = 0.0\overline{9}$, $0.01\overline{0} = 0.00\overline{9}$, $0.001\overline{0} = 0.000\overline{9}, \ldots$.

Exercise Set 7.1

1. Express each rational number in decimal form.
   (a) $\frac{6}{7}$
   (b) $\frac{3}{8}$
   (c) $\frac{5}{11}$
   (d) $\frac{13}{14}$
   (e) $\frac{9}{17}$
   (f) $\frac{1}{101}$

2. Express each decimal as the quotient of two integers in lowest terms.
   (a) 1.2
   (b) $1.\overline{2}$
   (c) $1.2\overline{3}$
   (d) $1.\overline{23}$
   (e) $1.23\overline{4}$
   (f) $1.23\overline{456}$

3. Prove that each real number is irrational by exhibiting a polynomial with integral coefficients for which it is a root, in accordance with Proposition 7.3.

(a) $\sqrt[5]{3}$
(b) $1 - \sqrt{6}$
(c) $\sqrt{6} + 1$
(d) $\sqrt{2} + \sqrt{5}$

4. *(a)* Let p be a prime number. Prove that $\sqrt{p}$ is an irrational number by using a technique similar to that used in the proof of Proposition 7.1.
(b) If one attempted to use the proof technique of Proposition 7.1 to prove that $\sqrt{4}$ was irrational (which obviously is false), where would the technique fail?

5. *(a)* Prove that $\sqrt{2}$ is an irrational number by using proof by contradiction as in the proof of Proposition 7.1 but without the assumption $(a, b) = 1$. (*Hint*: Prove that $2b^2 = a^2$ as in Proposition 7.1. Now use the Fundamental Theorem of Arithmetic to obtain a contradiction.)
(b) Let $p_1, p_2, \ldots, p_n$ be distinct prime numbers. Prove that $\sqrt{p_1 p_2 \cdots p_n}$ is an irrational number. [*Hint*: Parallel the proof of part (a) above.]
(c) Let n be a positive integer. Prove that if $\sqrt{n}$ is a rational number, then, in fact, $\sqrt{n}$ is an integer. Conclude that if n is a positive integer, then $\sqrt{n}$ is either integral or irrational.

6. Let d be a positive integer that is not a perfect square and let $w, x, y, z \in \mathbf{Q}$. If

$$w + x\sqrt{d} = y + z\sqrt{d}$$

prove that $w = y$ and $x = z$.

7. (This exercise requires a knowledge of Chapter 5.) Let m be a positive integer with $(m, 10) = 1$. Prove that the period length of the decimal representation of $\frac{1}{m}$ is $m - 1$ if and only if m is a prime number and 10 is a primitive root modulo m.

8. (The following exercise proves that e is irrational.)
(a) Assume that $e = \frac{a}{b}$ where (without loss of generality) a and b are positive integers. Let $n \in \mathbf{Z}$ with $b \le n$ and put

$$\alpha = n!\left(e - 1 - \frac{1}{1!} - \frac{1}{2!} - \cdots - \frac{1}{n!}\right)$$

Prove that α is a positive integer. (*Hint*: Recall from the theory of infinite series that $e = \sum_{n=0}^{\infty} \frac{1}{n!}$.)
(b) Prove that

$$\alpha = \frac{1}{n + 1} + \frac{1}{(n + 1)(n + 2)} + \frac{1}{(n + 1)(n + 2)(n + 3)} + \cdots$$

Deduce that $\alpha < \frac{1}{n}$. [*Hint*: For the deduction, note that the second term in the expansion is less than $1/(n + 1)^2$, the third term in the expansion is less than $1/(n + 1)^3$, etc.]
(c) Use the results of parts (a) and (b) above to obtain a contradiction to the opening assumption of part (a) that e is rational.

7.2

Finite Continued Fractions

We begin with a motivating example.

Example 6:

By Example 14 of Chapter 1, we have

$$803 = 154 \cdot 5 + 33$$
$$154 = 33 \cdot 4 + 22$$
$$33 = 22 \cdot 1 + 11$$
$$22 = 11 \cdot 2 + 0$$

By dividing each equation by the first factor of the product immediately to the right of the equal sign, we obtain

$$\frac{803}{154} = 5 + \frac{33}{154} \tag{1}$$

$$\frac{154}{33} = 4 + \frac{22}{33} \tag{2}$$

$$\frac{33}{22} = 1 + \frac{11}{22} \tag{3}$$

$$\frac{22}{11} = 2 \tag{4}$$

and

$$\frac{803}{154} = 5 + \frac{1}{\left(\frac{154}{33}\right)} \quad \text{[by (1)]}$$

$$= 5 + \cfrac{1}{4 + \cfrac{1}{\left(\frac{33}{22}\right)}} \quad \text{[by (2)]}$$

$$= 5 + \cfrac{1}{4 + \cfrac{1}{1 + \cfrac{1}{\left(\frac{22}{11}\right)}}} \quad \text{[by (3)]}$$

$$= 5 + \cfrac{1}{4 + \cfrac{1}{1 + \cfrac{1}{2}}} \quad \text{[by (4)]}$$

In sum, manipulation of the steps in the Euclidean algorithm used to obtain (803, 154) allows the expression of the rational number $\frac{803}{154}$ as a somewhat curious-looking more complex fraction called a *continued fraction* (the formal definition is below). You are no doubt wondering, "What is to be gained by the expression of a number as a continued fraction?" By the end of this chapter, you will have many answers.

We now present the formal definition of a (finite) continued fraction.

Definition 3: An expression of the form

$$a_0 + \cfrac{1}{a_1 + \cfrac{1}{a_2 + \cfrac{}{\ddots \cfrac{1}{a_{n-1} + \cfrac{1}{a_n}}}}}$$

where $a_0, a_1, a_2, \ldots, a_n \in \mathbf{R}$ and $a_1, a_2, \ldots, a_n > 0$ is said to be a *finite continued fraction* and is denoted concisely by $[a_0, a_1, a_2, \ldots, a_n]$. A finite continued fraction is said to be *simple* if $a_0, a_1, a_2, \ldots, a_n \in \mathbf{Z}$.

The remainder of this chapter deals exclusively with *simple* continued fractions. In the notation of Definition 3 above, we see that Example 6 expresses the rational number $\frac{803}{154}$ as the finite simple continued fraction $[5, 4, 1, 2]$. (This notation will cause no confusion with the notation for least common multiple, since least common multiples will not be used in this chapter.) We now give another example.

Example 7:

Express the number 0.7 as a finite simple continued fraction if possible. We have $0.7 = \frac{7}{10}$. We now repeatedly apply the division algorithm:

$$7 = 10 \cdot 0 + 7$$
$$10 = 7 \cdot 1 + 3$$
$$7 = 3 \cdot 2 + 1$$
$$3 = 1 \cdot 3 + 0$$

Note the first step here! (Beginning with the second step would ultimately express $\frac{10}{7}$ as a finite simple continued fraction.) Considerations exactly as in Example 6 now yield

$$\frac{7}{10} = 0 + \cfrac{1}{1 + \cfrac{1}{2 + \cfrac{1}{3}}}$$

or, more concisely, $\frac{7}{10} = [0, 1, 2, 3]$ as desired. In case you have not noticed, the expansion $[0, 1, 2, 3]$ may be obtained directly from the division algorithm steps above — look at the second factor in each product immediately to the right of each equal sign. This observation applies equally well to Example 6.

In view of Examples 6 and 7, it should be reasonably clear that all rational numbers are expressible as finite simple continued fractions. In fact, it is also true that every finite simple continued fraction represents a rational number. These two statements combine to provide us with a characterization of the rational numbers in terms of continued fractions, as contained in the following proposition.

Proposition 7.5: Let $\alpha \in \mathbf{R}$. Then $\alpha \in \mathbf{Q}$ if and only if α is expressible as a finite simple continued fraction.

Proof: ($\Rightarrow$) Assume that $\alpha \in \mathbf{Q}$. Then $\alpha = \frac{a}{b}$ for some $a, b \in \mathbf{Z}$ with $b \neq 0$. Without loss of generality, we may assume that $b > 0$. By the division algorithm, we obtain the following sequence of equations:

$$a = bq_0 + r_0, \quad 0 < r_0 < b \tag{5}$$

$$b = r_0 q_1 + r_1, \quad 0 < r_1 < r_0 \tag{6}$$

$$r_0 = r_1 q_2 + r_2, \quad 0 < r_2 < r_1$$

$$\vdots$$

$$r_{n-3} = r_{n-2} q_{n-1} + r_{n-1}, \quad 0 < r_{n-1} < r_{n-2}$$

$$r_{n-2} = r_{n-1} q_n$$

Note that $q_0, q_1, q_2, \ldots, q_n \in \mathbf{Z}$ and $q_1, q_2, \ldots, q_n > 0$. Now, dividing both sides of (5) above by b, we obtain

$$\frac{a}{b} = q_0 + \frac{r_0}{b} = q_0 + \frac{1}{\left(\frac{b}{r_0}\right)} \tag{7}$$

Dividing both sides of (6) above by r_0, we obtain

$$\frac{b}{r_0} = q_1 + \frac{r_1}{r_0} = q_1 + \frac{1}{\left(\frac{r_0}{r_1}\right)} \tag{8}$$

Continuing in such a manner, we obtain the equations

$$\frac{r_0}{r_1} = q_2 + \frac{r_2}{r_1} = q_2 + \cfrac{1}{\left(\frac{r_1}{r_2}\right)} \tag{9}$$

$$\vdots$$

$$\frac{r_{n-3}}{r_{n-2}} = q_{n-1} + \frac{r_{n-1}}{r_{n-2}} = q_{n-1} + \cfrac{1}{\left(\frac{r_{n-2}}{r_{n-1}}\right)}$$

$$\frac{r_{n-2}}{r_{n-1}} = q_n$$

Substituting the expression for b/r_0 from (8) into (7), we obtain

$$\frac{a}{b} = q_0 + \cfrac{1}{q_1 + \cfrac{1}{\left(\frac{r_0}{r_1}\right)}} \tag{10}$$

Substituting the expression for r_0/r_1 from (9) into (10), we obtain

$$\frac{a}{b} = q_0 + \cfrac{1}{q_1 + \cfrac{1}{q_2 + \cfrac{1}{\left(\frac{r_1}{r_2}\right)}}}.$$

Continuing in such a manner, we obtain

$$\frac{a}{b} = q_0 + \cfrac{1}{q_1 + \cfrac{1}{q_2 + \ddots + \cfrac{1}{q_{n-1} + \cfrac{1}{q_n}}}}$$

$$= [q_0, q_1, q_2, \ldots, q_n]$$

as desired.

($\Leftarrow$) Let α be expressible as a finite simple continued fraction, say $\alpha = [a_0, a_1, a_2, \ldots, a_n]$ where $a_0, a_1, a_2, \ldots, a_n \in \mathbf{Z}$ and $a_1, a_2, \ldots, a_n > 0$. We use induction on n. If $n = 0$, we have

$$\alpha = a_0 \in \mathbf{Z} \subseteq \mathbf{Q}$$

If $n = 1$, we have

$$\alpha = [a_0, a_1] = a_0 + \frac{1}{a_1} = \frac{a_0 a_1 + 1}{a_1} \in \mathbf{Q}$$

(Note that $a_1 \neq 0$.) Assume that $k \geq 1$ and that the desired result holds for

$n = k$ so that any finite simple continued fraction of length $k + 1$ is a rational number. We must show that any finite simple continued fraction of length $k + 2$ is a rational number so that the desired result holds for $n = k + 1$. Consider the finite simple continued fraction of length $k + 2$ given by $[a_0, a_1, a_2, \ldots, a_{k+1}]$ where $a_0, a_1, a_2, \ldots, a_{k+1} \in \mathbf{Z}$ and $a_1, a_2, \ldots, a_{k+1} > 0$. Now

$$[a_0, a_1, a_2, \ldots, a_{k+1}] = a_0 + \frac{1}{[a_1, a_2, \ldots, a_{k+1}]}$$

By the induction hypothesis, $[a_1, a_2, \ldots, a_{k+1}] \in \mathbf{Q}$, so

$$[a_1, a_2, \ldots, a_{k+1}] = \frac{a}{b}$$

for some $a, b \in \mathbf{Z}$ with $a, b \neq 0$. (Note that $a \neq 0$ here s[illegible] $a_1, a_2, \ldots, a_{k+1} > 0$.) Then

$$[a_0, a_1, a_2, \ldots, a_{k+1}] = a_0 + \frac{1}{\left(\frac{a}{b}\right)} = \frac{a_0 a + b}{a} \in \mathbf{Q}$$

as desired. ■

The finite simple continued fraction representation of a rational number is not unique since

$$[a_0, a_1, a_2, \ldots, a_n] = [a_0, a_1, a_2, \ldots, a_{n-1}, a_n - 1, 1]$$

if $a_n > 1$ and

$$[a_0, a_1, a_2, \ldots, a_n] = [a_0, a_1, a_2, \ldots, a_{n-2}, a_{n-1} + 1]$$

if $a_n = 1$. The reader is invited to check, for example, that

$$\frac{803}{154} = [5, 4, 1, 2] = [5, 4, 1, 1, 1]$$

In fact, by virtue of the identities above and Proposition 7.5, it can be shown that every rational number is expressible as a finite simple continued fraction in exactly two ways (see Exercise 13).

Exercise Set 7.2

9. Express each rational number as a finite simple continued fraction.

(a) $\frac{43}{30}$

(b) $\frac{30}{43}$

(c) $\frac{55}{34}$
(d) $\frac{37}{60}$
(e) $\frac{156}{49}$
(f) $\frac{64}{391}$

10. Repeat Exercise 9 above for the negative of each given rational number.

11. Express each finite simple continued fraction as a quotient of two integers.
(a) $[4, 3, 2, 1]$
(b) $[0, 4, 3, 2, 1]$
(c) $[1, 1, 1, 1, 1, 1]$
(d) $[5, 4, 6, 3, 7]$
(e) $[0, 2, 4, 6, 8]$
(f) $[9, 9, 9, 9]$

12. Repeat Exercise 11 for each given finite simple continued fraction where each first component has been replaced with its negative.

13. Prove that every rational number is expressible as a finite simple continued fraction in exactly two ways.

14. Let $\alpha \in \mathbf{Q}$ with $\alpha > 1$. Prove that if $\alpha = [a_0, a_1, a_2, \ldots, a_n]$, then

$$\frac{1}{\alpha} = [0, a_0, a_1, a_2, \ldots, a_n]$$

15. Let n be a positive integer and let f_n be the nth Fibonacci number (see Example 3 in Appendix A for a definition of the Fibonacci numbers). What is the finite simple continued fraction expansion of the rational number f_{n+1}/f_n? Prove your assertion. [Hint: Part (c) of Exercises 9 and 11 above may be helpful.]

16. Using the absolute least remainder algorithm in Exercise 15(c) of Chapter 1, parallel the development of finite simple continued fractions in this section.

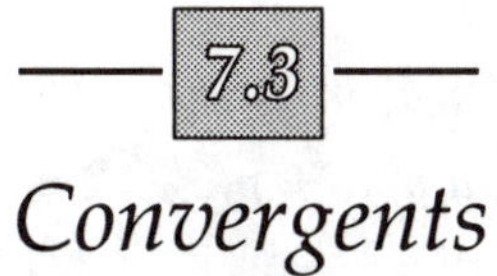

Convergents

Our ultimate goal is to be able to grasp the concept of an infinite simple continued fraction as a limit of finite simple continued fractions and to characterize those real numbers expressible as certain types of infinite simple continued fractions. The following definition is crucial in this regard.

Definition 4: Let $\alpha = [a_0, a_1, a_2, \ldots, a_n]$ be a finite continued fraction. The finite continued fraction $C_i = [a_0, a_1, a_2, \ldots, a_i]$, $0 \le i \le n$, is said to be the ith *convergent* of α.

Note, in particular, that the nth convergent of $\alpha = [a_0, a_1, a_2, \ldots, a_n]$ is equal to α. The computation of convergents is straightforward, as illustrated by the example below.

Example 8:

Consider the finite continued fraction $\alpha = [4, 3, 2, 1]$. The four convergents of α are as follows:

$$C_0 = [4] = 4$$

$$C_1 = [4, 3] = 4 + \frac{1}{3} = \frac{13}{3}$$

$$C_2 = [4, 3, 2] = 4 + \cfrac{1}{3 + \cfrac{1}{2}} = \frac{30}{7}$$

$$C_3 = [4, 3, 2, 1] = 4 + \cfrac{1}{3 + \cfrac{1}{2 + \cfrac{1}{1}}} = \frac{43}{10}$$

Note that $\alpha = C_3 = \frac{43}{10}$. Note also that the convergents of α are indeed converging (in the usual sense) to α; C_1 is closer to α than C_0 is, C_2 is closer to α than C_1 is, and C_3 is closer to α than C_2 is. (The observant reader will notice an interesting order property of these convergents, which will be dealt with more fully later.)

The convergents of a finite continued fraction are more easily computed by using certain recurrence relations given by the proposition below. These relations will also be useful in establishing needed properties of convergents.

Proposition 7.6: Let $\alpha = [a_0, a_1, a_2, \ldots, a_n]$ be a finite continued fraction. Define $p_0, p_1, p_2, \ldots, p_n$ and $q_0, q_1, q_2, \ldots, q_n$ by the following recurrence relations:

$$p_0 = a_0$$

$$p_1 = a_0 a_1 + 1$$

$$p_i = a_i p_{i-1} + p_{i-2}, \quad 2 \leq i \leq n$$

$$q_0 = 1$$

$$q_1 = a_1$$

$$q_i = a_i q_{i-1} + q_{i-2}, \quad 2 \leq i \leq n$$

Then

$$C_i = \frac{p_i}{q_i}, \quad 0 \leq i \leq n$$

Proof: We use induction on i. If $i = 0$, we have

$$C_0 = [a_0] = a_0 = \frac{a_0}{1} = \frac{p_0}{q_0}$$

If $i = 1$, we have

$$C_1 = [a_0, a_1] = a_0 + \frac{1}{a_1} = \frac{a_0 a_1 + 1}{a_1} = \frac{p_1}{q_1}$$

If $i = 2$, we have

$$C_2 = [a_0, a_1, a_2] = a_0 + \cfrac{1}{a_1 + \cfrac{1}{a_2}} = \frac{a_2(a_0 a_1 + 1) + a_0}{a_2 a_1 + 1} = \frac{a_2 p_1 + p_0}{a_2 q_1 + q_0} = \frac{p_2}{q_2}$$

Assume that $k \geq 2$ and that the desired result holds for $i = k$ so that

$$C_k = [a_0, a_1, a_2, \ldots, a_k] = \frac{p_k}{q_k} = \frac{a_k p_{k-1} + p_{k-2}}{a_k q_{k-1} + q_{k-2}}$$

We must show that

$$C_{k+1} = [a_0, a_1, a_2, \ldots, a_k, a_{k+1}] = \frac{p_{k+1}}{q_{k+1}} = \frac{a_{k+1} p_k + p_{k-1}}{a_{k+1} q_k + q_{k-1}}$$

so that the desired result holds for $i = k + 1$. We have

$$\begin{aligned}
C_{k+1} &= [a_0, a_1, a_2, \ldots, a_k, a_{k+1}] \\
&= \left[a_0, a_1, a_2, \ldots, a_{k-1}, a_k + \frac{1}{a_{k+1}}\right] \\
&= \frac{\left(a_k + \frac{1}{a_{k+1}}\right) p_{k-1} + p_{k-2}}{\left(a_k + \frac{1}{a_{k+1}}\right) q_{k-1} + q_{k-2}} \quad \text{(by the induction hypothesis)} \\
&= \frac{a_{k+1}(a_k p_{k-1} + p_{k-2}) + p_{k-1}}{a_{k+1}(a_k q_{k-1} + q_{k-2}) + q_{k-1}} \\
&= \frac{a_{k+1} p_k + p_{k-1}}{a_{k+1} q_k + q_{k-1}} \quad \text{(by the induction hypothesis)} \\
&= \frac{p_{k+1}}{q_{k+1}}
\end{aligned}$$

as desired. ■

Example 9:

We compute the four convergents of the finite continued fraction [5, 4, 1, 2] (which is equal to $\frac{803}{154}$ by Example 7) by using Proposition 7.6. In the notation of Proposition 7.6, we have $a_0 = 5$, $a_1 = 4$, $a_2 = 1$, and $a_3 = 2$, from which (by the same proposition)

$$p_0 = 5$$
$$p_1 = (5)(4) + 1 = 21$$
$$p_2 = (1)(21) + 5 = 26$$
$$p_3 = (2)(26) + 21 = 73$$

and

$$q_0 = 1$$
$$q_1 = 4$$
$$q_2 = (1)(4) + 1 = 5$$
$$q_3 = (2)(5) + 4 = 14$$

So

$$C_0 = \frac{p_0}{q_0} = \frac{5}{1}$$
$$C_1 = \frac{p_1}{q_1} = \frac{21}{4}$$
$$C_2 = \frac{p_2}{q_2} = \frac{26}{5}$$
$$C_3 = \frac{p_3}{q_3} = \frac{73}{14} \left(= \frac{803}{154}\right)$$

The interested reader is invited to use Proposition 7.6 to compute the convergents of the finite continued fraction [4, 3, 2, 1] and compare the results with Example 8.

We see immediately that the convergents in Example 9 above are (rational numbers) in lowest terms. This is true in general for convergents of finite simple continued fractions computed via the recurrence relations of Proposition 7.6; such a fact is an easy corollary of the following proposition.

Proposition 7.7: Let $\alpha = [a_0, a_1, a_2, \ldots, a_n]$ be a finite continued fraction and let all notation be as in Proposition 7.6. Then

$$p_i q_{i-1} - p_{i-1} q_i = (-1)^{i-1}, \quad 1 \le i \le n$$

Proof: We use induction on i. If $i = 1$, we have

$$p_1 q_0 - p_0 q_1 = (a_0 a_1 + 1)1 - a_0 a_1 = 1 = (-1)^{1-1}$$

Assume that $k \ge 1$ and the desired result holds for $i = k$ so that

$$p_k q_{k-1} - p_{k-1} q_k = (-1)^{k-1}$$

We must show that

$$p_{k+1}q_k - p_kq_{k+1} = (-1)^k$$

so that the desired result holds for $i = k + 1$. We have

$$\begin{aligned} p_{k+1}q_k - p_kq_{k+1} &= (a_{k+1}p_k + p_{k-1})q_k - p_k(a_{k+1}q_k + q_{k-1}) \\ &= p_{k-1}q_k - p_kq_{k-1} \\ &= -(p_kq_{k-1} - p_{k-1}q_k) \\ &= -(-1)^{k-1} \quad \text{(by the induction hypothesis)} \\ &= (-1)^k \end{aligned}$$

as desired. ■

The reader is invited to verify the truth of Proposition 7.7 in Example 9.

Proposition 7.7 has two important corollaries. The first corollary has already been alluded to and asserts that the convergents of a finite simple continued fraction are in lowest terms.

Corollary 7.8: Let $\alpha = [a_0, a_1, a_2, \ldots, a_n]$ be a finite simple continued fraction and let all notation be as in Proposition 7.6. Then $(p_i, q_i) = 1$, $0 \le i \le n$.

Proof: The fact that $(p_0, q_0) = 1$ is obvious. Let $d_i = (p_i, q_i)$, $1 \le i \le n$. Then, for all i with $1 \le i \le n$, we have $d_i \mid p_iq_{i-1}$ and $d_i \mid p_{i-1}q_i$. By Proposition 1.2, we obtain

$$d_i \mid p_iq_{i-1} - p_{i-1}q_i, \quad 1 \le i \le n$$

Proposition 7.7 then gives $d_i \mid (-1)^{i-1}$, $1 \le i \le n$, from which $d_i = 1$, $1 \le i \le n$. ■

The second corollary of Proposition 7.7 will be important in the extension of finite simple continued fractions to infinite simple continued fractions.

Corollary 7.9: Let $\alpha = [a_0, a_1, a_2, \ldots, a_n]$ be a finite simple continued fraction and let all notation be as in Proposition 7.6. Then

$$C_i - C_{i-1} = \frac{(-1)^{i-1}}{q_iq_{i-1}}, \quad 1 \le i \le n$$

and

$$C_i - C_{i-2} = \frac{(-1)^i a_i}{q_iq_{i-2}}, \quad 2 \le i \le n$$

Proof: For the first identity, we have

$$\begin{aligned} C_i - C_{i-1} &= \frac{p_i}{q_i} - \frac{p_{i-1}}{q_{i-1}} \\ &= \frac{p_i q_{i-1} - p_{i-1} q_i}{q_i q_{i-1}} \\ &= \frac{(-1)^{i-1}}{q_i q_{i-1}}, \quad 1 \le i \le n \quad \text{(by Proposition 7.7)} \end{aligned}$$

For the second identity, we have

$$\begin{aligned} C_i - C_{i-2} &= \frac{p_i}{q_i} - \frac{p_{i-2}}{q_{i-2}} \\ &= \frac{p_i q_{i-2} - p_{i-2} q_i}{q_i q_{i-2}} \\ &= \frac{(a_i p_{i-1} - p_{i-2}) q_{i-2} - p_{i-2}(a_i q_{i-1} - q_{i-2})}{q_i q_{i-2}} \\ &= \frac{a_i (p_{i-1} q_{i-2} - p_{i-2} q_{i-1})}{q_i q_{i-2}} \\ &= \frac{a_i (-1)^{i-2}}{q_i q_{i-2}} \quad \text{(by Proposition 7.7)} \\ &= \frac{(-1)^i a_i}{q_i q_{i-2}}, \quad 2 \le i \le n \ \blacksquare \end{aligned}$$

The convergents of a finite simple continued fraction possess an interesting order property: They alternate between being less than and greater than the rational number represented by the continued fraction with each successive convergent being closer to this rational number and, of course, the final convergent being equal to this rational number. (Perhaps you noticed this property in Example 8.) We now state this property precisely.

Proposition 7.10: Let $\alpha = [a_0, a_1, a_2, \ldots, a_n]$ be a finite simple continued fraction and let all notation be as in Proposition 7.6. Then

$$C_0 < C_2 < C_4 < \cdots < C_5 < C_3 < C_1$$

In other words, the even-subscripted convergents $C_0, C_2, C_4, \ldots$ form a strictly increasing sequence of rational numbers, the odd-subscripted convergents $C_1, C_3, C_5, \ldots$ form a strictly decreasing sequence of rational numbers, and every even-subscripted convergent is less than every odd-subscripted convergent.

Proof: By Corollary 7.9, we have

$$C_i - C_{i-2} = \frac{(-1)^i a_i}{q_i q_{i-2}}, \quad 2 \le i \le n$$

If i is even with $2 \le i \le n$, we have that $C_i - C_{i-2} > 0$, from which $C_{i-2} < C_i$

and

$$C_0 < C_2 < C_4 < \cdots \tag{11}$$

If i is odd with $2 \le i \le n$, we have that $C_i - C_{i-2} < 0$, from which $C_i < C_{i-2}$ and

$$\cdots < C_5 < C_3 < C_1 \tag{12}$$

It remains to show that every even-subscripted convergent is less than every odd-subscripted convergent. By Corollary 7.9, we also have

$$C_i - C_{i-1} = \frac{(-1)^{i-1}}{q_i q_{i-1}}, \quad 1 \le i \le n$$

If n is even, we have that $C_n - C_{n-1} < 0$, from which $C_n < C_{n-1}$; (11) and (12) now give

$$C_0 < C_2 < C_4 < \cdots < C_n < C_{n-1} < \cdots < C_5 < C_3 < C_1$$

as desired. If n is odd, we have that $C_n - C_{n-1} > 0$, from which $C_{n-1} < C_n$; (11) and (12) now give

$$C_0 < C_2 < C_4 < \cdots < C_{n-1} < C_n < \cdots < C_5 < C_3 < C_1$$

as desired. ■

The reader should verify that the convergents C_0, C_1, C_2, and C_3 of the finite simple continued fractions $[4, 3, 2, 1]$ (in Example 8) and $[5, 4, 1, 2]$ (in Example 9) satisfy $C_0 < C_2 < C_3 < C_1$ in accordance with Proposition 7.10 above.

Exercise Set 7.3

17. Find all convergents of each finite simple continued fraction and verify the ordering of these convergents as given by Proposition 7.10.

(a) $[1, 2, 3, 4]$
(b) $[0, 1, 2, 3, 4]$
(c) $[1, 1, 1, 1, 1, 1, 1, 1, 1]$
(d) $[0, 1, 1, 1, 1, 1, 1, 4]$
(e) $[3, 5, 2, 4]$
(f) $[0, 6, 9, 7]$

18. Let $\alpha = [a_0, a_1, a_2, \ldots, a_n]$ be a finite continued fraction with $a_0 > 0$ and let $C_i = p_i/q_i$ be the ith convergent of α. If $i \ge 1$, prove that

$$\frac{p_i}{p_{i-1}} = [a_i, a_{i-1}, a_{i-2}, \ldots, a_0]$$

and

$$\frac{q_i}{q_{i-1}} = [a_i, a_{i-1}, a_{i-2}, \ldots, a_1]$$

19. Let $\alpha = [a_0, a_1, a_2, \ldots, a_n]$ be a finite simple continued fraction with $a_0 > 0$ and let $C_i = \frac{p_i}{q_i}$ be the ith convergent of α.

(a) If $p_{n-1} = q_n$, prove that $q_n^2 + (-1)^{n-1} \equiv 0 \bmod p_n$.
(b) If $q_n^2 + (-1)^{n-1} \equiv 0 \bmod p_n$ and $p_n > q_n$, prove that $p_{n-1} = q_n$.
(c) Deduce a necessary and sufficient condition for the finite simple continued fraction expansion of α to be *palindromic* (read the same backward as forward). (*Hint*: Figure out the significance of the condition $p_{n-1} = q_n$ by using Exercise 18 above.)

Infinite Continued Fractions

We wish to define an infinite simple continued fraction as the limit of finite simple continued fractions. Some care must be taken to ensure that this limit exists. We first need an easy preliminary lemma that will be used several times in forthcoming results.

Lemma 7.11: Let $\alpha = [a_0, a_1, a_2, \ldots, a_n]$ be a finite simple continued fraction and let all notation be as in Proposition 7.6. Then $q_i \geq i$, $0 \leq i \leq n$.

Proof: We use induction on i. If $i = 0$, we have

$$q_0 = 1 \geq 0$$

If $i = 1$, we have

$$q_1 = a_1 \geq 1 \quad (\text{since } a_1 \in \mathbf{Z} \text{ and } a_1 > 0)$$

Assume that $k \geq 1$ and that the desired result holds for $i = k$ so that $q_k \geq k$. We must show that $q_{k+1} \geq k + 1$ so that the desired result holds for $i = k + 1$. We have

$$\begin{aligned} q_{k+1} &= a_{k+1}q_k + q_{k-1} \\ &\geq (1)(k) + 1 \quad (\text{why?}) \\ &= k + 1 \end{aligned}$$

as desired. ■

We may now define the concept of an infinite simple continued fraction. We record this definition as part of the proposition below that guarantees that the value of this infinite simple continued fraction exists. The proof below assumes that you have some familiarity with the theory of limits and the theory of sequences; the theory may be found in any introductory analysis book.

Proposition 7.12: Let $a_0, a_1, a_2, \ldots \in \mathbf{Z}$ with $a_1, a_2, \ldots > 0$ and let $C_i = [a_0, a_1, a_2, \ldots, a_i]$, $i \geq 0$. Then

$$\lim_{i \to \infty} C_i$$

exists. This limit is said to be the value of the *infinite simple continued fraction* $[a_0, a_1, a_2, \ldots]$; C_i is said to be the ith *convergent* of this infinite simple continued fraction.

Proof: The sequence of even-subscripted convergents $C_0, C_2, C_4, \ldots$ is strictly increasing and bounded above (for example, by C_1) by Proposition 7.10; so

$$\lim_{i \to \infty} C_{2i}$$

exists. Similarly, the sequence of odd-subscripted convergents $C_1, C_3, C_5, \ldots$ is strictly decreasing and bounded below (for example, by C_0) by Proposition 7.10; so

$$\lim_{i \to \infty} C_{2i+1}$$

exists. It remains to show that

$$\lim_{i \to \infty} C_{2i} = \lim_{i \to \infty} C_{2i+1}$$

We have

$$\begin{aligned} C_{2i+1} - C_{2i} &= \frac{1}{q_{2i+1}q_{2i}} \quad \text{(by Corollary 7.9)} \\ &\leq \frac{1}{(2i+1)(2i)} \quad \text{(by Lemma 7.11)} \end{aligned}$$

Now

$$0 \leq \lim_{i \to \infty} (C_{2i+1} - C_{2i}) \leq \lim_{i \to \infty} \frac{1}{(2i+1)(2i)} = 0$$

implies that

$$\lim_{i \to \infty} (C_{2i+1} - C_{2i}) = 0$$

from which

$$\lim_{i \to \infty} C_{2i+1} - \lim_{i \to \infty} C_{2i} = 0$$

and

$$\lim_{i \to \infty} C_{2i} = \lim_{i \to \infty} C_{2i+1}$$

as desired. ■

We have seen in Proposition 7.5 that those real numbers expressible as *finite* simple continued fractions are precisely the *rational* numbers. The following proposition asserts that those real numbers expressible as *infinite* simple continued fractions are precisely the *irrational* numbers. So *all* real numbers have representations as simple continued fractions; the rational numbers are characterized by having finite representations, and the irrational numbers are characterized by having infinite representations. In addition, the first half of the proposition below will give us a procedure for computing an infinite simple continued fraction expansion of a given irrational number.

Proposition 7.13: Let $\alpha \in \mathbf{R}$. Then $\alpha \in \mathbf{R} - \mathbf{Q}$ if and only if α is expressible as an infinite simple continued fraction.

Proof: ($\Rightarrow$) Let $\alpha = \alpha_0 \in \mathbf{R} - \mathbf{Q}$. Define $a_0, a_1, a_2, \ldots$ and $\alpha_1, \alpha_2, \ldots$ by the following recurrence relations:

$$a_i = [\alpha_i], \quad i \geq 0$$

$$\alpha_{i+1} = \frac{1}{\alpha_i - a_i}, \quad i \geq 0$$

Here $[\cdot]$ denotes the greatest integer function. Clearly, $a_0, a_1, a_2, \ldots \in \mathbf{Z}$. An easy induction establishes that $\alpha_0, \alpha_1, \alpha_2, \ldots \in \mathbf{R} - \mathbf{Q}$, from which $a_i < \alpha_i < a_i + 1$ for all $i \geq 0$. (Why?) So, for all i with $i \geq 0$, we have that $0 < \alpha_i - a_i < 1$, which implies that

$$\alpha_{i+1} = \frac{1}{\alpha_i - a_i} > 1$$

and

$$a_{i+1} = [\alpha_{i+1}] \geq 1$$

So $a_1, a_2, \ldots > 0$. We show that $\alpha = [a_0, a_1, a_2, \ldots]$. Rewriting $\alpha_{i+1} = 1/(\alpha_i - a_i)$ as

$$\alpha_i = a_i + \frac{1}{\alpha_{i+1}}$$

for all i with $i \geq 0$, we have

$$\begin{aligned}
\alpha = \alpha_0 &= a_0 + \frac{1}{\alpha_1} \\
&= a_0 + \cfrac{1}{a_1 + \cfrac{1}{\alpha_2}} \\
&\vdots \\
&= a_0 + \cfrac{1}{a_1 + \cfrac{1}{a_2 + \ddots + \cfrac{}{a_i + \cfrac{1}{\alpha_{i+1}}}}} \\
&= [a_0, a_1, a_2, \ldots, a_i, \alpha_{i+1}] \\
&= \frac{\alpha_{i+1} p_i + p_{i-1}}{\alpha_{i+1} q_i + q_{i-1}}, \quad i > 0
\end{aligned}$$

where the last equality follows from Proposition 7.6 with p_{i-1}/q_{i-1} and p_i/q_i being the $(i - 1)$th and ith convergents of $[a_0, a_1, a_2, \ldots, a_i]$, respectively.

Now

$$
\begin{aligned}
|\alpha - C_i| &= \left|\frac{\alpha_{i+1}p_i + p_{i-1}}{\alpha_{i+1}q_i + q_{i-1}} - \frac{p_i}{q_i}\right| \\
&= \left|\frac{-(p_i q_{i-1} - p_{i-1}q_i)}{(\alpha_{i+1}q_i + q_{i-1})q_i}\right| \\
&= \left|\frac{-(-1)^{i-1}}{(\alpha_{i+1}q_i + q_{i-1})q_i}\right| \quad \text{(by Proposition 7.7)} \\
&= \frac{1}{(\alpha_{i+1}q_i + q_{i-1})q_i} \\
&< \frac{1}{(a_{i+1}q_i + q_{i-1})q_i} \quad (\text{since } a_{i+1} < \alpha_{i+1}) \\
&= \frac{1}{q_{i+1}q_i} \\
&\leq \frac{1}{(i+1)(i)}, \quad i > 0 \quad \text{(by Lemma 7.11)}
\end{aligned}
$$

implies that

$$\lim_{i \to \infty} (\alpha - C_i) = 0$$

from which

$$\lim_{i \to \infty} \alpha - \lim_{i \to \infty} C_i = 0$$

and

$$\alpha = \lim_{i \to \infty} \alpha = \lim_{i \to \infty} C_i = [a_0, a_1, a_2, \ldots]$$

as desired.

($\Leftarrow$) Let α be expressible as an infinite simple continued fraction, say $\alpha = [a_0, a_1, a_2, \ldots]$ with $a_0, a_1, a_2, \ldots \in \mathbf{Z}$ and $a_1, a_2, \ldots > 0$. Assume, by way of contradiction, that $\alpha \in \mathbf{Q}$, say $\alpha = \frac{a}{b}$ with $a, b \in \mathbf{Z}$ and, without loss of generality, $b > 0$. Let $C_i = [a_0, a_1, a_2, \ldots, a_i] = p_i/q_i$, $i \geq 0$. By Proposition 7.10, we have that

$$C_{2i} < \alpha < C_{2i+1}, \quad i \geq 0$$

Subtracting C_{2i} from all terms of this inequality, we obtain

$$0 < \alpha - C_{2i} < C_{2i+1} - C_{2i}, \quad i \geq 0$$

Now, by Corollary 7.9, we have that

$$0 < \alpha - \frac{p_{2i}}{q_{2i}} < \frac{1}{q_{2i+1}q_{2i}}, \quad i \geq 0$$

Multiplying all terms of this inequality by q_{2i} gives

$$0 < \alpha q_{2i} - p_{2i} < \frac{1}{q_{2i+1}}, \quad i \geq 0$$

Substituting $\frac{a}{b}$ for α and multiplying all terms of the inequality above by b

yields

$$0 < aq_{2i} - bp_{2i} < \frac{b}{q_{2i+1}}, \quad i \geq 0$$

Choose i such that $2i + 1 > b$. Then we have

$$0 < aq_{2i} - bp_{2i} < \frac{b}{q_{2i+1}} \leq \frac{b}{2i+1} \quad \text{(by Lemma 7.11)}$$

and, by our choice of i, we obtain $0 < aq_{2i} - bp_{2i} < 1$, a contradiction since $aq_{2i} - bp_{2i} \in \mathbf{Z}$. So $\alpha \in \mathbf{R} - \mathbf{Q}$ and the proof is complete. ■

Before finding infinite simple continued fraction expansions for several irrational numbers, we prove that unlike finite simple continued fraction expansions, infinite simple continued fraction expansions are unique. Hereafter, we may then speak of *the* infinite simple continued fraction expansion of a given irrational number.

Proposition 7.14: Let $\alpha \in \mathbf{R} - \mathbf{Q}$. Then the expression of α as an infinite simple continued fraction is unique.

Proof: Before proving uniqueness, we need two preliminary results. Let $\alpha = [a_0, a_1, a_2, \ldots]$ be an expression of α as an infinite simple continued fraction. First, by Proposition 7.10 we have

$$C_0 < \alpha < C_1$$

or, equivalently,

$$a_0 < \alpha < a_0 + \frac{1}{a_1}$$

Inasmuch as $a_1 \geq 1$, we have

$$[\alpha] = a_0$$

(Here $[\cdot]$ denotes the greatest integer function.) Second, we have

$$\begin{aligned}
\alpha &= \lim_{i\to\infty} [a_0, a_1, a_2, \ldots, a_i] \\
&= \lim_{i\to\infty} \left(a_0 + \frac{1}{[a_1, a_2, \ldots, a_i]} \right) \\
&= a_0 + \frac{1}{\lim_{i\to\infty} [a_1, a_2, \ldots, a_i]} \\
&= a_0 + \frac{1}{[a_1, a_2, \ldots]}
\end{aligned}$$

We are now able to prove uniqueness. Let

$$\alpha = [a_0, a_1, a_2, \ldots] = [b_0, b_1, b_2, \ldots]$$

be two expressions of α as infinite simple continued fractions. We must show that $a_i = b_i$ for all $i \geq 0$. From the first result, we have

$$[\alpha] = a_0 = b_0$$

Now the second result implies

$$a_0 + \frac{1}{[a_1, a_2, \ldots]} = b_0 + \frac{1}{[b_1, b_2, \ldots]}$$

so that

$$[a_1, a_2, \ldots] = [b_1, b_2, \ldots]$$

So $a_0 = b_0$ and $[a_1, a_2, \ldots] = [b_1, b_2, \ldots]$. Now assume that $a_k = b_k$ with $k \geq 0$ and that $[a_{k+1}, a_{k+2}, \ldots] = [b_{k+1}, b_{k+2}, \ldots]$. The same reasoning as above yields $a_{k+1} = b_{k+1}$ and $[a_{k+2}, a_{k+3}, \ldots] = [b_{k+2}, b_{k+3}, \ldots]$. So, by induction, $a_i = b_i$ for all $i \geq 0$, and the proof is complete. ■

As stated previously, the procedure for computing the infinite simple continued fraction expansion of an irrational number is given by the first half of the proof of Proposition 7.13. We illustrate this procedure with several examples.

Example 10:

Find the infinite simple continued fraction expansion of $\sqrt{2}$.
We use the procedure and notation in the first half of the proof of Proposition 7.13. We have

$$\alpha = \alpha_0 = \sqrt{2} = 1.41421\ldots$$

$$a_0 = [\alpha_0] = 1 \qquad \alpha_1 = \frac{1}{\alpha_0 - a_0} = 2.41421\ldots$$

$$a_1 = [\alpha_1] = 2 \qquad \alpha_2 = \frac{1}{\alpha_1 - a_1} = 2.41421\ldots$$

$$a_2 = [\alpha_2] = 2 \qquad \alpha_3 = \frac{1}{\alpha_2 - a_2} = 2.41421\ldots$$

$$a_3 = [\alpha_3] = 2 \qquad \alpha_4 = \frac{1}{\alpha_3 - a_3} = 2.41421\ldots$$

$$\vdots$$

Exercise 27 asks you to establish that the pattern above continues by proving that $\alpha_i = \sqrt{2} + 1$ if $i \geq 1$. So the infinite simple continued fraction expansion of $\sqrt{2}$ is $[1, 2, 2, 2, \ldots]$ where the 2's repeat ad infinitum.

Example 11:

Find the infinite simple continued fraction expansion of $\sqrt{3}$.
We use the procedure and notation in the first half of the proof of

Proposition 7.13. We have

$$\alpha = \alpha_0 = \sqrt{3} = 1.73205\ldots$$

$$a_0 = [\alpha_0] = 1 \qquad \alpha_1 = \frac{1}{\alpha_0 - a_0} = 1.36602\ldots$$

$$a_1 = [\alpha_1] = 1 \qquad \alpha_2 = \frac{1}{\alpha_1 + a_1} = 2.73205\ldots$$

$$a_2 = [\alpha_2] = 2 \qquad \alpha_3 = \frac{1}{\alpha_2 - a_2} = 1.36602\ldots$$

$$a_3 = [\alpha_3] = 1 \qquad \alpha_4 = \frac{1}{\alpha_3 - a_3} = 2.73205\ldots$$

$$a_4 = [\alpha_4] = 2 \qquad \alpha_5 = \frac{1}{\alpha_4 - a_4} = 1.36602\ldots$$

$$\vdots$$

Exercise 28 asks you to establish that the pattern above continues by proving that $\alpha_i = (\sqrt{3} + 1)/2$ if $i \geq 1$ and i is odd and that $\alpha_i = \sqrt{3} + 1$ if $i \geq 1$ and i is even. So the infinite simple continued fraction expansion of $\sqrt{3}$ is $[1, 1, 2, 1, 2, \ldots]$ where the 1, 2 blocks repeat ad infinitum.

Example 12:

Find the infinite simple continued fraction expansion of e.
We use the procedure and notation in the first half of the proof of Proposition 7.13. We have

$$\alpha = \alpha_0 = e = 2.71828\ldots$$

$$a_0 = [\alpha_0] = 2 \qquad \alpha_1 = \frac{1}{\alpha_0 - a_0} = 1.39221\ldots$$

$$a_1 = [\alpha_1] = 1 \qquad \alpha_2 = \frac{1}{\alpha_1 - a_1} = 2.54964\ldots$$

$$a_2 = [\alpha_2] = 2 \qquad \alpha_3 = \frac{1}{\alpha_2 - a_2} = 1.81935\ldots$$

$$a_3 = [\alpha_3] = 1 \qquad \alpha_4 = \frac{1}{\alpha_3 - a_3} = 1.22047\ldots$$

$$a_4 = [\alpha_4] = 1 \qquad \alpha_5 = \frac{1}{\alpha_4 - a_4} = 4.53557\ldots$$

$$\vdots$$

So the infinite simple continued fraction expansion of e is $[2, 1, 2, 1, 1, \ldots]$. Note that on the basis of our computations so far, this expansion does not appear to have any repeating pattern. (Of course, one cannot say for sure on the slight bit of numerical evidence above!)

We will further discuss the distinctions between Examples 10 and 11 as compared to Example 12 in Section 7.5.

All of us, at some time or other in our mathematical lives, have probably used the rational number $\frac{22}{7}$ as an approximation for π. A natural question then arises: How good is this approximation? Is precision necessarily gained by allowing a greater numerator and denominator as in $\frac{314}{100} = \frac{157}{50}$? More generally, if our desire is to keep rational approximations to irrational numbers "as simple as possible" in terms of the sizes of the numerators and denominators, what approximations should we use? Given an irrational number α, the convergents of the infinite simple continued fraction expansion of α provide the answers. These convergents give a sequence of best rational approximations to α in the following sense: The ith convergent p_i/q_i of α is the closest rational number to α among all rational numbers with positive denominator q_i or less. To prove this fact, we need a preliminary technical proposition.

Proposition 7.15: Let $\alpha \in \mathbf{R} - \mathbf{Q}$ and let p_i/q_i, $i = 0, 1, 2, \ldots$, be the convergents of the infinite simple continued fraction expansion of α. If $a, b \in \mathbf{Z}$ and $0 < b < q_{i+1}$, then

$$|q_i\alpha - p_i| \leq |b\alpha - a|$$

Proof: Consider the system of equations given by

$$\begin{aligned} p_i x + p_{i+1} y &= a \\ q_i x + q_{i+1} y &= b \end{aligned}$$

By solving the system simultaneously and invoking Proposition 7.7, we obtain

$$\begin{aligned} x &= (-1)^i(bp_{i+1} - aq_{i+1}) \\ y &= (-1)^i(aq_i - bp_i) \end{aligned}$$

(Verify this!) We now show that $x \neq 0$. For, if $x = 0$, we have $bp_{i+1} = aq_{i+1}$; since $(p_{i+1}, q_{i+1}) = 1$ by Corollary 7.8, we obtain $q_{i+1} \mid b$, which contradicts $0 < b < q_{i+1}$. If $y = 0$, then we have $a = p_i x$ and $b = q_i x$; since $|x| \geq 1$, we obtain

$$|b\alpha - a| = |x|\,|q_i\alpha - p_i| \geq |q_i\alpha - p_i|$$

and the desired inequality is obviously true. So, we assume hereafter that $x \neq 0$ and $y \neq 0$. Under this assumption, we now show that x and y have opposite signs. If $y < 0$, then the equation $q_i x = b - q_{i+1}y$ implies that $q_i x > 0$, from which $x > 0$. If $y > 0$, then $b < q_{i+1}$ implies that $b < q_{i+1}y$, from which the equation $q_i x = b - q_{i+1}y$ implies that $q_i x < 0$ and $x < 0$. Now, since α lies between the successive convergents p_i/q_i and p_{i+1}/q_{i+1} by Proposition 7.10, we have

$$\frac{p_i}{q_i} < \alpha < \frac{p_{i+1}}{q_{i+1}}$$

or

$$\frac{p_{i+1}}{q_{i+1}} < \alpha < \frac{p_i}{q_i}$$

In either case, we obtain the fact that $q_i\alpha - p_i$ and $q_{i+1}\alpha - p_{i+1}$ have opposite signs. Hence, the quantities $x(q_i\alpha - p_i)$ and $y(q_{i+1}\alpha - p_{i+1})$ have the same sign (why?), and we have

$$\begin{aligned}|b\alpha - a| &= |(q_ix + q_{i+1}y)\alpha - (p_ix + p_{i+1}y)| \\ &= |x(q_i\alpha - p_i) + y(q_{i+1}\alpha - p_{i+1})| \\ &= |x(q_i\alpha - p_i)| + |y(q_{i+1}\alpha - p_{i+1})| \quad \text{(why?)} \\ &= |x|\,|q_i\alpha - p_i| + |y|\,|q_{i+1}\alpha - p_{i+1}| \\ &> |x|\,|q_i\alpha - p_i| \quad \text{(why?)} \\ &\geq |q_i\alpha - p_i| \quad \text{(why?)}\end{aligned}$$

which implies the desired inequality. ■

Our result on the convergents of infinite simple continued fractions being best rational approximations of irrational numbers now follows from Proposition 7.15 above as an easy corollary.

Corollary 7.16: Let $\alpha \in \mathbf{R} - \mathbf{Q}$ and let p_i/q_i, $i = 0, 1, 2, \ldots,$ be the convergents of the infinite simple continued fraction expansion of α. If $a, b \in \mathbf{Z}$ and $1 \leq b \leq q_i$, then

$$\left|\alpha - \frac{p_i}{q_i}\right| \leq \left|\alpha - \frac{a}{b}\right|$$

Proof: Assume, by way of contradiction, that $|\alpha - p_i/q_i| > |\alpha - \frac{a}{b}|$. Then

$$|q_i\alpha - p_i| = q_i\left|\alpha - \frac{p_i}{q_i}\right| > b\left|\alpha - \frac{a}{b}\right| = |b\alpha - a|$$

which contradicts Proposition 7.15. ■

The importance of Corollary 7.16 is seen in the following example.

Example 13:

From Example 12, we obtain the infinite simple continued fraction expansion of e as $[2, 1, 2, 1, 1, \ldots]$ (with no discernible repeating pattern evident). The tenth convergent of e (as the interested reader may verify) is $\frac{2721}{1001}$. By Corollary 7.16, then, no rational number with a denominator less than or equal to 1001 comes closer to e than $\frac{2721}{1001}$. So any rational approximation closer to e than $\frac{2721}{1001}$ must have a denominator greater than 1001.

Our final result in this section shows that if a rational number is "sufficiently close" to an irrational number, then it must be a convergent of the infinite simple continued fraction expansion of that irrational number.

Proposition 7.17: Let $a, b \in \mathbf{Z}$ with $(a, b) = 1$ and $b > 0$. If α is an irrational number and

$$\left|\alpha - \frac{a}{b}\right| < \frac{1}{2b^2}$$

then $\frac{a}{b}$ is a convergent of the infinite simple continued fraction expansion of α.

Proof: Let p_i/q_i, $i = 0, 1, 2, \ldots$, be the convergents of the infinite simple continued fraction expansion of α and assume, by way of contradiction, that $\frac{a}{b}$ is not such a convergent. Since the denominators of the convergents form a nondecreasing sequence of integers, there exists a unique integer k such that $q_k \leq b < q_{k+1}$. Then

$$\begin{aligned} |q_k\alpha - p_k| &\leq |b\alpha - a| \quad \text{(by Proposition 7 15)} \\ &= b\left|\alpha - \frac{a}{b}\right| \\ &< \frac{1}{2b} \quad \text{(by hypothesis)} \end{aligned}$$

and we have that

$$\left|\alpha - \frac{p_k}{q_k}\right| < \frac{1}{2bq_k}$$

Now $\frac{a}{b} \neq p_k/q_k$ implies that $bp_k - aq_k \neq 0$, from which $|bp_k - aq_k| \geq 1$, and we obtain

$$\begin{aligned} \frac{1}{bq_k} &\leq \frac{|bp_k - aq_k|}{bq_k} \\ &= \left|\frac{p_k}{q_k} - \frac{a}{b}\right| \\ &\leq \left|\frac{p_k}{q_k} - \alpha\right| + \left|\alpha - \frac{a}{b}\right| \\ &< \frac{1}{2bq_k} + \frac{1}{2b^2} \end{aligned}$$

which is equivalent to $b < q_k$ (verify this!), yielding a contradiction (why?), and completing the proof. ■

Example 14:

Use Proposition 7.17 to prove that the rational number $\frac{3363}{2378}$ is a convergent of the infinite simple continued fraction expansion of $\sqrt{2}$.

A calculator easily verifies that

$$\left|\sqrt{2} - \frac{3363}{2378}\right| \approx 0.000\,000\,063$$

while

$$\frac{1}{2(2378)^2} \approx 0.000\,000\,088$$

Since the first quantity is less than the second quantity, Proposition 7.17 implies that $\frac{3363}{2378}$ is a convergent of the infinite simple continued fraction expansion of $\sqrt{2}$ as desired.

The converse of Proposition 7.17 is false; to wit, given an irrational number α, it is not necessarily the case that $|\alpha - \frac{a}{b}| < 1/(2b^2)$ for *every* convergent $\frac{a}{b}$ of the infinite simple continued fraction expansion of α. We ask you for an appropriate counterexample in Exercise 25.

Exercise Set 7.4

20. Express each real number as an infinite simple continued fraction.
(a) $\sqrt{7}$
(b) $\sqrt{11}$
(c) $\sqrt{13}$
(d) $\sqrt{17}$
(e) $\sqrt{19}$
(f) $\dfrac{1+\sqrt{3}}{2}$
(g) $\dfrac{1+\sqrt{7}}{2}$

21. Find the first five convergents of the simple continued fraction expansion of each real number.
(a) $\sqrt[3]{2}$
(b) π
(c) e^2
(d) πe

22. Find the best rational approximation to π with denominator less than or equal to each integer.
(a) 10 (*Hint*: Corollary 7.16 guarantees that no rational number with a denominator less than or equal to 7 comes closer to π than $\frac{22}{7}$. Now check to see whether there are any rational numbers with denominators of 8, 9, or 10 that come closer to π than $\frac{22}{7}$.)
(b) 110
(c) 115

23. Use Proposition 7.17 to prove that the rational number $\frac{49171}{18089}$ is a convergent of the infinite simple continued fraction expansion of e.

24. Use Proposition 7.17 to prove that the rational number $\frac{1146408}{364913}$ is a convergent of the infinite simple continued fraction expansion of π.

25. Find a counterexample to the converse of Proposition 7.17.

26. On the basis of numerical evidence in Examples 10–12 and Exercises 20 and 21 above, formulate a conjecture as to the form taken by those irrational numbers having infinite simple continued fraction expansions with some block of digits repeating ad infinitum.

27. In Example 10, prove that $\alpha_i = \sqrt{2} + 1$ if $i \geq 1$.

28. In Example 11, prove that $\alpha_1 = \frac{\sqrt{3}+1}{2}$ if $i \geq 1$ and i is odd and that $\alpha_i = \sqrt{3} + 1$ if $i \geq 1$ and i is even.

29. Let $\alpha \in \mathbf{R} - \mathbf{Q}$ with $\alpha > 1$. Prove that the ith convergent of the infinite simple continued fraction expansion of α is the reciprocal of the $(i + 1)$th convergent of the infinite simple continued fraction expansion of $\frac{1}{\alpha}$. (*Hint*: Does the result of Exercise 14 extend to *irrational* numbers and *infinite* simple continued fractions?)

30. Let $\alpha \in \mathbf{R} - \mathbf{Q}$ and let p_i/q_i, $i = 0, 1, 2, \ldots$, be the convergents of the infinite simple continued fraction expansion of α. Prove that

$$\left|\alpha - \frac{p_i}{q_i}\right| < \frac{1}{q_i^2}$$

(*Hint*: Use Corollary 7.9.)

31. Let $\alpha \in \mathbf{R} - \mathbf{Q}$. Prove that of any two consecutive convergents of the infinite simple continued fraction expansion of α, at least one, say p_i/q_i, satisfies

$$\left|\alpha - \frac{p_i}{q_i}\right| < \frac{1}{2q_i^2}$$

(*Hint*: First prove that

$$\frac{1}{q_i q_{i+1}} = \left|\frac{p_{i+1}}{q_{i+1}} - \frac{p_i}{q_i}\right| = \left|\alpha - \frac{p_{i+1}}{q_{i+1}}\right| + \left|\alpha - \frac{p_i}{q_i}\right|\Bigg)$$

32. *(a)* Let $\alpha \in \mathbf{R} - \mathbf{Q}$. Prove that of any three consecutive convergents of the infinite simple continued fraction expansion of α, at least one, say p_i/q_i, satisfies

$$\left|\alpha - \frac{p_i}{q_i}\right| < \frac{1}{\sqrt{5}\, q_i^2}$$

(b) Let $\alpha \in \mathbf{R} - \mathbf{Q}$. Prove that there are infinitely many distinct rational numbers $\frac{a}{b}$ where $a, b \in \mathbf{Z}$ with $b \neq 0$ such that

$$\left|\alpha - \frac{a}{b}\right| < \frac{1}{\sqrt{5}\, b^2}$$

(This is a famous theorem from A. Hurwitz, which was proved in 1891. See also Student Project 6.)

7.5

Eventually Periodic Continued Fractions

In the previous section, Example 10 established that the infinite simple continued fraction expansion of $\sqrt{2}$ is $[1, 2, 2, 2, \ldots]$ where the 2's repeat ad infinitum, while Example 11 established that the infinite simple continued fraction expansion of $\sqrt{3}$ is $[1, 1, 2, 1, 2, \ldots]$ where the 1, 2 blocks repeat ad infinitum. The repeating behavior exhibited in these two examples motivates the following definition.

Definition 5: Let $\alpha \in \mathbf{R} - \mathbf{Q}$ and let $\alpha = [a_0, a_1, a_2, \ldots]$ be the expression of α as an infinite simple continued fraction. If there exist nonnegative integers ρ and N such that $a_n = a_{n+\rho}$ for all $n \geq N$, then α is said to be *eventually periodic*; the sequence $a_N, a_{N+1}, \ldots, a_{N+(\rho-1)}$ with ρ minimal is said to be the *period* of α, and ρ is said to be the *period length*. If N can be taken to be 1, then α is said to be *periodic*. An eventually periodic infinite simple continued fraction

$$\alpha = [a_0, a_1, a_2, \ldots, a_{N-1}, a_N, a_{N+1}, \ldots, a_{N+(\rho-1)}, a_N, a_{N+1}, \ldots, a_{N+(\rho-1)}, \ldots]$$

is denoted $\alpha = [a_0, a_1, a_2, \ldots, a_{N-1}, \overline{a_N, a_{N+1}, \ldots, a_{N+(\rho-1)}}]$.

If the wording of Definition 5 seems familiar (and it should!), compare it with that of Definition 2. Two examples that illustrate the content of Definition 5 immediately follow; a third is provided by a previous exercise.

Example 15:

(a) By Example 10, we have that $\sqrt{2} = [1, \overline{2}]$ is eventually periodic, with period 2 and period length 1.

(b) By Example 11, we have that $\sqrt{3} = [1, \overline{1, 2}]$ is eventually periodic, with period 1, 2 and period length 2.

(c) By Exercise 20(f), we have that $\frac{1+\sqrt{3}}{2} = [\overline{1, 2}]$ is periodic, with period 1, 2 and period length 2.

(In each of these examples, the eventually periodic continued fraction expansion was computed from a given irrational number via the procedure in the first half of the proof of Proposition 7.13. A method for computing the irrational number represented by a given eventually periodic infinite simple continued fraction will be given in Example 18 at the end of this section.)

What, if anything, characterizes those irrational numbers having eventually periodic infinite simple continued fraction expansions? (See Exercise 26 from the previous section.) We begin our search for the answer to this question with a definition.

Definition 6: Let $\alpha \in \mathbf{R} - \mathbf{Q}$. Then α is said to be a *quadratic irrational number* if α is a root of a (quadratic) polynomial $Ax^2 + Bx + C$ with $A, B, C \in \mathbf{Z}$ and $A \neq 0$.

Example 16:

(a) If a is a positive integer that is not a perfect square, then $\sqrt{a}$ is a quadratic irrational number by Example 2(a).
(b) The real number $2 + \sqrt{7}$ is a quadratic irrational number by Example 2(b).

Quadratic irrational numbers are characterized by a nice form as given by the proposition below.

Proposition 7.18: Let $\alpha \in \mathbf{R} - \mathbf{Q}$. Then α is a quadratic irrational number if and only if

$$\alpha = \frac{a + \sqrt{b}}{c}$$

where $a, b, c \in \mathbf{Z}$, $b > 0$, b is not a perfect square, and $c \neq 0$.

Proof: ($\Rightarrow$) Assume that α is a quadratic irrational number. Then there exist $A, B, C \in \mathbf{Z}$ with $A \neq 0$ such that

$$A\alpha^2 + B\alpha + C = 0$$

By the quadratic formula, we have

$$\alpha = \frac{-B \pm \sqrt{B^2 - 4AC}}{2A}$$

Note that $B^2 - 4AC$ is a nonsquare positive integer since α is irrational. If the plus sign is taken above, put $a = -B$, $b = B^2 - 4AC$, and $c = 2A$; if the minus sign is taken above, put $a = B$, $b = B^2 - 4AC$, and $c = -2A$. In any event, $a, b, c \in \mathbf{Z}$, $b > 0$, b is not a perfect square, and $c \neq 0$ as desired.
($\Leftarrow$) Assume that

$$\alpha = \frac{a + \sqrt{b}}{c}$$

where $a, b, c \in \mathbf{Z}$, $b > 0$, b is not a perfect square, and $c \neq 0$. Then $c^2 \neq 0$, and it is easy to check that α is a root of the quadratic polynomial

$$c^2x^2 - 2acx + (a^2 - b)$$

Put $A = c^2$, $B = -2ac$, and $C = a^2 - b$. Then α is a root of the quadratic polynomial $Ax^2 + Bx + C$ with $A, B, C \in \mathbf{Z}$ and $A \neq 0$ as desired. ■

Example 17:

$\frac{1+\sqrt{5}}{2}$ is a quadratic irrational number. (Simply let $a = 1$, $b = 5$, and $c = 2$ in Proposition 7.18 above.) The quantity $\frac{1+\sqrt{5}}{2}$ is called the *golden ratio* since ancient civilizations regarded geometric figures in which this number appeared to be aesthetically pleasing to the eye. The interested reader is invited to find a quadratic polynomial for which the golden ratio is a root in accordance with Definition 6.

The following (quite lengthy) sequence of results will culminate in the characterization of those irrational numbers that possess eventually periodic infinite simple continued fraction expansions. (Can you guess the desired characterization?) We begin with a lemma that characterizes quadratic irrational numbers by a form slightly more specialized than that given in Proposition 7.18.

Lemma 7.19: Let $\alpha \in \mathbf{R} - \mathbf{Q}$. Then α is a quadratic irrational number if and only if

$$\alpha = \frac{P + \sqrt{d}}{Q}$$

where $P, d, Q \in \mathbf{Z}$, $d > 0$, d is not a perfect square, $Q \neq 0$, and $Q \mid d - P^2$.

Proof: ($\Rightarrow$) Assume that α is a quadratic irrational number. Then, by Proposition 7.18,

$$\alpha = \frac{a + \sqrt{b}}{c}$$

where $a, b, c \in \mathbf{Z}$, $b > 0$, b is not a perfect square, and $c \neq 0$. Multiplying both numerator and denominator of α by $|c|$, we obtain

$$\alpha = \frac{a\,|c| + \sqrt{bc^2}}{c\,|c|}$$

Putting $P = a\,|c|$, $d = bc^2$, and $Q = c\,|c|$ yields $P, d, Q \in \mathbf{Z}$, $d > 0$, d not a perfect square, $Q \neq 0$, and $Q \mid d - P^2$ as desired (these verifications are left as easy exercises for the reader).
($\Leftarrow$) This is immediate by Proposition 7.18. ■

Recall that the procedure in the first half of the proof of Proposition 7.13

will find the infinite simple continued fraction expansion of any irrational number. We now use the form given by Lemma 7.19 to obtain an alternate procedure for finding such expansions *for quadratic irrational numbers.*

Proposition 7.20: In accordance with Lemma 7.19, let

$$\alpha = \frac{P_0 + \sqrt{d}}{Q_0}$$

be a quadratic irrational number where $P_0, d, Q_0 \in \mathbf{Z}$, $d > 0$, d is not a perfect square, $Q_0 \neq 0$, and $Q_0 \mid d - P_0^2$. Define $\alpha_0, \alpha_1, \alpha_2, \ldots, a_0, a_1, a_2, \ldots, P_1, P_2, \ldots$, and $Q_1, Q_2, \ldots$ by the following recurrence relations:

$$\alpha_i = \frac{P_i + \sqrt{d}}{Q_i}, \quad i \geq 0$$

$$a_i = [\alpha_i], \quad i \geq 0$$

$$P_{i+1} = a_i Q_i - P_i, \quad i \geq 0$$

$$Q_{i+1} = \frac{d - P_{i+1}^2}{Q_i}, \quad i \geq 0$$

(Here $[\cdot]$ denotes the greatest integer function.) Then $\alpha = [a_0, a_1, a_2, \ldots]$.

Proof: We first show that $P_i, Q_i \in \mathbf{Z}$, $Q_i \neq 0$, and $Q_i \mid d - P_i^2$ for all $i \geq 0$. We use induction on i. If $i = 0$, the conditions hold by hypothesis. Assume that $k \geq 0$ and that the conditions hold for $i = k$ so that $P_k, Q_k \in \mathbf{Z}$, $Q_k \neq 0$, and $Q_k \mid d - P_k^2$. We must show that $P_{k+1}, Q_{k+1} \in \mathbf{Z}$, $Q_{k+1} \neq 0$, and $Q_{k+1} \mid d - P_{k+1}^2$ so that the conditions hold for $i = k + 1$. Since

$$P_{k+1} = a_k Q_k - P_k$$

and $P_k, Q_k \in \mathbf{Z}$ by the induction hypothesis, we have $P_{k+1} \in \mathbf{Z}$. Now

$$\begin{aligned} Q_{k+1} &= \frac{d - P_{k+1}^2}{Q_k} \\ &= \frac{d - (a_k Q_k - P_k)^2}{Q_k} \\ &= \frac{d - P_k^2}{Q_k} + 2a_k P_k - a_k^2 Q_k \end{aligned} \tag{13}$$

Since $Q_k \mid d - P_k^2$ by the induction hypothesis, we have $Q_{k+1} \in \mathbf{Z}$. Furthermore, since d is not a perfect square, we have $d - P_{k+1}^2 \neq 0$ in (13) and so $Q_{k+1} \neq 0$. Finally, (13) yields

$$Q_k = \frac{d - P_{k+1}^2}{Q_{k+1}}$$

Inasmuch as $Q_k \in \mathbf{Z}$ by the induction hypothesis, we have $Q_{k+1} \mid d - P_{k+1}^2$, and the induction is complete. It remains to show that $\alpha = [a_0, a_1, a_2, \ldots]$. By the first half of the proof of Proposition 7.13, it suffices to prove that

$$\alpha_{i+1} = \frac{1}{\alpha_i - a_i}, \quad i \geq 0$$

We have

$$\begin{aligned}
\alpha_i - a_i &= \frac{P_i + \sqrt{d}}{Q_i} - a_i \\
&= \frac{\sqrt{d} - (a_i Q_i - P_i)}{Q_i} \\
&= \frac{\sqrt{d} - P_{i+1}}{Q_i} \\
&= \frac{(\sqrt{d} - P_{i+1})(\sqrt{d} + P_{i+1})}{Q_i(\sqrt{d} + P_{i+1})} \\
&= \frac{d - P_{i+1}^2}{Q_i(\sqrt{d} + P_{i+1})} \\
&= \frac{Q_i Q_{i+1}}{Q_i(\sqrt{d} + P_{i+1})} \\
&= \frac{Q_{i+1}}{\sqrt{d} + P_{i+1}} \\
&= \frac{1}{\alpha_{i+1}}
\end{aligned}$$

This is equivalent to what was to be proven. So $\alpha = [a_0, a_1, a_2, \ldots]$, and the proof is complete. ■

We are nearly ready to state and prove the main result of this section. The following lemma, the proof of which is left as an exercise for the reader, will be needed.

Lemma 7.21: Let α be a quadratic irrational number and let $a, b, c, d \in \mathbf{Z}$. Then

$$\frac{a\alpha + b}{c\alpha + d}$$

is either a rational number or a quadratic irrational number.

Proof: Exercise 36. ■

We will also need to consider one final concept, that of the conjugate of a quadratic irrational number. Conjugates will be crucial in Section 7.6.

Definition 7: Let

$$\alpha = \frac{a + \sqrt{b}}{c}$$

be a quadratic irrational number where $a, b, c \in \mathbf{Z}$ and $c \neq 0$. The *conjugate* of

α, denoted α', is

$$\alpha' = \frac{a - \sqrt{b}}{c}$$

Several arithmetic properties of conjugates are recorded in the lemma below; proofs of these properties are left as exercises for the reader.

Lemma 7.22: Let

$$\alpha_1 = \frac{a_1 + \sqrt{b}}{c_1}$$

and

$$\alpha_2 = \frac{a_2 + \sqrt{b}}{c_2}$$

where $a_1, a_2, b, c_1, c_2 \in \mathbf{Z}$, and $c_1, c_2 \neq 0$.

(a) $(\alpha_1 + \alpha_2)' = \alpha_1' + \alpha_2'$
(b) $(\alpha_1 - \alpha_2)' = \alpha_1' - \alpha_2'$
(c) $(\alpha_1\alpha_2)' = \alpha_1'\alpha_2'$
(d) $\left(\dfrac{\alpha_1}{\alpha_2}\right)' = \dfrac{\alpha_1'}{\alpha_2'}$, provided $\alpha_2 \neq 0$

Proof: Exercise 37. ■

Are you ready for this next result? Do you remember the question that we are trying to answer? What are we attempting to characterize?

Theorem 7.23: Let $\alpha \in \mathbf{R} - \mathbf{Q}$. Then the expression of α as an infinite simple continued fraction is eventually periodic if and only if α is a quadratic irrational number.

Proof: ($\Rightarrow$) Assume that the expression of α as an infinite simple continued fraction is eventually periodic. Then there exist nonnegative integers p and N such that

$$\alpha = [a_0, a_1, \ldots, a_{N-1}, \overline{a_N, a_{N+1}, \ldots, a_{N+(p-1)}}]$$

Put $\beta = [\overline{a_N, a_{N+1}, \ldots, a_{N+(p-1)}}]$. Then

$$\beta = [a_N, a_{N+1}, \ldots, a_{N+(p-1)}, \beta]$$

and, by Proposition 7.6, we have

$$\beta = \frac{\beta P_{p-1} + P_{p-2}}{\beta Q_{p-1} + Q_{p-2}} \tag{14}$$

where P_{p-2}/Q_{p-2} and P_{p-1}/Q_{p-1} are convergents of $[\overline{a_N, a_{N+1}, \ldots, a_{N+(p-1)}}]$.

Since the expression of β as a simple continued fraction is infinite, β is irrational; furthermore (14) gives that β is a root of the quadratic polynomial

$$Q_{p-1}x^2 + (Q_{p-2} - P_{p-1})x - P_{p-2}$$

and so β is a quadratic irrational number. Now

$$\alpha = [a_0, a_1, \ldots, a_{N-1}, \beta]$$

By Proposition 7.6, we have

$$\alpha = \frac{\beta p_{N-1} + p_{N-2}}{\beta q_{N-1} + q_{N-2}} \tag{15}$$

where p_{N-2}/q_{N-2} and p_{N-1}/q_{N-1} are convergents of $[a_0, a_1, \ldots, a_{N-1}]$. Since the expression of α as a simple continued fraction is infinite, α is irrational. Since β is a quadratic irrational number, Lemma 7.21 and (15) show that α is a quadratic irrational number as desired.

($\Leftarrow$) Assume that α is a quadratic irrational number. By Proposition 7.20, we have

$$\alpha = [a_0, a_1, a_2, \ldots]$$

where $a_0, a_1, a_2, \ldots$ are defined by the following recurrence relations:

$$\alpha = \frac{P_0 + \sqrt{d}}{Q_0}$$

$$\alpha_i = \frac{P_i + \sqrt{d}}{Q_i}, \quad i \geq 0$$

$$a_i = [\alpha_i], \quad i \geq 0$$

$$P_{i+1} = a_i Q_i - P_i, \quad i \geq 0$$

$$Q_{i+1} = \frac{d - P_{i+1}^2}{Q_i}, \quad i \geq 0$$

Now $\alpha = [a_0, a_1, a_2, \ldots, a_{i-1}, \alpha_i]$; by Proposition 7.6, we have

$$\alpha = \frac{\alpha_i p_{i-1} + p_{i-2}}{\alpha_i q_{i-1} + q_{i-2}}$$

By Lemma 7.22, we then have

$$\alpha' = \frac{\alpha_i' p_{i-1} + p_{i-2}}{\alpha_i' q_{i-1} + q_{i-2}}$$

Solving for α_i', we obtain

$$\alpha_i' = \frac{-q_{i-2}}{q_{i-1}} \left(\frac{\alpha' - \dfrac{p_{i-2}}{q_{i-2}}}{\alpha' - \dfrac{p_{i-1}}{q_{i-1}}} \right)$$

Now the convergents p_{i-2}/q_{i-2} and p_{i-1}/q_{i-1} tend to α as i tends to ∞, from which

$$\left(\frac{\alpha' - \dfrac{p_{i-2}}{q_{i-2}}}{\alpha' - \dfrac{p_{i-1}}{q_{i-1}}}\right)$$

tends to 1 as i tends to ∞. So there exists a positive integer I such that $\alpha_i' < 0$ if $i \geq I$. Since $\alpha_i > 0$ if $i \geq 1$, we have

$$\alpha_i - \alpha_i' > 0, \quad i \geq I$$

or, equivalently,

$$\frac{P_i + \sqrt{d}}{Q_i} - \frac{P_i - \sqrt{d}}{Q_i} > 0, \quad i \geq I$$

Simplifying, we obtain

$$\frac{2\sqrt{d}}{Q_i} > 0, \quad i \geq I$$

and so

$$Q_i > 0, \quad i \geq I \tag{16}$$

Now

$$Q_i \leq Q_i Q_{i+1} = d - P_{k+1}^2 \leq d, \quad i \geq I \tag{17}$$

We combine (16) and (17) to get

$$0 < Q_i \leq d, \quad i \geq I \tag{18}$$

Also,

$$P_{i+1}^2 \leq P_{i+1}^2 + Q_i Q_{i+1} = d, \quad i \geq I$$

implies that

$$-\sqrt{d} < P_{i+1} < \sqrt{d}, \quad i \geq I \tag{19}$$

Now (18) and (19) imply that for $i > I$, the pair of integers P_i, Q_i assumes only finitely many possible pairs of values. So $P_N = P_M$ and $Q_N = Q_M$ for some $N < M$. If $p = M - N$, then the defining recurrence relation for α_i gives $\alpha_n = \alpha_{n+p}$ for all $n \geq N$. Now the defining recurrence relation for a_i gives $a_n = a_{n+p}$ for all $n \geq N$, and the infinite continued fraction expansion of α is eventually periodic as desired. ■

Given a quadratic irrational number, one can find the eventually periodic infinite simple continued fraction expansion of this number by using either the procedure in the first half of the proof of Proposition 7.13 or the procedure of Proposition 7.20. We conclude this section by providing an example that illustrates how to find the quadratic irrational number represented by a given eventually periodic infinite simple continued fraction expansion.

Example 18:

Find the quadratic irrational number represented by the eventually periodic infinite simple continued fraction $[3, \overline{2, 1}]$.

We first find the quadratic irrational number represented by the periodic infinite simple continued fraction $\alpha = [\overline{2, 1}]$. Note that we have

$$\alpha = [2, 1, \overline{2, 1}] = [2, 1, \alpha]$$

so that

$$\alpha = 2 + \cfrac{1}{1 + \cfrac{1}{\alpha}}$$

Simplifying the equation above, we obtain $\alpha^2 - 2\alpha - 2 = 0$, from which, by the quadratic formula, we have $\alpha = 1 + \sqrt{3}$. (The negative root $1 - \sqrt{3}$ is rejected since $\alpha > 0$.) So

$$\alpha = [\overline{2, 1}] = 1 + \sqrt{3}$$

Now

$$\begin{aligned} [3, \overline{2, 1}] &= [3, \alpha] \\ &= [3, 1 + \sqrt{3}] \\ &= 3 + \frac{1}{1 + \sqrt{3}} \\ &= \frac{4 + 3\sqrt{3}}{1 + \sqrt{3}} \\ &= \frac{5 + \sqrt{3}}{2} \end{aligned}$$

and hence $[3, \overline{2, 1}] = \frac{5+\sqrt{3}}{2}$.

Exercise Set 7.5

33. Find the quadratic irrational number associated with each eventually periodic infinite simple continued fraction expansion.
   (a) $[\overline{1}]$
   (b) $[3, \overline{1}]$
   (c) $[2, \overline{4}]$
   (d) $[1, \overline{3, 2}]$
   (e) $[2, \overline{3, 4}]$
   (f) $[1, 2, \overline{3, 4}]$

34. Express each of the quadratic irrational numbers of Exercise 20 as an eventually periodic infinite simple continued fraction by using the procedure of Proposition 7.20.

35. On the basis of numerical evidence in Exercises 33 and 34, formulate a conjecture as to the form taken by those quadratic irrational numbers having *periodic* infinite simple continued fraction expansions. [*Note*: This is not so easy to guess. (But try anyway!) We will discuss periodic continued fractions in Section 7.6.]

36. Prove Lemma 7.21.

37. Prove Lemma 7.22.

38. *(a)* Let n be a positive integer. Prove that the infinite simple continued fraction expansion of $\sqrt{n^2+1}$ is $[n, \overline{2n}]$.

(b) Let n be a positive integer with $n \geq 2$. Prove that the infinite simple continued fraction expansion of $\sqrt{n^2-1}$ is $[n-1, \overline{1, 2n-2}]$.

(c) Find the infinite simple continued fraction expansions of $\sqrt{730}$ and $\sqrt{440}$.

39. *(a)* Let n be a positive integer. Prove that the infinite simple continued fraction expansion of $\sqrt{n^2+2}$ is $[n, \overline{n, 2n}]$.

(b) Let n be a positive integer with $n \geq 3$. Prove that the infinite simple continued fraction expansion of $\sqrt{n^2-2}$ is $[n-1, \overline{1, n-2, 1, 2n-2}]$

(c) Find the infinite simple continued fraction expansions of $\sqrt{531}$ and $\sqrt{674}$.

40. Let n be a positive integer. Prove that the period length of the infinite simple continued fraction expansion of $\sqrt{n}$ is 1 if and only if $n = m^2 + 1$ for some integer m.

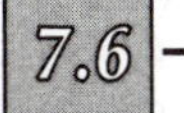

Periodic Continued Fractions

In Section 7.5, we established that quadratic irrational numbers are precisely those numbers that are expressible as *eventually periodic* infinite simple continued fractions. In this section, we wish to obtain a characterization of those quadratic irrational numbers that are expressible as *periodic* infinite simple continued fractions. [Do you have your own conjecture? (See Exercise 35.)] Such a characterization is immediately given by the following theorem.

Theorem 7.24: Let α be a quadratic irrational number. Then the expression of α as an infinite simple continued fraction is periodic if and only if $\alpha > 1$ and $-1 < \alpha' < 0$, where α' denotes the conjugate of α. Such quadratic irrational numbers are said to be *reduced.*

We precede the proof of Theorem 7.24 with an example.

Example 19:

(a) Let $\alpha = \frac{1+\sqrt{3}}{2}$. Then $\alpha' = \frac{1-\sqrt{3}}{2}$, and it is easily verified that $\alpha > 1$ and $-1 < \alpha' < 0$. So α is reduced, and, by Theorem 7.24, α has a periodic infinite simple continued fraction expansion. Indeed, $\alpha = [\overline{1, 2}]$ by Exercise 20(f). Similarly, $\alpha = \frac{1+\sqrt{7}}{2}$ has a periodic infinite simple continued fraction expansion [see Exercise 20(g)].

(b) Let $\alpha = \frac{1+\sqrt{11}}{2}$. Then $\alpha' = \frac{1-\sqrt{11}}{2}$; although $\alpha > 1$, it is easily verified that $\alpha' < -1$. So α is not reduced and, by Theorem 7.24, α does not have a

periodic infinite simple continued fraction expansion. (Note, however, that Theorem 7.23 guarantees an eventually periodic expansion. What is this expansion?)

(c) Let d be a positive integer that is not a perfect square and consider $\alpha = \sqrt{d}$. Then $\alpha' = -\sqrt{d}$, and, since $\alpha > 1$, we have that $\alpha' = -\alpha < -1$. So α is not reduced, and, by Theorem 7.24, α does not have a periodic infinite simple continued fraction expansion. The eventually periodic infinite simple continued fraction expansion of $\alpha = \sqrt{d}$ does take a nice form, however; we will conclude the section with a description of this form. The reader may wish to review Exercise 20(a)–(e) with the hope of formulating a conjecture in this regard.

We now prove Theorem 7.24.

Proof: (of Theorem 7.24) ($\Leftarrow$) Assume that $\alpha = \alpha_0$ is a reduced quadratic irrational number. Then, by the first half of the proof of Proposition 7.13, we have that $\alpha = [a_0, a_1, a_2, \ldots]$ where

$$a_i = [\alpha_i], \quad i \geq 0$$

$$\alpha_{i+1} = \frac{1}{\alpha_i - a_i}, \quad i \geq 0$$

The second of these equations gives $\alpha_i = a_i + 1/\alpha_{i+1}$; Lemma 7.22 immediately yields

$$\alpha_i' = a_i + \frac{1}{\alpha_{i+1}'}$$

for all i. Since $\alpha = \alpha_0$ is reduced, we have that $-1 < \alpha_0' < 0$. Mathematical induction now establishes that $-1 < \alpha_i' < 0$ for all i [see Exercise 44(a)]; so $-1 < a_i + 1/\alpha_{i+1}' < 0$ or

$$-1 - \frac{1}{\alpha_{i+1}'} < a_i < \frac{-1}{\alpha_{i+1}'}$$

for all i. Consequently,

$$a_i = \left[\frac{-1}{\alpha_{i+1}'}\right], \quad i \geq 0$$

Since α is a quadratic irrational number, the proof of Theorem 7.23 implies that $\alpha_j = \alpha_k$ for integers j and k with $0 < j < k$. So $\alpha_j' = \alpha_k'$ and

$$a_{j-1} = \left[\frac{-1}{\alpha_j'}\right] = \left[\frac{-1}{\alpha_k'}\right] = a_{k-1}$$

Furthermore,

$$\alpha_{j-1} = a_{j-1} + \frac{1}{\alpha_j} = a_{k-1} + \frac{1}{\alpha_k} = \alpha_{k-1}$$

so that $\alpha_j = \alpha_k$ implies that $\alpha_{j-1} = \alpha_{k-1}$. Iterating this implication j times yields $\alpha_0 = \alpha_{k-j}$, from which

$$\begin{aligned}\alpha = \alpha_0 &= [a_0, a_1, a_2, \ldots, a_{k-j-1}, \alpha_{k-j}] \\ &= [a_0, a_1, a_2, \ldots, a_{k-j-1}, \alpha_0] \\ &= [\overline{a_0, a_1, a_2, \ldots, a_{k-j-1}}]\end{aligned}$$

and the infinite simple continued fraction expansion is periodic as desired.

($\Rightarrow$) Assume that the expression of α as an infinite simple continued fraction is periodic with period length ρ so that $\alpha = [\overline{a_0, a_1, a_2, \ldots, a_{\rho-1}}]$ where $a_0, a_1, a_2, \ldots, a_{\rho-1}$ are positive integers. Clearly, $\alpha > a_0 \geq 1$. It remains to show that $-1 < \alpha' < 0$. We give the proof for $\rho \geq 2$; the (easier) proof for $\rho = 1$ is similar and appears as Exercise 44(b). Assume that $\rho \geq 2$. We have

$$\begin{aligned}\alpha &= [a_0, a_1, a_2, \ldots, a_{\rho-1}, \alpha] \\ &= \frac{\alpha p_{\rho-1} + p_{\rho-2}}{\alpha q_{\rho-1} + q_{\rho-2}}\end{aligned} \tag{20}$$

where the last equality follows from Proposition 7.6 with $p_{\rho-2}/q_{\rho-2}$ and $p_{\rho-1}/q_{\rho-1}$ being the $(\rho - 2)$th and $(\rho - 1)$th convergents of $[a_0, a_1, a_2, \ldots, a_{\rho-1}]$, respectively. Lemma 7.22 then yields

$$\alpha' = \frac{\alpha' p_{\rho-1} + p_{\rho-2}}{\alpha' q_{\rho-1} + q_{\rho-2}} \tag{21}$$

Algebraic manipulations of equations (20) and (21) show (respectively) that α and α' are roots of the quadratic polynomial

$$f(x) = q_{\rho-1}x^2 + (q_{\rho-2} - p_{\rho-1})x - p_{\rho-2}$$

Since $f(x)$ has at most two roots and we have that $\alpha > 1$, it suffices to show that $f(x)$ has a root strictly between -1 and 0; indeed, this root must be α', and we will have $-1 < \alpha' < 0$ as desired. Accordingly, we prove that $f(-1) > 0$ and $f(0) < 0$. We have

$$f(-1) = (q_{\rho-1} - q_{\rho-2}) + (p_{\rho-1} - p_{\rho-2})$$

Inasmuch as Proposition 7.6 implies that $p_{\rho-1} > p_{\rho-2}$ and $q_{\rho-1} \geq q_{\rho-2}$, we have $f(-1) > 0$. Also, Proposition 7.6 implies that $p_{\rho-2} > 0$, from which

$$f(0) = -p_{\rho-2} < 0$$

and the proof is complete. ■

Let d be a positive integer that is not a perfect square. As mentioned in Example 19(c), even though the infinite continued fraction expansion of $\sqrt{d}$ is not periodic, the eventually periodic expansion takes a nice form. (Have you formulated a conjecture?) We conclude this section with a description of this form. Part of this description is given immediately by the following proposition.

Proposition 7.25: Let d be a positive integer that is not a perfect square.

Then the infinite simple continued fraction expansion of $\sqrt{d}$ takes the form

$$[a_0, \overline{a_1, a_2, \ldots, a_{\rho-1}, 2a_0}]$$

where ρ is the period length and $a_0 = [\sqrt{d}]$. Here $[\cdot]$ denotes the greatest integer function.

In other words, the period of the infinite simple continued fraction expansion of $\sqrt{d}$ will always begin immediately after the first term, and the final term in the period will always be exactly twice this first term. Please check your answers in Exercise 20(a)–(e) for this behavior!

Proof: (of Proposition 7.25) Consider $\alpha = [\sqrt{d}] + \sqrt{d}$. It is easily verified that α is reduced; by Theorem 7.24, the infinite simple continued fraction expansion of α is periodic, say

$$\alpha = [\overline{a_0, a_1, a_2, \ldots, a_{\rho-1}}]$$

where ρ is the period length. Noting that $a_0 = [[\sqrt{d}] + \sqrt{d}] = 2[\sqrt{d}]$, we have that

$$\begin{aligned}
\sqrt{d} = ([\sqrt{d}] + \sqrt{d}) - [\sqrt{d}] &= \alpha - [\sqrt{d}] \\
&= [\overline{a_0, a_1, a_2, \ldots, a_{\rho-1}}] - [\sqrt{d}] \\
&= [a_0, \overline{a_1, a_2, \ldots, a_{\rho-1}, a_0}] - [\sqrt{d}] \\
&= [2[\sqrt{d}], \overline{a_1, a_2, \ldots, a_{\rho-1}, 2[\sqrt{d}]}] - [\sqrt{d}] \\
&= [[\sqrt{d}], \overline{a_1, a_2, \ldots, a_{\rho-1}, 2[\sqrt{d}]}]
\end{aligned}$$

as desired. ■

The proof of Proposition 7.25 above shows striking similarities between the infinite simple continued fraction expansions of $\sqrt{d}$ and $[\sqrt{d}] + \sqrt{d}$. These similarities are important enough to be highlighted separately in a corollary. In addition, we will find the last statement of this corollary useful in Section 8.4.

Corollary 7.26: Let d be a positive integer that is not a perfect square. Then the infinite simple continued fraction expansions of $\sqrt{d}$ and $[\sqrt{d}] + \sqrt{d}$ differ only in the first component (with the first component of the latter being twice the first component of the former), and the period lengths are equal. Furthermore, the values generated in Proposition 7.20 by

$$\alpha = \alpha_0 = \sqrt{d}, \qquad P_0 = 0, \qquad Q_0 = 1$$

differ from those generated by

$$\alpha = \alpha_0 = [\sqrt{d}] + \sqrt{d}, \qquad P_0 = [\sqrt{d}], \qquad Q_0 = 1$$

precisely in P_0 and a_0.

Proof: Both claims in the next to the last statement follow directly from the proof of Proposition 7.25. The claim in the last statement is a straightforward computation using the recurrence relations of Proposition 7.20. ■

More is true about the infinite simple continued fraction expansion of $\sqrt{d}$

given by $[a_0, \overline{a_1, a_2, \ldots, a_{\rho-1}, 2a_0}]$ in Proposition 7.25. It is a fact that the sequence $a_1, a_2, \ldots, a_{\rho-1}$ is *palindromic*; in other words, the sequence $a_1, a_2, \ldots, a_{\rho-1}$ reads the same backward as forward. The proof of this fact is motivated in Exercise 45. We illustrate with an example.

Example 20:

(a) From Exercise 20(a), we have that $\sqrt{7} = [2, \overline{1, 1, 1, 4}]$; clearly, the sequence 1, 1, 1 is palindromic.
(b) From Exercise 20(b), we have that $\sqrt{11} = [3, \overline{3, 6}]$; clearly, the sequence 3 is palindromic.
(c) From Exercise 20(c), we have that $\sqrt{13} = [3, \overline{1, 1, 1, 1, 6}]$; clearly, the sequence 1, 1, 1, 1 is palindromic.
(d) From Exercise 20(d), we have that $\sqrt{17} = [4, \overline{8}]$; here the palindromic sequence is empty.
(e) From Exercise 20(e), we have that $\sqrt{19} = [4, \overline{2, 1, 3, 1, 2, 8}]$; clearly, the sequence 2, 1, 3, 1, 2 is palindromic.

The eventually periodic infinite simple continued fraction expansions of $\sqrt{d}$ for nonsquare integers d with $2 \leq d \leq 99$ are given in Table 4 of Appendix E.

Exercise Set 7.6

41. Use Theorem 7.24 to determine whether each of the following quadratic irrational numbers has a periodic infinite simple continued fraction expansion.
(a) $2 + \sqrt{10}$
(b) $\dfrac{1 + \sqrt{11}}{5}$
(c) $3 + \sqrt{13}$
(d) $\dfrac{-3 + \sqrt{17}}{8}$
(e) $\dfrac{7 + \sqrt{19}}{2}$
(f) $\dfrac{4 + \sqrt{23}}{7}$

42. Given that the infinite simple continued fraction expansions of $4 + \sqrt{23}$ is $[\overline{8, 1, 3, 1}]$, find the infinite simple continued fraction expansion of $\sqrt{23}$.

43. Given that the infinite simple continued fraction expansion of $\sqrt{29}$ is $[5, \overline{2, 1, 1, 2, 10}]$, find the infinite simple continued fraction expansion of $5 + \sqrt{29}$.

44. This exercise will complete the proof of Theorem 7.24.
(a) Use mathematical induction to establish that $-1 < \alpha_i' < 0$ for all i in the proof of ($\Leftarrow$).
(b) Prove the case $\rho = 1$ in ($\Rightarrow$).

45. (Given the expansion of $\sqrt{d}$ as $[a_0, \overline{a_1, a_2, \ldots, a_{\rho-1}, 2a_0}]$ in Proposition

7.25, the following exercise establishes that the sequence $a_1, a_2, \ldots, a_{\rho-1}$ is palindromic.)

(a) Prove that if the infinite simple continued fraction expansion of α is

$$[\overline{a_0, a_1, a_2, \ldots, a_{\rho-1}}]$$

then the infinite simple continued fraction expansion of $\frac{-1}{\alpha'}$ is

$$[\overline{a_{\rho-1}, a_{\rho-2}, a_{\rho-3}, \ldots, a_0}]$$

Here ρ is the period length in both cases. [*Hint*: Let $\beta = [\overline{a_{\rho-1}, a_{\rho-2}, a_{\rho-3}, \ldots, a_0}]$. Show that $\frac{-1}{\beta}$ satisfies the quadratic polynomial $f(x)$ in the proof of Theorem 7.24. (Use a procedure similar to that used in the proof of Theorem 7.24 along with Exercise 18.) It then follows that $\alpha' = \frac{-1}{\beta}$.]

(b) Let d be a positive integer that is not a perfect square. By Proposition 7.25, let the infinite simple continued fraction expansion of $\sqrt{d}$ be $[a_0, \overline{a_1, a_2, \ldots, a_{\rho-1}, 2a_0}]$, where ρ is the period length. Prove that the sequence $a_1, a_2, \ldots, a_{\rho-1}$ is palindromic. [*Hint*: By Corollary 7.26, write the infinite simple continued fraction expansion of $[\sqrt{d}] + \sqrt{d}$ as $[\overline{2a_0, a_1, a_2, \ldots, a_{\rho-1}}]$. Now express $1/(\sqrt{d} - [\sqrt{d}])$ as an infinite simple continued fraction in two ways: (i) by writing $\sqrt{d} - [\sqrt{d}] = [0, \overline{a_1, a_2, \ldots, a_{\rho-1}, 2a_0}]$ and using the infinite analogue of Exercise 14 and (ii) by using part (a) above.]

Concluding Remarks

In Example 6 of Section 7.2, we posed the question "What is to be gained by the expression of a number as a continued fraction?" Numerous answers to this question are now possible. We have seen several elegant characterizations of numbers in terms of their continued fraction expansions: rational numbers in terms of finite simple continued fractions (Proposition 7.5), irrational numbers in terms of infinite simple continued fractions (Proposition 7.13), quadratic irrational numbers in terms of eventually periodic infinite simple continued fractions (Theorem 7.23), and reduced quadratic irrational numbers in terms of periodic infinite simple continued fractions (Theorem 7.24). As such, continued fractions introduce unexpected order into the set of real numbers. In addition, the convergents of an infinite simple continued fraction can be used to obtain "best" rational approximations to the irrational number represented by the continued fraction in the sense of Corollary 7.16. As we will see in Section 8.4, continued fractions can also be used to solve certain Diophantine equations. For these reasons (as well as the usefulness of the techniques of this chapter in advanced topics in number theory), the study of continued fractions is of substantial significance in elementary number theory.

Student Projects

1. (Programming project for calculator or computer)
Given a real number α, compute the simple continued fraction expansion of α.

[Exercises 2–5 are motivated by the following article: Ian Richards, "Continued Fractions Without Tears," *Mathematics Magazine, 54* (1981), 163–171.]

2. Farey approximations (named after J. Farey, the discoverer of some of the properties associated with them) are rational number approximations to irrational numbers (for example, $\pi \approx \frac{22}{7}$, etc.). Two algorithms for obtaining such approximations will be developed in Student Projects 3 and 4. The second algorithm is closely connected with continued fractions. In what follows, all lowercase variables are assumed to be integral and nonnegative.
Definition: A pair of distinct nonnegative rational numbers $\frac{a}{b}, \frac{c}{d}$ is said to be a *Farey pair if* $bc - ad = 1$. The *mediant* of a Farey pair $\frac{a}{b}, \frac{c}{d}$ is $\frac{a+c}{b+d}$. The closed interval $\left[\frac{a}{b}, \frac{c}{d}\right]$ is said to be a *Farey interval* if $\frac{a}{b}, \frac{c}{d}$ is a Farey pair. Let $\left[\frac{a}{b}, \frac{c}{d}\right]$ be a Farey interval.

(a) Prove that the length of the interval $\left[\frac{a}{b}, \frac{c}{d}\right]$ is $\frac{1}{bd}$.

(b) Prove that $\frac{a+c}{b+d} \in \left(\frac{a}{b}, \frac{c}{d}\right)$. [Here $\left(\frac{a}{b}, \frac{c}{d}\right)$ denotes the open interval with endpoints $\frac{a}{b}$ and $\frac{c}{d}$.]

(c) Prove that $\left[\frac{a}{b}, \frac{a+c}{b+d}\right]$ and $\left[\frac{a+c}{b+d}, \frac{c}{d}\right]$ are Farey intervals.

(d) Let $\frac{x}{y} \in \left(\frac{a}{b}, \frac{c}{d}\right)$ with $\frac{x}{y} \neq \frac{a+c}{b+d}$. Prove that $y > b + d$. Conclude that every rational number in the open interval $\left(\frac{a}{b}, \frac{c}{d}\right)$ has a greater denominator than the denominator of the mediant of $\frac{a}{b}, \frac{c}{d}$.

3. *Farey Algorithm 1*: Approximation of $\alpha \in \mathbf{R} - \mathbf{Q}$, $\alpha > 0$, by rational numbers

(i) Start with the Farey interval $\left[\frac{n}{1}, \frac{n+1}{1}\right]$ where $n \in \mathbf{Z}$ and $n < \alpha < n + 1$.

(ii) Compute the mediant of the corresponding Farey pair.

(iii) Of the two Farey subintervals created, keep the interval containing α and go to (ii).

This algorithm produces a sequence of mediants (which are rational numbers) converging to α.

(a) Use 10 iterations of Farey Algorithm 1 to approximate π with rational numbers. Compare the sequence of mediants so obtained with the sequence of convergents obtained in Exercise 21(b) of this chapter. With π as a reference point, illustrate the sequence of mediants obtained on a number line. What is the error in using the tenth iteration as an approximation for π?

(b) Prove that if Farey Algorithm 1 is applied to $\alpha \in \mathbf{Q}$, then some mediant in the produced sequence of mediants will equal α.

4. *Farey Algorithm 2*: Approximation of $\alpha \in \mathbf{R} - \mathbf{Q}$, $\alpha > 0$, by rational numbers

(i) Put $a = 0$, $b = 1$, $c = 1$, $d = 0$, $\gamma = \alpha$, and $i = 0$.

(ii) Put $s_i = [\gamma]$, $p_i = a + s_i c$, and $q_i = b + s_i d$. (Here $[\cdot]$ denotes the greatest integer function.)

(iii) Put $a \leftarrow c$, $b \leftarrow d$, $c \leftarrow p_i$, $d \leftarrow q_i$, $\gamma \leftarrow 1/(\gamma - s_i)$, and $i \leftarrow i + 1$. (Here $a \leftarrow c$ means that the current value of c is assigned to a; the other assignments proceed similarly.) Go to (ii).

This algorithm produces a sequence of rational numbers p_i/q_i.

(a) At each stage of Farey Algorithm 1, one endpoint of the Farey interval changes (approaching α), and the other endpoint does not change. Assume that the left-hand endpoint changes s_0 times in succession, followed by the right-hand endpoint changing s_1 times in succession, followed by the left-hand endpoint changing s_2 times in succession, and so on. Prove that these s_i are precisely the s_i in Farey Algorithm 2. Deduce that Farey Algorithm 1 computes all mediants in the succession of mediants while Farey Algorithm 2 computes only the s_0th, $(s_0 + s_1)$th, $(s_0 + s_1 + s_2)$th, ... mediants in the succession.

(b) Prove that Farey Algorithm 2 is equivalent to the continued fraction procedure given by the first half of the proof of Proposition 7.13 of this chapter.

(c) Use five iterations of Farey Algorithm 2 to approximate π with rational numbers. Compare the sequence of rational numbers so obtained with the sequence of convergents obtained in Exercise 21(b) of this chapter. With π as a reference point, illustrate the sequence of rational numbers obtained on a number line. What is the error in using the fifth iteration as an approximation for π?

5. Which Farey Algorithm seems to be the most efficient? Why? Explain in what sense this algorithm finds "best" rational number approximations to irrational numbers.
6. Prove that the result of Exercise 32(b) does not hold if $\sqrt{5}$ is replaced with any larger constant. [In other words, the constant $\sqrt{5}$ in the result of Exercise 32(b) is the "best" possible.]
7. Prove or disprove the following conjecture.
   Conjecture: Let $n \in \mathbf{Z}$ with $n \geq 2$. Then there exist positive integers a and b with $n = a + b$ such that the finite simple continued fraction expansion of $\frac{a}{b}$ contains only 1's and 2's.

8

A Few Applications

In this final chapter, we collect a few interesting applications of the previous material in this book. The first application shows how the modular arithmetic of Chapter 2 can be used to ascertain winning strategies in a simple pencil-and-paper game. The second application involves cryptography — loosely, the making and breaking of secret codes. We will study a very important procedure used to encode and decode information called the RSA encryption system and discuss the number theory behind this procedure. The discussion only requires a knowledge of Chapters 1 through 3. The third application involves the development of two primality tests using concepts from Chapters 4 and 5. (An additional test is developed in an exercise.) Along with Student Project 5, which develops yet another primality test, this will bring the number of primality tests in this book to six. (What were the other two primality tests?) The fourth and final application concerns Pell's Equation, a Diophantine equation of considerable significance in number theory. An analysis of the solutions of Pell's Equation in a special case will require a heavy usage of material in Chapter 7; as such, this application provides a link between Chapters 6 and 7.

One additional remark is in order. Much of the material in this chapter involves computations with rather large numbers; as such, the reader will find a calculator absolutely necessary. Furthermore, the use of computation software with infinite precision such as *Mathematica*® or MAPLE is recommended if available and, in fact, is necessary for a few of the exercises.

8.1

A Recreational Application

Number theory arises in many "recreational" contexts, from the theory of magic squares [see Ball (1944)] to the analysis of various ways to shuffle a standard deck of playing cards [see Gardner (1975)]. The word *recreational* implies a certain amount of entertainment value in the application; after the previous seven chapters, it is perhaps appropriate that we take a break from our theoretical discussions and relax a bit. In this section, we describe a very simple game that, when studied, yields an application of modular arithmetic. For lack of a better name, we call this game Square-Off.

Square-Off is played by two players on a 6×6 grid as shown below:

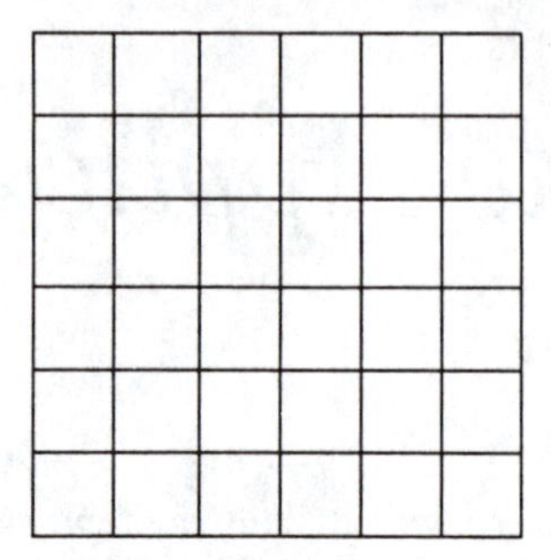

Players alternate taking turns. On any turn, a player may cross off 1, 2, or 3 of the 36 small squares. [The squares need not be contiguous or connected in any way (but see Student Project 2).] The player forced to cross off the last small square loses. It is particularly instructive at this point for the reader to find a partner and play several games of Square-Off. Players should alternate who starts. Let Player 1 be the player to begin and Player 2 be the player going second. As you play, try to figure out which player, if any, is favored in Square-Off. Is there a strategy whereby one player can consistently win the game?

After playing several games of Square-Off, you begin to see a winning strategy for Player 1. Obviously, if Player 1 leaves Player 2 with only one square on any move, Player 1 wins. If Player 1 responds to Player 2's crossing off x squares on a move by crossing off $4 - x$ squares on the next move (making a total of 4 squares crossed off on the pair of moves), we see that Player 1 can win by leaving Player 2 with 5 squares on any move. Similar reasoning yields the fact that Player 1 can win by leaving Player 2 with 9 squares on any move, or 13 squares on any move, or, in fact, any number of squares congruent to 1 modulo 4 on any move. So if Player 1 crosses off 3 squares on the first move, Player 1 leaves Player 2 with 33 squares (which is congruent to 1 modulo 4), and Player 1 can win by responding to each of Player 2's subsequent moves as described above. Modular arithmetic has successfully described a winning strategy for Player 1!

Many related questions instantly present themselves. Does Square-Off on a 5×5 grid still favor the player who moves first? In general, how does the size of the grid affect winning strategies? What if a move consists of crossing off one, two, three, *or four* squares? In general, how does the number of choices in any move affect winning strategies? Again, it is instructive for the reader to pause and experiment with Square-Off on grids of differing sizes with varying numbers of choices in any move. We hope that you will indeed find such experimentation to be recreational and fun. The answers to these questions are motivated in the exercises below.

Exercise Set 8.1

1. (a) Experiment with Square-Off on grids of sizes 1×1 (trivial!), 2×2, 3×3, and 4×4 with varying numbers of choices in any move.
   (b) Compile charts as to how varying numbers of choices in any move affect the player favored on a grid of fixed size.
   (c) Compile charts as to how differing grid sizes affect the player favored in games with a fixed number of choices in any move.
2. Prove that Player 2 has a winning strategy in Square-Off on a 5×5 grid where a move consists of crossing off one, two, or three squares.
3. Prove that Player 2 has a winning strategy in Square-Off on a 6×6 grid where a move consists of crossing off one, two, three, or four squares.
4. Prove that Player 1 has a winning strategy in Square-Off on a 5×5 grid where a move consists of crossing off one, two, three, or four squares.
5. Let k and n be positive integers. Consider Square-Off on an $n \times n$ grid where a move consists of crossing off any number of squares from 1 to k inclusive. Prove that Player 1 has a winning strategy if and only if $n^2 \not\equiv 1 \bmod (k + 1)$ (with Player 2 having a winning strategy otherwise).

8.2

Cryptography; The RSA Encryption System

Cryptography is that branch of science concerned with the development and analysis of procedures used to encode and decode information. The word *cryptography* is derived from the Greek words *kryptos,* meaning "hidden," and *graphein,* meaning "to write." Quite literally, then, the object of cryptography is to render communications unintelligible to all persons except the intended recipients. Historically, the need for cryptography has been primarily in the areas of political and military intelligence, but with the advent of high-speed digital computers in many areas of business, the need for the security of information has increased rapidly in the commercial sector. Cryptography appears to be the only practical way of protecting information transmitted via public communication networks such as telephone lines and microwave links. While not providing absolute protection, the coding of information frequently makes the penetration of sensitive systems by unwarranted persons costly and dangerous.

The information to be coded in cryptography is termed *plaintext,* while the coded information is termed *ciphertext.* The conversion process from plaintext to ciphertext is called *encryption* (or *enciphering*), and the conversion process from ciphertext back to plaintext is called *decryption* (or *deciphering*). To facilitate the encryption of information, all desired plaintext symbols are given numerical equivalents. In this section, we will encipher messages involving only the capital letters A–Z and use the numerical equivalents in Table 8.1.

Table 8.1

Letter	Numerical Equivalent	Letter	Numerical Equivalent
A	00	N	13
B	01	O	14
C	02	P	15
D	03	Q	16
E	04	R	17
F	05	S	18
G	06	T	19
H	07	U	20
I	08	V	21
J	09	W	22
K	10	X	23
L	11	Y	24
M	12	Z	25

The phrase NUMBER THEORY thus would be converted to the numerical string 132012010417 190704141724 before encryption. In practice, one might also have numerical equivalents for the small letters a–z, the digits 0–9, punctuation marks, and so on.

We now discuss a particular encryption system that was developed in 1977 by Ronald Rivest, Adi Shamir, and Len Adelman called the *RSA encryption system* (R for Rivest, S for Shamir, and A for Adelman, in case you were wondering). The RSA encryption system is an example of a *public-key encryption system.* In such a system, each member of a network of n communicants may encode and send a message to any other member of the network. This is acheived by publishing a directory of *encryption keys*— $e_1, e_2, \ldots, e_n$—one for each communicant in the network. If individual i wishes to encode and send a message to individual j, then encryption key e_j is used by individual i. Individual j then uses his or her *decryption key* d_j to recover the plaintext. The key d_j is known only to individual j; indeed, in a public-key encryption system, d_j cannot be found in a realistic time by anyone in the network other than individual j, even though the corresponding encryption key e_j is public. So only the intended recipient of any message may decode the message. This feature of the system prevents unauthorized persons from gaining access to a message *even though any member can encode and send a message to any other member.* The RSA encryption system is thus ideal for the large information-processing environments so prevalent in today's society.

We now describe the mechanics of the RSA encryption system in detail. Concentrate only on these mechanics for now—we will deal with the theory behind the mechanics later. The encryption process below includes a stage called *formatting,* which puts the plaintext message in a form suitable for the RSA encryption scheme.

The RSA Encryption Scheme

(0) The Public Encryption Key
Let p and q be distinct odd prime numbers (typically, very large), let $m = pq$, and let e be a positive integer such that $(e, \phi(m)) = 1$. The ordered pair (e, m) is the public encryption key.
[The component prime numbers p and q of m are *not* publicized; neither is $\phi(m)$.]

(1) Formatting
Translate each letter of the plaintext into its two-digit numerical equivalent from Table 8.1. Format these numerical equivalents into blocks of maximal even length such that each block of digits, viewed as a single positive integer, is less than m. If the final block contains fewer than the required even number of digits, then "dummy" digits, say 23's (for X's) are added so that this block has sufficient digits.

(2) The Encryption Scheme
Encipher each block P, when viewed as a single positive integer, via

$$P^e \equiv C \bmod m, 0 \leq C < m$$

to obtain a new block C, which is viewed as a single positive integer.

We now give a (presumably much needed) example of the RSA encryption scheme.

Example 1:

Use the RSA encryption scheme to encipher the message I AM HAVING FUN NOW with public encryption key (11, 3127). (Here 3127 is the product of two distinct prime numbers; can you find them?)

(1) Formatting
Translating each letter of the plaintext into its two-digit numerical equivalent, we obtain 08 0012 070021081306 052013 131422. We now format this numerical string into blocks of length four:

$$0800\ 1207\ 0021\ 0813\ 0605\ 2013\ 1314\ 2223 \qquad \textbf{(1)}$$

Four is chosen as the block length since it is the maximal even number with the property that all blocks of this length, viewed as single positive integers, are less than 3127. Note that a 23 (for X) has been added to the final block so that it also has length four. In essence, then, we are about to encipher the message IA MH AV IN GF UN NO WX, which is the original message formatted into blocks of length two with a "dummy" letter X added at the end.

(2) The Encryption Scheme
We now encipher each block P of length four in (1) above, when viewed as a single positive integer, via $P^{11} \equiv C \bmod 3127$, $0 \leq C < 3127$, to obtain a new block C, which is viewed as a single positive integer. Somewhat tedious

computations (see Student Project 1 of Chapter 2) reveal that

$$0800^{11} \equiv 2882 \bmod 3127$$
$$1207^{11} \equiv 2913 \bmod 3127$$
$$0021^{11} \equiv 1900 \bmod 3127$$
$$0813^{11} \equiv 1479 \bmod 3127$$
$$0605^{11} \equiv 0344 \bmod 3127$$
$$2013^{11} \equiv 2808 \bmod 3127$$
$$1314^{11} \equiv 2219 \bmod 3127$$
$$2223^{11} \equiv 2099 \bmod 3127$$

Hence, the enciphered message is 2882 2913 1900 1479 344 2808 2219 2099.

We now describe the RSA decryption scheme. Again, concentrate only on the mechanics of the scheme for now; the relevant theory will be discussed later. The decryption process includes a stage called *deformatting,* which recovers the original plaintext message from the deciphered numerical values.

The RSA Decryption Scheme

(0) The Private Decryption Key
Let (e, m) be the public encryption key. The ordered pair (d, m) is the private decryption key where d is the inverse of e modulo $\phi(m)$.

(1) The Decryption Scheme
Decipher each block C, which is viewed as a single positive integer, via

$$C^d \equiv P \bmod m, 0 \leq P < m$$

to obtain a new block P of maximal even length such that each block of digits, when viewed as a single positive integer, is less than m.

(2) Deformatting
Replace each two-digit block with its alphabetical equivalent from Table 8.1. Deformat the resulting blocks of letters into a meaningful message, truncating all "dummy" digits.

We now give a concrete example of the RSA decryption scheme.

Example 2:

Use the RSA decryption scheme to decipher the enciphered message 2882 2913 1900 1479 344 2808 2219 2099 of Example 1. (Of course, we know the answer; the point here is how to get it!)

(0) The Private Decryption Key
The inverse of 11 modulo $\phi(3127)$ is 1371. (This is a nontrivial calculation unless the two component prime numbers of 3127 are known. We will discuss this more fully later.) So the private decryption key is (1371, 3127).

(1) The Decryption Scheme
We now decipher each block C, which is viewed as a single positive integer, via $C^{1371} \equiv P \bmod 3127$, $0 \leq P < 3127$, to obtain a new block P of length four, when viewed as a single positive integer. Very(!) tedious computations reveal that

$$2882^{1371} \equiv 0800 \bmod 3127$$

$$2913^{1371} \equiv 1207 \bmod 3127$$

$$1900^{1371} \equiv 0021 \bmod 3127$$

$$1479^{1371} \equiv 0813 \bmod 3127$$

$$344^{1371} \equiv 0605 \bmod 3127$$

$$2808^{1371} \equiv 2013 \bmod 3127$$

$$2219^{1371} \equiv 1314 \bmod 3127$$

$$2099^{1371} \equiv 2223 \bmod 3127$$

So the deciphered message is

$$0800\ 1207\ 0021\ 0813\ 0605\ 2013\ 1314\ 2223 \qquad \textbf{(2)}$$

(2) Deformatting
Replacing each two-digit block of the deciphered message (2) above with its alphabetical equivalent, we obtain IA MH AV IN GF UN NO WX; deformatting these blocks of letters into a meaningful message, we obtain I AM HAVING FUN NOW (the X is clearly a "dummy" letter and is truncated).

We now discuss the theory behind the RSA encryption system. We assume all notation presented in the encryption and decryption schemes above. We first give the derivation of the RSA decryption scheme $C^d \equiv P \bmod m$ from the RSA encryption scheme $P^e \equiv C \bmod m$. Assume that we are given an enciphered message in the RSA system so that a block P has been encrypted into a new block C via $P^e \equiv C \bmod m$, $0 \leq C < m$. Since d is the inverse of e modulo $\phi(m)$, we have $ed \equiv 1 \bmod \phi(m)$ or, equivalently,

$$ed = k\phi(m) + 1$$

for some integer k. Then if $(P, m) = 1$, we obtain

$$C^d \equiv (P^e)^d \equiv P^{ed} \equiv P^{k\phi(m)+1} \equiv (P^{\phi(m)})^k P \equiv P \bmod m$$

where the last congruence is by Euler's Theorem (Theorem 2.17). So we have $C^d \equiv P \bmod m$ as desired. The case where $(P, m) \neq 1$ is similar and is motivated in Exercise 10. This completes the desired derivation.

Perhaps the most intriguing feature of the RSA encryption system is the fact that knowledge of the public encryption key (e, m) does not lead to the knowledge of the corresponding decryption key (d, m) in a realistic period of time. Since d is the inverse of e modulo $\phi(m)$, the computation of the decryption key d requires the computation of $\phi(m)$. If the two component prime numbers p and q of m are known, this computation of $\phi(m)$ is ridiculously easy $[\phi(m) = \phi(pq) = \phi(p)\phi(q) = (p - 1)(q - 1)]$. However, the component prime numbers of m are known only to the intended recipient of the message; persons other than this recipient would have to compute $\phi(m)$ given m only, and this computation may be prohibitive. For example, if each component prime number of m has 150 digits (remember that p and q are taken to be *large* prime numbers!), then m has approximately 300 digits. The computation of $\phi(m)$ essentially requires a factorization of m into p and q; assuming one computer operation per nanosecond (or 10^9 computer operations per second, roughly the speed of an average supercomputer today), the time required by the most efficient algorithm for factoring an integer of this size is approximately $(4.9)10^{12}$ years! As a (much easier) illustration of the difficulties involved here, the reader is invited to actually compute $\phi(3127)$ in Example 2. A general method for factoring m is illustrated in Student Project 1.

Exercise Set 8.2

6. Consider the RSA encryption scheme with public encryption key (9, 3127).
(a) Encipher the message SEND ENVOY TODAY.
(b) Decipher the message 2490 769 2502 978 428 1142 1210 2417 2778.

7. Consider the RSA encryption scheme with public encryption key (11, 2623).
(a) Encipher the message PATIENCE IS A VIRTUE.
(b) Decipher the message 284 926 2489 445 662 2445 926 178.

8. Consider the RSA encryption scheme with public encryption key (13, 2419).
(a) Encipher the message PLEASE REPLY SOON.
(b) Decipher the message 132 1985 2188 1013 726 26 1450 343.

The following exercise points out a subtlety in the RSA Encryption System. See the Hints and Answers section for further details.

9. Consider the RSA encryption scheme with public encryption key (13, 2419) as in Exercise 8 above.
(a) Encipher the message PLEASE SEND REPLY.
(b) Decipher the message 2259 966 1785 1380 299 1380 2209 1785.

10. Assume all notation in the RSA encryption and decryption schemes. Derive the decryption scheme $C^d \equiv P \bmod m$ from the encryption scheme $P^e \equiv C \bmod m$ in the case where $(P, m) \neq 1$. (*Hint*: If $m = pq$, prove that $C^d \equiv P \bmod p$ and $C^d \equiv P \bmod q$.)

11. Assume all notation in the RSA encryption and decryption schemes.
(a) Prove that the component primes p and q of m are easily found if m and $\phi(m)$ are known.
(b) Find p and q if $m = 176399$ and $\phi(m) = 175560$.
(c) Find p and q if $m = 551923$ and $\phi(m) = 550368$.

12. Assume all notation in the RSA encryption scheme. Let d' be the inverse of e modulo $[p - 1, q - 1]$ (here $[\cdot,\cdot]$ denotes the least common multiple). Prove that (d', m) acts as a decryption key.

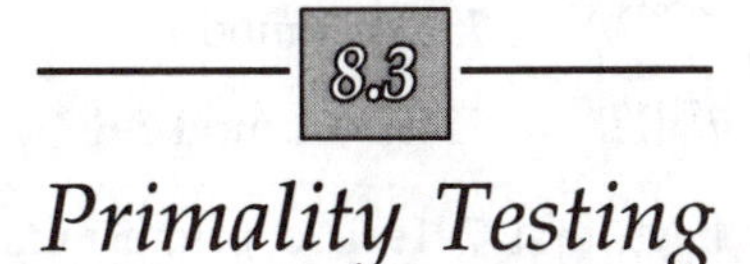

Primality Testing

Recall that a *primality test* is an algorithm that determines whether a given positive integer is prime. In view of the need for large prime numbers as discussed in Section 8.2, primality tests are not only important but also necessary. So far, we have seen two examples of such tests, namely the sieve of Eratosthenes (Example 6 of Chapter 1) and the converse of Wilson's Theorem (Proposition 2.12). Both tests are highly inefficient; in this section, we present two primality tests that are extremely useful for numbers taking a certain form. We will conclude this section with a comment on the practicality of these tests; see if you can anticipate the discussion! The first test is given immediately by the following theorem.

Theorem 8.1: Let n be a positive integer. If an integer a exists such that

$$a^{n-1} \equiv 1 \bmod n$$

and

$$a^{(n-1)/q} \not\equiv 1 \bmod n$$

for all prime divisors q of $n - 1$, then n is a prime number.

Proof: Since $a^{n-1} \equiv 1 \bmod n$, we have $\text{ord}_n a \mid n - 1$ by Proposition 5.1, and so there exists a positive integer k such that $n - 1 = k(\text{ord}_n a)$. We show that $\text{ord}_n a = n - 1$. Assume, by way of contradiction, that $\text{ord}_n a \neq n - 1$. Then $k > 1$; let q be a prime divisor of k. Then q is a prime divisor of $n - 1$ and

$$a^{(n-1)/q} = a^{k(\text{ord}_n a)/q} = (a^{\text{ord}_n a})^{k/q} \equiv 1 \bmod n$$

which contradicts the hypotheses of the theorem. So $\text{ord}_n a = n - 1$ as desired. Now

$$\text{ord}_n a \leq \phi(n) \leq n - 1$$

Since $\text{ord}_n a = n - 1$, we have that $\phi(n) = n - 1$. By Exercise 13(b) of Chapter 3, n is a prime number. ■

Example 3:

Let $n = 7919$. We show that $a = 7$ satisfies the hypotheses of Theorem 8.1. A straightforward verification establishes that $7^{7918} \equiv 1 \bmod 7919$. The prime divisors of 7918 are 2, 37, and 107; straightforward verifications establish that

$$7^{7918/2} \equiv 7918 \not\equiv 1 \bmod 7919$$

$$7^{7918/37} \equiv 755 \not\equiv 1 \bmod 7919$$

$$7^{7918/107} \equiv 5549 \not\equiv 1 \bmod 7919$$

By Theorem 8.1, we have that 7919 is a prime number. (In fact, it is the 1000th prime number!)

The following corollary gives a slightly improved primality test.

Corollary 8.2: Let n be an odd positive integer. If an integer a exists such that

$$a^{(n-1)/2} \equiv -1 \bmod n$$

and

$$a^{(n-1)/q} \not\equiv 1 \bmod n$$

for all odd prime divisors q of $n - 1$, then n is a prime number.

Proof: Since

$$a^{n-1} = (a^{(n-1)/2})^2 \equiv (-1)^2 \equiv 1 \bmod n$$

all hypotheses of Theorem 8.1 are satisfied, from which n is a prime number. ■

Example 4:

Let $n = 48611$. We show that $a = 2$ satisfies the hypotheses of Corollary 8.2. The odd prime divisors of 48610 are 5 and 4861 [Exercise 13(a) asks you to test 4861 for primality]; straightforward verifications establish that

$$2^{48610/2} \equiv -1 \bmod 48611$$

$$2^{48610/5} \equiv 43432 \not\equiv 1 \bmod 48611$$

$$2^{48610/4861} \equiv 1024 \not\equiv 1 \bmod 48611$$

By Corollary 8.2, we have that 48611 is a prime number. (In fact, it is the 5000th prime number!)

We now comment on the practicality of the primality tests given by Theorem 8.1 and Corollary 8.2 above. Note that these primality tests require the prime factorization of $n - 1$ in order to determine the primality (or nonprimality) of n. Inasmuch as finding the prime factorization of an integer may be extremely time-consuming, the primality tests in this section are practical only if we know information about the prime factorization of $n - 1$ beforehand. Such is the case for the special class of numbers of the form $2^{2^n} + 1$ with n a nonnegative integer, for example; these numbers are known as

Fermat numbers. [A Fermat prime (see Definition 8 of Chapter 1) is simply a prime Fermat number.] Exercise 14 develops a primality test for Fermat numbers. See also Student Project 5.

Exercise Set 8.3

13. Use Theorem 8.1 or Corollary 8.2 to test the following positive integers for primality.
(a) 4861
(b) 5059
(c) 5067
(d) 6263
(e) 6941
(f) 7457

The following exercise develops a characterization of Fermat primes.

14. Let n be a nonnegative integer and let $F_n = 2^{2^n} + 1$ be a Fermat number.
(a) Prove that if $3^{(F_n-1)/2} \equiv -1 \bmod F_n$, then F_n is a prime number. (*Note*: This yields a primality test known as *Pepin's Test.*)
(b) Prove that if F_n is a prime number, then either $n = 0$ or $3^{(F_n-1)/2} \equiv -1 \bmod F_n$. [*Hint*: If $n \geq 1$, use the law of quadratic reciprocity to evaluate the Legendre symbol $(3/F_n)$. Now use Euler's Criterion (Theorem 4.4).]
(c) Use the characterization given by parts (a) and (b) to determine whether the Fermat numbers F_4, F_5, and F_6 are prime numbers.

Pell's Equation

In this section, we study the solvability of the Diophantine equation $x^2 - dy^2 = n$ where d and n are fixed integers. Through a mistake by Euler, this equation has come to be known as *Pell's Equation,* even though English mathematician John Pell never studied it. Fermat was the first mathematician to deal systematically with Pell's Equation, followed by Euler and Lagrange. The equation arises quite naturally in more advanced areas of number theory, one of which is motivated in Student Project 6.

We begin our discussion of Pell's Equation with a few observations. In the case where $d < 0$ and $n < 0$, the equation $x^2 - dy^2 = n$ has no solutions since $x^2 - dy^2$ is nonnegative and n is negative. Consider the case where $d < 0$ and $n \geq 0$. Then $x^2 - dy^2 = n$ implies that $x^2 = n + dy^2$; since $x^2 \geq 0$, we have that $n + dy^2 \geq 0$, from which it follows that

$$|y| \leq \sqrt{\frac{n}{|d|}}$$

A symmetric consideration yields $|x| \leq \sqrt{n}$, from which Pell's Equation can have at most finitely many solutions in this case. Note further that if d is a perfect square, say $d = m^2$, then $x^2 - dy^2 = n$ becomes

$$x^2 - m^2y^2 = (x + my)(x - my) = n$$

Any factorization of n into two integers, say $n = ab$, yields at most one solution of Pell's Equation since any such solution must satisfy the simultaneous system

$$x + my = a$$

$$x - my = b$$

Since there are finitely many factorizations of n into two integers, Pell's Equation has at most finitely many solutions if d is a perfect square. In what follows, then, we will restrict d to be a positive integer that is not a perfect square. In addition, we will describe only positive integral solutions of Pell's Equation. All other solutions are obtainable by changing the sign of one or both components of a solution (since x and y are both squared) and by considering the cases where $x = 0$ or $y = 0$.

The theorem below shows the relationship between Pell's Equation and the infinite simple continued fraction expansion of $\sqrt{d}$.

Theorem 8.3: Let d and n be integers such that $d > 0$, d is not a perfect square, and $0 < |n| < \sqrt{d}$. If x, y is a positive solution of Pell's Equation $x^2 - dy^2 = n$, then $\frac{x}{y}$ is a convergent of the infinite simple continued fraction expansion of $\sqrt{d}$.

Proof: Assume that $n > 0$. Then $x^2 - dy^2 = n$ becomes

$$(x + y\sqrt{d})(x - y\sqrt{d}) = n$$

and we have that $x - y\sqrt{d} > 0$ or, equivalently, $\frac{x}{y} - \sqrt{d} > 0$. Now

$$\begin{aligned}
\frac{x}{y} - \sqrt{d} &= \frac{x - y\sqrt{d}}{y} \\
&= \frac{x^2 - dy^2}{y(x + y\sqrt{d})} \\
&< \frac{n}{y(2y\sqrt{d})} \quad \text{(why?)} \\
&< \frac{\sqrt{d}}{2y^2\sqrt{d}} \quad \text{(why?)} \\
&= \frac{1}{2y^2}
\end{aligned}$$

So we have $0 < \frac{x}{y} - \sqrt{d} < 1/(2y^2)$; Proposition 7.17 then implies that $\frac{x}{y}$ is a convergent of the infinite simple continued fraction expansion of $\sqrt{d}$ as desired. Now assume that $n < 0$. Dividing both sides of the equation

$x^2 - dy^2 = n$ by $-d$, we obtain

$$y^2 - \left(\frac{1}{d}\right)x^2 = \frac{-n}{d}$$

An argument similar to that used for $n > 0$ above establishes that $\frac{y}{x}$ is a convergent of the infinite simple continued fraction expansion of $\frac{1}{\sqrt{d}}$. Exercise 29 of Chapter 7 then implies that $\frac{x}{y}$ is a convergent of the infinite simple continued fraction expansion of $\sqrt{d}$ as desired. ■

For the remainder of this section, we restrict to the special case of Pell's Equation given by $x^2 - dy^2 = 1$. (Exercise 18 investigates the related special case given by $x^2 - dy^2 = -1$.) Our goal is to obtain a complete characterization of the positive solutions of $x^2 - dy^2 = 1$ in terms of the convergents of the infinite simple continued fraction expansion of $\sqrt{d}$. This characterization requires two preliminary rather technical lemmata, both relying heavily on Proposition 7.20. You may wish to refamiliarize yourself with this proposition now.

Lemma 8.4: Let d be a positive integer that is not a perfect square and let p_i/q_i denote the ith convergent of the infinite simple continued fraction expansion of $\sqrt{d}$. Then

$$p_i^2 - dq_i^2 = (-1)^{i-1}Q_{i+1}, \quad i \geq 0$$

where $Q_1, Q_2, \ldots$ are defined as in Proposition 7.20.

Proof: Let all notation be as in Proposition 7.20 with $\alpha_0 = \sqrt{d}$. Since

$$\sqrt{d} = \alpha_0 = [a_0, a_1, a_2, \ldots, a_i, \alpha_{i+1}], \quad i > 0$$

Proposition 7.6 yields

$$\sqrt{d} = \frac{\alpha_{i+1}p_i + p_{i-1}}{\alpha_{i+1}q_i + q_{i-1}}, \quad i > 0$$

By the definition of α_{i+1} in Proposition 7.20, we obtain immediately (verify!) that

$$\sqrt{d} = \frac{(P_{i+1} + \sqrt{d})p_i + Q_{i+1}p_{i-1}}{(P_{i+1} + \sqrt{d})q_i + Q_{i+1}q_{i-1}}, \quad i > 0$$

and so

$$dq_i + (P_{i+1}q_i + Q_{i+1}q_{i-1})\sqrt{d} = (P_{i+1}p_i + Q_{i+1}p_{i-1}) + p_i\sqrt{d}, \quad i > 0$$

Now Exercise 6 of Chapter 7 implies that $dq_i = P_{i+1}p_i + Q_{i+1}p_{i-1}$ and $P_{i+1}q_i + Q_{i+1}q_{i-1} = p_i$. Multiplying the first of these equations by q_i and the second by p_i, subtracting the first equation from the second equation, and simplifying yields

$$p_i^2 - dq_i^2 = (p_iq_{i-1} - p_{i-1}q_i)Q_{i+1}, \quad i > 0$$

Proposition 7.7 now gives the desired result if $i > 0$. The easy verification of the desired result for $i = 0$ is left as an exercise for the reader. ■

Lemma 8.5: Let d be a positive integer that is not a perfect square, let ρ be the period length of the infinite simple continued fraction expansion of $\sqrt{d}$, and let all notation be as in Proposition 7.20. If $\alpha = \alpha_0 = \sqrt{d}$, $P_0 = 0$, and $Q_0 = 1$, then $Q_i = 1$ if and only if $\rho \mid i$; moreover, $Q_i = -1$ holds for no i.

Proof: In view of Corollary 7.26, it suffices to prove the desired results given $\alpha = \alpha_0 = [\sqrt{d}] + \sqrt{d}$, $P_0 = [\sqrt{d}]$, and $Q_0 = 1$ in Proposition 7.20; hereafter, we assume these values for α_0, P_0, and Q_0. Now α_0 is reduced (as noted in the proof of Proposition 7.25) and, by Theorem 7.24, we have that the infinite simple continued fraction expansion of α_0 is periodic, say

$$\alpha_0 = [\overline{a_0, a_1, a_2, \ldots, a_{\rho-1}}]$$

where ρ is the period length. If i is a nonnegative integer and $\alpha_i = [a_i, a_{i+1}, a_{i+2}, \ldots]$, we have that $\alpha_0 = \alpha_\rho = \alpha_{2\rho} = \ldots$. Furthermore, $\alpha_i = \alpha_0$ implies that $\rho \mid i$, or otherwise the period length would be strictly less than ρ. (Why?) So $\alpha_i = \alpha_0$ if and only if $\rho \mid i$. We are now ready to prove the next to the last statement of the lemma. If $\rho \mid i$, we have that

$$\frac{P_i + \sqrt{d}}{Q_i} = \alpha_i = \alpha_0 = [\sqrt{d}] + \sqrt{d}$$

or, equivalently,

$$P_i - Q_i[\sqrt{d}] = (Q_i - 1)\sqrt{d}$$

Inasmuch as $Q_i \neq 1$ implies that the left side of this equation is rational and the right side of this equation is irrational, we have $Q_i = 1$. Conversely, if $Q_i = 1$, we have that $\alpha_i = P_i + \sqrt{d}$; since α_i is periodic (why?), we have that $-1 < \alpha_i' < 0$ by Theorem 7.24. So

$$-1 < P_i - \sqrt{d} < 0$$

or, equivalently,

$$\sqrt{d} - 1 < P_i < \sqrt{d}$$

from which it follows that $P_i = [\sqrt{d}]$. Then $\alpha_i = \alpha_0$ and $\rho \mid i$ as desired. It remains to prove the last statement of the lemma. Assume, by way of contradiction, that $Q_i = -1$ for some i. Then $\alpha_i = -P_i - \sqrt{d}$ and, by Theorem 7.24, we have that $\alpha_i > 1$ and $-1 < \alpha_i' < 0$. These two inequalities imply that $\sqrt{d} < P_i < -\sqrt{d} - 1$ (as the reader will verify); this contradiction completes the proof. ■

We now characterize the positive solutions of the Pell Equation $x^2 - dy^2 = 1$ in terms of the convergents of the infinite simple continued fraction expansion of $\sqrt{d}$.

Theorem 8.6: Let d be a positive integer that is not a perfect square. Let p_i/q_i denote the ith convergent of the eventually periodic simple continued fraction expansion of $\sqrt{d}$ and let ρ be the period length of this expansion. If ρ is even, the positive solutions of the Pell Equation $x^2 - dy^2 = 1$ are given precisely by $x = p_{n\rho-1}$ and $y = q_{n\rho-1}$ where n is a positive integer; if ρ is odd, the positive solutions of the Pell Equation $x^2 - dy^2 = 1$ are given precisely by $x = p_{2n\rho-1}$ and $y = q_{2n\rho-1}$ where n is a positive integer.

Proof: By Lemma 8.4, we have that

$$p_i^2 - dq_i^2 = (-1)^{i-1}Q_{i+1}, \quad i \geq 0 \tag{3}$$

Note first that the right-hand side of (3) is 1 only if $\rho \mid i + 1$ by Lemma 8.5. If $\rho \mid i + 1$, then (3) becomes

$$p_{n\rho-1}^2 - dq_{n\rho-1}^2 = (-1)^{n\rho-2}$$

where n is a positive integer. So if ρ is even, the positive integers $x = p_{n\rho-1}$ and $y = q_{n\rho-1}$ solve $x^2 - dy^2 = 1$ and, if ρ is odd, the positive integers $x = p_{2n\rho-1}$ and $y = q_{2n\rho-1}$ solve $x^2 - dy^2 = 1$. These solutions are the only positive solutions of $x^2 - dy^2 = 1$ since any such solution must be a convergent of the infinite simple continued fraction expansion of $\sqrt{d}$ by Theorem 8.3. ∎

Example 5:

Find all positive solutions of the Pell Equation $x^2 - 7y^2 = 1$.
We use Theorem 8.6 above and assume all notation contained therein. By Exercise 20(a) of Chapter 7, the infinite simple continued fraction expansion of $\sqrt{7}$ is $[2, \overline{1, 1, 1, 4}]$. So $\rho = 4$ and, by Theorem 8.6, the desired solutions are given by $x = p_{4n-1}$ and $y = q_{4n-1}$ where n is a positive integer. The least positive solution (obtained by putting $n = 1$) is $x = p_3 = 8$ and $y = q_3 = 3$.

Example 6:

Find all positive solutions of the Pell Equation $x^2 - 13y^2 = 1$.
We use Theorem 8.6 above and assume all notation contained therein. By Exercise 20(c) of Chapter 7, the infinite simple continued fraction expansion of $\sqrt{13}$ is $[3, \overline{1, 1, 1, 1, 6}]$. So $\rho = 5$ and, by Theorem 8.6, the desired solutions are given by $x = p_{10n-1}$ and $y = q_{10n-1}$ where n is a positive integer. The least positive solution (obtained by putting $n = 1$) is $x = p_9 = 649$ and $y = q_9 = 180$.

The least positive solution of the Pell Equation $x^2 - dy^2 = 1$ is said to be the *fundamental solution.* Examples 5 and 6 above show that the "size" of the fundamental solution of the Pell Equation $x^2 - dy^2 = 1$ will vary depending on d. As a further illustration, the interested reader is invited to verify that the fundamental solution of the Pell Equation $x^2 - 61y^2 = 1$ is $x = 1766319049$ and $y = 226153980$. Before you think that the size of the fundamental solution strictly increases with increasing d, you only need to consider the equation $x^2 - 62y^2 = 1$, which has fundamental solution $x = 63$ and $y = 8$. The term *fundamental* suggests that some importance is attached to these least positive solutions; indeed, all positive solutions are expressible in terms of the fundamental solution as established by the following theorem.

Theorem 8.7: Let d be a positive integer that is not a perfect square and let x_1, y_1 be the fundamental solution of the Pell Equation $x^2 - dy^2 = 1$. Then all positive solutions of this equation are given precisely by x_n, y_n where

$$x_n + y_n\sqrt{d} = (x_1 + y_1\sqrt{d})^n$$

and n is a positive integer.

Proof: We first show that x_n, y_n is a solution of the Pell Equation $x^2 - dy^2 = 1$. We have

$$\begin{aligned} x_n^2 - dy_n^2 &= (x_n + y_n\sqrt{d})(x_n - y_n\sqrt{d}) \\ &= (x_1 + y_1\sqrt{d})^n(x_1 - y_1\sqrt{d})^n \quad \text{(why?)} \\ &= (x_1^2 - dy_1^2)^n \\ &= 1^n \\ &= 1 \end{aligned}$$

as desired. We now show that every positive solution of $x^2 - dy^2 = 1$ is x_n, y_n for some n. Assume, by way of contradiction, that X, Y is a positive solution of $x^2 - dy^2 = 1$ not of the form x_n, y_n for any n. Then there exists a positive integer k such that

$$(x_1 + y_1\sqrt{d})^k < X + Y\sqrt{d} < (x_1 + y_1\sqrt{d})^{k+1} \quad \text{(why?)}$$

Multiplying through by the quantity $(x_1 + y_1\sqrt{d})^{-k}$, we obtain

$$1 < (X + Y\sqrt{d})(x_1 + y_1\sqrt{d})^{-k} < x_1 + y_1\sqrt{d}$$

or, equivalently,

$$1 < (X + Y\sqrt{d})(x_1 - y_1\sqrt{d})^k < x_1 + y_1\sqrt{d} \quad \text{(why?)}$$

Now let r and s be integers such that

$$r + s\sqrt{d} = (X + Y\sqrt{d})(x_1 - y_1\sqrt{d})^k$$

Then

$$1 < r + s\sqrt{d} < x_1 + y_1\sqrt{d} \tag{4}$$

We also have

$$\begin{aligned} r^2 - ds^2 &= (r + s\sqrt{d})(r - s\sqrt{d}) \\ &= (X + Y\sqrt{d})(x_1 - y_1\sqrt{d})^k(X - Y\sqrt{d})(x_1 + y_1\sqrt{d})^k \\ &= (X^2 - dY^2)(x_1^2 - dy_1^2)^k \\ &= 1 \end{aligned}$$

So r, s is a solution of the Pell Equation $x^2 - dy^2 = 1$. Since $r + s\sqrt{d} > 1$, we have that $0 < r - s\sqrt{d} < 1$. (Why?) Consequently,

$$2r = (r + s\sqrt{d}) + (r - s\sqrt{d}) > 1 + 0 > 0$$

and

$$2s\sqrt{d} = (r + s\sqrt{d}) - (r - s\sqrt{d}) > 1 - 1 = 0$$

These latter inequalities imply that r and s are positive, from which r, s is a positive solution of the Pell Equation $x^2 - dy^2 = 1$. Since x_1, y_1 is the fundamental solution of $x^2 - dy^2 = 1$, we have that $x_1 \le r$ and $y_1 \le s$, contradicting (4). So all positive solutions of the Pell Equation $x^2 - dy^2 = 1$ are as desired, and the proof is complete. ■

Note that once enough convergents of $\sqrt{d}$ have been computed to determine the fundamental solution of $x^2 - dy^2 = 1$, Theorem 8.7 frees us from having to compute additional convergents in order to obtain subsequent solutions: All subsequent solutions may be obtained directly from the fundamental solution! The terminology is now clear: The "fundamental" solution is that positive solution upon which all other positive solutions are built in the sense of Theorem 8.7. We illustrate with two examples.

Example 7:

By Example 5, the fundamental solution of the Pell Equation $x^2 - 7y^2 = 1$ is $x_1 = 8$ and $y_1 = 3$. Theorem 8.7 establishes that all positive solutions of this equation are given precisely by x_n, y_n where

$$x_n + y_n\sqrt{7} = (8 + 3\sqrt{7})^n$$

and n is a positive integer. In particular, the next to the least positive solution of $x^2 - 7y^2 = 1$, obtainable by putting $n = 2$ in the formula above and expanding $(8 + 3\sqrt{7})^2$, is $x_2 = 127$ and $y_2 = 48$.

Example 8:

By Example 6, the fundamental solution of the Pell Equation $x^2 - 13y^2 = 1$ is $x_1 = 649$ and $y_1 = 180$. Theorem 8.7 establishes that all positive solutions of this equation are given precisely by x_n, y_n where

$$x_n + y_n\sqrt{13} = (649 + 180\sqrt{13})^n$$

and n is a positive integer. In particular, the next to the least positive solution of $x^2 - 13y^2 = 1$, obtainable by putting $n = 2$ in the formula above and expanding $(649 + 180\sqrt{13})^2$, is $x_2 = 842401$ and $y_2 = 233640$.

Analyses of the general Pell Equation $x^2 - dy^2 = n$ are not as complete as those for the special cases $x^2 - dy^2 = 1$ (above) and $x^2 - dy^2 = -1$ (see Exercises 18–20). For a discussion of results pertaining to the general Pell Equation, the interested reader may consult Niven, Zuckerman, and Montgomery (1991). See also Exercise 21.

Exercise Set 8.4

15. Find all integral solutions of each Pell Equation.

(a) $x^2 - y^2 = -3$
(b) $x^2 + 2y^2 = 6$
(c) $x^2 - 2y^2 = -6$
(d) $x^2 + 3y^2 = -4$
(e) $x^2 - 4y^2 = 3$
(f) $x^2 + 4y^2 = 11$

(g) $x^2 + 5y^2 = 9$
(h) $x^2 - 9y^2 = 7$

16. Find all positive solutions of each Pell Equation in the sense of Theorem 8.6.
(a) $x^2 - 21y^2 = 1$
(b) $x^2 - 29y^2 = 1$
(c) $x^2 - 47y^2 = 1$
(d) $x^2 - 58y^2 = 1$

17. Find the fundamental solution of each Pell Equation in Exercise 16 and use Theorem 8.7 to find the next to the least positive solution in each case.

18. Let d be a positive integer that is not a perfect square. Characterize the positive solutions of the Pell Equation $x^2 - dy^2 = -1$ in the spirit of Theorem 8.6.

19. Find all positive solutions of each Pell Equation.
(a) $x^2 - 21y^2 = -1$
(b) $x^2 - 29y^2 = -1$
(c) $x^2 - 47y^2 = -1$
(d) $x^2 - 58y^2 = -1$

20. Let $d \in \mathbf{Z}$ with $d \equiv 3 \bmod 4$. Prove that the Pell Equation $x^2 - dy^2 = -1$ has no solutions.

21. Let d be a positive integer that is not a perfect square and let n be an integer.
(a) Let x_1, y_1 be the fundamental solution of the Pell Equation $x^2 - dy^2 = 1$ and let r_0, s_0 be a positive solution of the Pell Equation $x^2 - dy^2 = n$. If the integers r_m, s_m are defined by
$$r_m + s_m\sqrt{d} = (r_0 + s_0\sqrt{d})(x_1 + y_1\sqrt{d})^m$$
where m is a positive integer, prove that r_m, s_m is a positive solution of the Pell Equation $x^2 - dy^2 = n$.
(b) Prove that the Pell Equation $x^2 - dy^2 = n$ has either no solutions or infinitely many solutions.

Concluding Remarks

The purpose of this chapter was to convey to the reader a sense of the applicability of the concepts of elementary number theory. You may not have noticed that the applications themselves were of three types. The game of Square-Off in Section 8.1 illustrated a *recreational application* of number theory. Number theory frequently aids in the solution and analysis of puzzles and games, often in unexpected ways. The RSA encryption system of Section 8.2 and the primality tests of Section 8.3 were *real-world applications* of number theory; these algorithms (or modifications) are actually being implemented and used by various organizations within today's society for which information security is paramount. The discussion of Pell's Equation in Section 8.4 was a *theoretical application* of number theory; we applied the theory of continued fractions to obtain a theoretical characterization of the solutions of a particular Diophantine equation.

Further applications of elementary number theory abound. Recreational applications may be found in Ball (1944) and Gardner (1975); cryptographic and primality testing applications may be found in Konheim (1981) and Bressoud (1990) respectively; theoretical applications may be found in Hardy and Wright (1979). As noted at the end of the Introduction, important applications of elementary number theory in the field of computer science may be found in Knuth (1981). The interested reader is referred to Student Project 7 for one such application. Another computer science application of elementary number theory is alluded to in Student Project 3.

Student Projects

1. (Programming project for calculator or computer)
The following procedure illustrates a method for factoring the m of the RSA encryption system into its component prime numbers p and q. This method, devised by Fermat to facilitate general factorization, is appropriately called the *Fermat factorization method.* Consider $m = 551923$ [from Exercise 11(c)].

Step 1: Find the smallest positive integer k for which $k^2 \geq m$, namely, $k = 743$.

Step 2: Consider the sequence

$$k^2 - m, (k + 1)^2 - m, (k + 2)^2 - m, \ldots$$

until a perfect square is encountered, say $x^2 - m = y^2$:

$$743^2 - 551923 = 126$$
$$744^2 - 551923 = 1613$$
$$745^2 - 551923 = 3102$$
$$\vdots$$
$$777^2 - 551923 = 51806$$
$$778^2 - 551923 = 53361 = 231^2$$

Here $x = 778$ and $y = 231$. [This sequence must terminate, since $((m + 1)/2)^2 - m = ((m - 1)/2)^2$.]

Step 3: Since $m = x^2 - y^2 = (x + y)(x - y)$, then m factors into $x + y$ and $x - y$:

$$\begin{aligned} 551923 &= 778^2 - 231^2 \\ &= (778 + 231)(778 - 231) \\ &= 1009 \cdot 547 \end{aligned}$$

So $p = 1009$ and $q = 547$.

Given a positive integer m as in the RSA encryption system, factor m into its component prime numbers p and q using the Fermat factorization method above. (Note that repeated applications of the method will completely factor any positive integer.)

2. Analyze the game of Square-Off (see Section 8.1), where the squares crossed off in any move must be contiguous.
3. (Euler, 1782) There are 36 officers, one from each of six ranks from each of six regiments. Is it possible to arrange these officers in a 6×6 square so that every row and every column of the square contains exactly one officer of each rank and exactly one officer of each regiment? [This "recreational" problem has connections to a field within computer science known as *coding theory*. For more on these connections, consult R. Hill, *A First Course in Coding Theory* (Oxford: Clarendon Press, 1986).]
4. Read the following article: R. Rivest, A. Shamir, and L. Adleman, "A Method for Obtaining Digital Signatures and Public-Key Cryptosystems," *Communications of the ACM, 21* (1978), 120–126.
5. *(a)* Let n be a positive integer and let A and B be positive integers such that $n - 1 = AB$, $A > \sqrt{n}$, and $(A, B) = 1$. If an integer a exists such that

$$a^{n-1} \equiv 1 \bmod n$$

and

$$(a^{(n-1)/q} - 1, n) = 1$$

for all prime divisors q of A, prove that n is a prime number.

(b) In what way does the primality test of part (a) above represent an improvement over the tests given by Theorem 8.1 and Corollary 8.2?
6. Let d be a positive integer that is not a perfect square and let

$$\mathbf{Z}[\sqrt{d}] = \{a + b\sqrt{d}: a, b \in \mathbf{Z}\}$$

Definition: An element $\alpha \in \mathbf{Z}[\sqrt{d}]$ is said to be a *unit* in $\mathbf{Z}[\sqrt{d}]$ if there exists $\beta \in \mathbf{Z}[\sqrt{d}]$ such that $\alpha\beta = 1$.

Definition: Let $\alpha \in \mathbf{Z}[\sqrt{d}]$. The *norm* of α, denoted $N(\alpha)$, is $N(\alpha) = \alpha\alpha'$ where α' denotes the conjugate of α.

It is a fact that $\alpha \in \mathbf{Z}[\sqrt{d}]$ is a unit in $\mathbf{Z}[\sqrt{d}]$ if and only if $N(\alpha) = \pm 1$.

(a) Explain the relationship of the concepts above to Pell's Equation.

(b) Describe all units in $\mathbf{Z}[\sqrt{2}]$.

(c) Describe all units in $\mathbf{Z}[\sqrt{3}]$.

[A detailed discussion of the concepts in this exercise may be found in Chapter 9 of Niven, Zuckerman, and Montgomery (1991).]
7. As noted in the Introduction and again in Chapter 8, many applications of elementary number theory in the field of computer science may be found in Knuth (1981). Using Knuth as a reference, read about the applications of primitive roots in the computer generation of pseudo-random numbers.

Appendices

Mathematical Induction

Mathematical induction is a powerful proof technique that is used quite frequently in number theory as well as in other areas of mathematics. In this appendix, we summarize the two essential forms of mathematical induction and provide examples of the use of each form. This summary is intended to be rather concise; more detailed discussions of the topics here may be found, for example, in Fletcher and Patty (1992).

A.1 The First Principle of Mathematical Induction

The first form of mathematical induction is called the *first principle of mathematical induction.*

The First Principle of Mathematical Induction: Let m be an integer and let S be a set of integers such that

(a) $m \in S$ and
(b) if $k \geq m$ and $k \in S$, then $k + 1 \in S$.

Then S contains all integers greater than or equal to m.

The first principle of mathematical induction can be proven by using the well-ordering property of the system of integers. The first principle of mathematical induction is useful for proving statements true for all integers greater than or equal to a specified integer m. Such a proof by induction requires two steps:

(a) Prove the desired statement true for $n = m$ (the *basis step*) and
(b) assuming that $k \geq m$ and that the desired statement is true for $n = k$ (the *induction hypothesis*), prove that the desired statement is true for $n = k + 1$ (the *inductive step*).

Then, by the first principle of mathematical induction, the set of integers for which the statement is true must include all integers greater than or equal to m. We illustrate the use of the first principle of mathematical induction with two examples.

Example 1:

(First Principle of Mathematical Induction) Prove that if n is a positive integer, then

$$1 + 2 + 3 + \cdots + n = \frac{n(n+1)}{2}$$

The proof is by induction. (Since we wish to show that the given statement is true for all integers greater than or equal to 1, we take $m = 1$ in the first principle of mathematical induction.) We first show that the desired statement is true for $n = 1$. If $n = 1$, the left-hand side of the desired statement becomes 1, while the right-hand side of the desired statement becomes $\frac{1(1+1)}{2}$; inasmuch as both of these quantities are equal to 1, we have that the desired statement is true for $n = 1$. Now assume that $k \geq 1$ and that the desired statement is true for $n = k$ so that

$$1 + 2 + 3 + \cdots + k = \frac{k(k+1)}{2} \quad \text{(the induction hypothesis)}$$

We must show that the desired statement is true for $n = k + 1$; in other words, we must show that

$$1 + 2 + 3 + \cdots + k + (k+1) = \frac{(k+1)((k+1)+1)}{2}$$

We have

$$\begin{aligned}
1 + 2 + 3 + \cdots + k + (k+1) &= \frac{k(k+1)}{2} + (k+1) \\
&\quad \text{(by the induction hypothesis)} \\
&= \frac{k(k+1)}{2} + \frac{2k+2}{2} \\
&= \frac{k(k+1) + 2k + 2}{2} \\
&= \frac{k^2 + 3k + 2}{2} \\
&= \frac{(k+1)(k+2)}{2} \\
&= \frac{(k+1)((k+1)+1))}{2}
\end{aligned}$$

So the desired statement is true for $n = k + 1$. By the first principle of mathematical induction, the desired statement is true for all positive integers, and the proof is complete.

Example 2:

(First Principle of Mathematical Induction) Let x be a nonzero real number with $x > -1$. Prove that if n is an integer with $n \geq 2$, then

$$(1 + x)^n > 1 + nx$$

(This result is known as *Bernoulli's Inequality.*)

The proof is by induction. (Since we wish to show that the given statement is true for all integers greater than or equal to 2, we take $m = 2$ in the first principle of mathematical induction.) We first show that the desired statement is true for $n = 2$. If $n = 2$, the left-hand side of the desired statement becomes $(1 + x)^2$, while the right-hand side of the desired statement becomes $1 + 2x$; inasmuch as

$$(1 + x)^2 = 1 + 2x + x^2 > 1 + 2x$$

since x is nonzero, we have that the desired statement is true for $n = 2$. Now assume that $k \geq 2$ and that the desired statement is true for $n = k$ so that

$$(1 + x)^k > 1 + kx \quad \text{(the induction hypothesis)}$$

We must show that the desired statement is true for $n = k + 1$; in other words, we must show that

$$(1 + x)^{k+1} > 1 + (k + 1)x$$

We have

$$\begin{aligned}(1 + x)^{k+1} &= (1 + x)^k(1 + x) \\ &> (1 + kx)(1 + x) \quad \text{(by the induction hypothesis since } x > -1) \\ &= 1 + kx + x + kx^2 \\ &= 1 + (k + 1)x + kx^2 \\ &> 1 + (k + 1)x \quad \text{(since } k \geq 2 \text{ and } x \text{ is nonzero)}\end{aligned}$$

So the desired statement is true for $n = k + 1$. By the first principle of mathematical induction, the desired statement is true for all integers greater than or equal to 2, and the proof is complete.

A.2 The Second Principle of Mathematical Induction

The second form of mathematical induction is called the *second principle of mathematical induction.*

The Second Principle of Mathematical Induction: Let m be an integer and let S be a set of integers such that (*a*) $m \in S$ and

(*b*) if $k \geq m$ and $\{m, m + 1, m + 2, \ldots, k\} \subseteq S$, then $k + 1 \in S$.

Then S contains all integers greater than or equal to m.

The second principle of mathematical induction is useful in problems involving *recurrence relations.* Such a proof by induction requires two steps:

(*a*) prove the desired statement true for $n = m$ (the *basis step*) and

(*b*) assuming that $k \geq m$ and that the desired statement is true for the integers $m, m + 1, m + 2, \ldots, k$ (the *induction hypothesis*), prove that the desired statement is true for $n = k + 1$ (the *inductive step*).

Then, by the second principle of mathematical induction, the set of integers for which the statement is true must include all integers greater than or equal to m. We illustrate the use of the second principle of mathematical induction with two examples involving recurrence relations. (Although the examples may appear difficult, they are actually quite straightforward. Persevere!)

Example 3:

(Second Principle of Mathematical Induction) The *Fibonacci numbers* are the elements of the sequence $f_1, f_2, f_3, \ldots$ where $f_1 = 1$, $f_2 = 1$, and, if n is a positive integer, then $f_{n+2} = f_n + f_{n+1}$. (Here, $f_{n+2} = f_n + f_{n+1}$ is the recurrence relation. This relation says that to get any Fibonacci number after the first two Fibonacci numbers, simply add the previous two Fibonacci numbers. The Fibonacci numbers are thus given by $1, 1, 2, 3, 5, 8, 13, 21, 34, \ldots$; note that each number of the sequence after the first two numbers is simply the sum of the two previous numbers.) Prove that if n is a positive integer, then

$$f_n = \frac{(1 + \sqrt{5})^n - (1 - \sqrt{5})^n}{2^n\sqrt{5}}$$

The proof is by induction. Since the first two Fibonacci numbers are defined explicitly, our basis step will consist of verification of the desired statement for $n = 1$ *and* $n = 2$. (We will take $m = 2$ in the second principle of mathematical induction.) If $n = 1$, the left-hand side of the desired statement becomes f_1, while the right-hand side of the desired statement becomes

$$\frac{(1 + \sqrt{5})^1 - (1 - \sqrt{5})^1}{2^1\sqrt{5}}$$

Inasmuch as both quantities equal 1, we have that the desired statement is true for $n = 1$. If $n = 2$, the left-hand side of the desired statement becomes f_2, while the right-hand side of the desired statement becomes

$$\frac{(1 + \sqrt{5})^2 - (1 - \sqrt{5})^2}{2^2\sqrt{5}}$$

Inasmuch as both quantities equal 1, we have that the desired statement is true for $n = 2$. Now assume that $k \geq 2$ and that the desired statement is true for the integers $2, 3, \ldots, k$ (the induction hypothesis). We must show that the desired statement is true for the integer $k + 1$; in other words, we must show that

$$f_{k+1} = \frac{(1 + \sqrt{5})^{k+1} - (1 - \sqrt{5})^{k+1}}{2^{k+1}\sqrt{5}}$$

Since $k \geq 2$, we have

$$\begin{aligned} f_{k+1} &= f_{k-1} + f_k \quad \text{(by the recurrence relation defining the Fibonacci numbers)} \\ &= \frac{(1 + \sqrt{5})^{k-1} - (1 - \sqrt{5})^{k-1}}{2^{k-1}\sqrt{5}} + \frac{(1 + \sqrt{5})^{k} - (1 - \sqrt{5})^{k}}{2^{k}\sqrt{5}} \\ &\quad \text{(by the induction hypothesis)} \\ &= \frac{4[(1 + \sqrt{5})^{k-1} - (1 - \sqrt{5})^{k-1}] + 2[(1 + \sqrt{5})^{k} - (1 - \sqrt{5})^{k}]}{2^{k+1}\sqrt{5}} \\ &= \frac{\begin{array}{l}[-(1 + \sqrt{5})(1 - \sqrt{5})][(1 + \sqrt{5})^{k-1} - (1 - \sqrt{5})^{k-1}] \\ \quad + [(1 + \sqrt{5}) + (1 - \sqrt{5})][(1 + \sqrt{5})^{k} - (1 - \sqrt{5})^{k}]\end{array}}{2^{k+1}\sqrt{5}} \\ &= \frac{(1 + \sqrt{5})^{k+1} - (1 - \sqrt{5})^{k+1}}{2^{k+1}\sqrt{5}} \end{aligned}$$

So the desired statement is true for $n = k + 1$. By the second principle of mathematical induction, the desired statement is true for all positive integers, and the proof is complete.

Example 4:

(Second Principle of Mathematical Induction) Let $f_1, f_2, f_3, \ldots$ be the Fibonacci numbers. Define a sequence $F_1, F_2, F_3, \ldots$ by $F_1 = 0$, $F_2 = 1$, and, if n is a positive integer, then $F_{n+2} = F_{n+1} + F_n + F_{n+1}F_n$. Prove that if n is an integer with $n \geq 2$, then

$$F_n = 2^{f_{n-1}} - 1$$

(Note immediately that the desired statement makes no sense for $n = 1$; if $n = 1$, the right-hand side of the desired statement involves f_0, which is undefined.) The proof is by induction. Since the first two elements of the sequence $F_1, F_2, F_3, \ldots$ are defined explicitly, our basis step will consist of verification of the desired statement for $n = 2$ *and* $n = 3$. (We will take $m = 3$ in the second principle of mathematical induction.) If $n = 2$, the left-hand side of the desired statement becomes F_2, while the right-hand side of the desired statement becomes $2^{f_1} - 1$; inasmuch as both quantities equal 1, we have that the desired statement is true for $n = 2$. If $n = 3$, the left-hand side of the desired statement becomes F_3, while the right-hand side of the desired statement becomes $2^{f_2} - 1$; inasmuch as both quantities equal 1 (using the recurrence relation to compute F_3), we have that the desired statement is true for $n = 3$. Now assume that $k \geq 3$ and that the desired statement is true for the integers, $3, 4, \ldots, k$ (the induction hypothesis). We must show that the desired statement is true for the integer $k + 1$; in other words, we must show that

$$F_{k+1} = 2^{f_k} - 1$$

Since $k \geq 3$, we have

$$\begin{aligned} F_{k+1} &= F_k + F_{k-1} + F_kF_{k-1} \\ &\quad \text{(by the recurrence relation defining } F_1, F_2, F_3, \ldots) \\ &= (2^{f_{k-1}} - 1) + (2^{f_{k-2}} - 1) + (2^{f_{k-1}} - 1)(2^{f_{k-2}} - 1) \\ &\quad \text{(by the induction hypothesis)} \\ &= 2^{f_{k-1}+f_{k-2}} - 1 \\ &= 2^{f_k} - 1 \quad \text{(by the recurrence relation defining the Fibonacci numbers)} \end{aligned}$$

So the desired statement is true for $n = k + 1$. By the second principle of mathematical induction, the desired statement is true for all integers greater than or equal to 2, and the proof is complete.

Two remarks are in order here. First, the first and second principles of mathematical induction are sometimes referred to as *weak induction* and *strong* (or *complete*) *induction*, respectively, from the fact that more is assumed in the induction hypothesis of the second principle than that of the first principle. Second, since the first and second principles of mathematical induction are both logically equivalent to the well-ordering principle (see Introduction), they are necessarily logically equivalent to each other. In this sense, the use of the words *weak* and *strong* may be somewhat misleading here.

Exercise Set A

1. Use the first principle of mathematical induction to prove each statement.

(a) If n is a positive integer, then

$$1^2 + 2^2 + 3^2 + \cdots + n^2 = \frac{n(n + 1)(2n + 1)}{6}$$

(b) If n is a positive integer, then

$$1^3 + 2^3 + 3^3 + \cdots + n^3 = \frac{n^2(n + 1)^2}{4}$$

(c) If n is an integer with $n \geq 5$, then

$$2^n > n^2$$

2. Use the second principle of mathematical induction to prove each statement.

(a) Define a sequence $x_1, x_2, x_3, \ldots$ by $x_1 = 1$, $x_2 = 3$, and, if n is a positive integer, then $x_{n+2} = 3x_{n+1} - 2x_n$. If n is a positive integer, then $x_n = 2^n - 1$.

(b) Let $f_1, f_2, f_3, \ldots$ be the Fibonacci numbers. If n is an integer with $n > 1$, then

$$f_n^2 + f_{n-1}^2 = f_{2n-1}$$

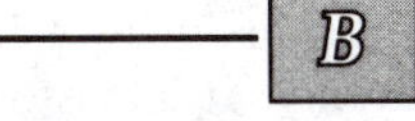

Equivalence Relations

The relation of congruence first discussed in Chapter 2 is an extremely important type of relation called an *equivalence relation* (see Proposition 2.1). In this appendix, we summarize the elementary theory of relations and equivalence relations. This summary is intended to be rather concise; more detailed discussions of the topics here may be found, for example, in Fletcher and Patty (1992).

B.1 Relations

Definition 1: Let A and B be sets. The *Cartesian product* of A and B, denoted A × B, is

$$A \times B = \{(a, b)\colon a \in A, b \in B\}$$

Here the element (a, b) of the Cartesian product is an *ordered* pair.

Example 1:

(a) Let $A = \{a, b\}$ and $B = \{1, 2, 3\}$. Then

$$A \times B = \{(a, 1), (a, 2), (a, 3), (b, 1), (b, 2), (b, 3)\}$$

Note that A × B is a set of *ordered* pairs. For example, $(a, 1) \in A \times B$, but $(1, a) \notin A \times B$.

(b) Let $A = B = \mathbf{R}$. Then $A \times B = \mathbf{R} \times \mathbf{R}$ (usually denoted $\mathbf{R}^2$), the set of all ordered pairs of real numbers. In other words, $\mathbf{R}^2$ is the Cartesian plane in the usual sense.

Definition 2: Let A and B be sets. If $R \subseteq A \times B$, then R is said to be a *relation from A to B*. A relation from A to A is said to be a *relation on A*.

So a relation from set A to set B is simply some set of ordered pairs whose first components come from A and whose second components come from B. If the sets A and B are equal, then we have a relation *on* the set A. (In the succeeding sections of this appendix, we will restrict to these latter types of relations.) If an ordered pair (a, b) is in a relation R, we sometimes use the notation $a \mathrm{R} b$ to show the relationship of a to b. The next example illustrates the usefulness of this alternate notation.

Example 2:

(a) Let $\mathrm{A} = \{-1, 2, 5\}$ and $\mathrm{B} = \{-2, 0, 3, 4\}$. If R is the relation from A to B commonly denoted $|$ (divides), then

$$\mathrm{R} = | = \{(-1, -2), (-1, 0), (-1, 3), (-1, 4), (2, -2), (2, 0), (2, 4), (5, 0)\}$$

R consists precisely of those ordered pairs in $\mathrm{A} \times \mathrm{B}$ whose first components divide their second components evenly. It is customary to express the fact that $(-1, -2) \in \mathrm{R}$ (for example) by writing $-1 \mid -2$.

(b) Let $\mathrm{A} = \mathrm{B} = \mathbf{R}$. If R is the relation on $\mathbf{R}$ commonly denoted $<$, then

$$\mathrm{R} = < = \{(x, y) \in \mathbf{R}^2 : x < y\}$$

R consists precisely of those ordered pairs in $\mathbf{R}^2$ whose first components are less than their second components. It is customary to express the fact that $(x, y) \in \mathrm{R}$ by writing $x < y$.

B.2 Equivalence Relations

Definition 3: Let A be a nonempty set and let R be a relation on A.

(a) R is said to be *reflexive on A* if $(a, a) \in \mathrm{R}$ for all $a \in \mathrm{A}$ (or, equivalently, $a \mathrm{R} a$ for all $a \in \mathrm{A}$.)

(b) R is said to be *symmetric* if $(a, b) \in \mathrm{R}$ implies that $(b, a) \in \mathrm{R}$ for all $a, b \in \mathrm{A}$ (or, equivalently, $a \mathrm{R} b$ implies that $b \mathrm{R} a$ for all $a, b \in \mathrm{A}$).

(c) R is said to be *transitive* if $(a, b) \in \mathrm{R}$ and $(b, c) \in \mathrm{R}$ imply that $(a, c) \in \mathrm{R}$ for all $a, b, c \in \mathrm{A}$ (or, equivalently, $a \mathrm{R} b$ and $b \mathrm{R} c$ imply that $a \mathrm{R} c$ for all $a, b, c \in \mathrm{A}$).

If R is reflexive on A, symmetric, and transitive, then R is said to be an *equivalence relation on A*.

Example 3:

(a) Let R be the relation of equality on $\mathbf{R}$. Then R is reflexive on $\mathbf{R}$ (since every real number is equal to itself), R is symmetric (since $a = b$ implies that $b = a$ for all $a, b \in \mathbf{R}$), and R is transitive (since $a = b$ and $b = c$ imply that $a = c$ for all $a, b, c \in \mathbf{R}$). So the relation of equality on $\mathbf{R}$ is an equivalence relation on $\mathbf{R}$.

(b) Let R be the relation of divides on $\mathbf{Z}$. Then R is reflexive on $\mathbf{Z}$ (since every integer divides itself), R is *not* symmetric (since $a \mid b$ does not necessarily imply that $b \mid a$; consider, for example, $a = 1$ and $b = 2$), and R is transitive (since $a \mid b$ and $b \mid c$ imply that $a \mid c$ for all $a, b, c \in \mathbf{Z}$; see Proposition 1.1). Since the relation of divides on $\mathbf{Z}$ is not symmetric, the relation of divides is not an equivalence relation on $\mathbf{Z}$.

(c) Let R be the relation on **Z** defined as follows: If $a, b \in \mathbf{Z}$, then $a \mathrm{R} b$ if and only if $a + b$ is even. Then R is reflexive on **Z** (since $a + a = 2a$ is even for every $a \in \mathbf{Z}$), R is symmetric (since $a + b$ being even implies that $b + a$ is even for all $a, b \in \mathbf{Z}$), and R is transitive [since $a + b$ and $b + c$ being even implies that $(a + b) + (b + c)$ is even, from which $(a + b) + (b + c) - 2b$ is even and $a + c$ is even for all $a, b, c \in \mathbf{Z}$]. So the relation R on **Z** is an equivalence relation on **Z**.

(d) Let R be the relation on **Z** defined as follows: If $a, b \in \mathbf{Z}$, then $a \mathrm{R} b$ if and only if $a + b$ is odd. Then R is *not* reflexive on **Z** (since $a + a = 2a$ is even for every $a \in \mathbf{Z}$), R is symmetric (since $a + b$ being odd implies that $b + a$ is odd for all $a, b \in \mathbf{Z}$), and R is *not* transitive (since $a + b$ and $b + c$ being odd implies that $a + c$ is *even* for all $a, b, c \in \mathbf{Z}$; consider, for example, $a = 1$, $b = 2$, and $c = 3$). Since the relation R on **Z** is neither reflexive nor transitive, the relation R is not an equivalence relation on **Z**.

Definition 4: Let R be an equivalence relation on a set A and let $a \in \mathrm{A}$. The *equivalence class containing a*, denoted $[a]$, is

$$[a] = \{b \in \mathrm{A} : b \mathrm{R} a\}$$

Example 4:

(a) Let R be the equivalence relation of equality on **R** [see Example 3(a)]. If $a \in \mathbf{R}$, then

$$[a] = \{b \in \mathbf{R} : b = a\} = \{a\}$$

So every equivalence class in **R** under the relation of equality consists of exactly one element.

(b) Let R be the equivalence relation on **Z** defined as in Example 3(c). If $a \in \mathbf{Z}$, then

$$[a] = \{b \in \mathbf{Z} : b + a \text{ is even}\}$$

If a is even, then $b + a$ is even if and only if b is even; so if a is even, we have

$$[a] = \{\ldots, -6, -4, -2, 0, 2, 4, 6, \ldots\}$$

Similarly, if a is odd, then $b + a$ is even if and only if b is odd; so if a is odd, we have

$$[a] = \{\ldots, -7, -5, -3, -1, 1, 3, 5, 7, \ldots\}$$

B.3 Partitions

Definition 5: Let A be a nonempty set. A set S of nonempty subsets of A is said to be a *partition* of A if every element of A is an element of exactly one element of S.

Note carefully that there are two requirements to be satisfied by a partition S of a set A. The first requirement is that the sets in the partition be nonempty. The second requirement is that every element of A appear in one and only one set of the partition. We pause to consider two examples of partitions.

Example 5:

(a) Let R be the equivalence relation of equality on **R** [see Example 3(a)]. Since every equivalence class in **R** under the relation of equality consists of exactly one element [see Example 4(a)], the set of distinct equivalence classes of **R** under equality forms a partition of **R**. In other words, one can view **R** as having been partitioned into singleton sets by the equivalence relation of equality.

(b) Let R be the equivalence relation on **Z** defined as in Example 3(c). There are two distinct equivalence classes in **Z** under the relation R: the set of even integers and the set of odd integers [see Example 4(b)]. Since every integer is either even or odd (and no integer is both!), the set of distinct equivalence classes of **Z** under R forms a partition of **Z**. In other words, one can view **Z** as having been partitioned into the set of even integers and the set of odd integers by the equivalence relation R.

Note the different behaviors exhibited in (a) and (b) of Example 5. In (a), an infinite set was partitioned into infinitely many finite sets, while in (b) an infinite set was partitioned into finitely many infinite sets. The fact that we obtained a partition in each case, however, is not mere coincidence. Even though the characteristics of partitions may vary widely, given any equivalence relation defined on a set A, it is a fact that the set of distinct equivalence classes forms a partition of A. This is precisely the content of the following theorem.

Theorem 1: Let R be an equivalence relation on a nonempty set A and let S be the set of all distinct equivalence classes under R. Then S is a partition of A.

A proof of Theorem 1 may be found in Fletcher and Patty (1992). More examples of equivalence relations and the resulting partitions arise from consideration of the congruence relation investigated in Chapter 2.

Exercise Set B

1. Let m be a positive integer. Let R be the relation on **Z** defined as follows: If $a, b \in \mathbf{Z}$, then $a \mathrm{R} b$ if and only if $m \mid a + b$. Prove that R is an equivalence relation on **Z** for precisely two values of m. What are the equivalence classes in **Z** under R in these two cases?
2. Let R be the relation on **R** defined as follows: If $x, y \in \mathbf{R}$, then $x \mathrm{R} y$ if and only if $x^2 - y^2 = 0$. Prove that R is an equivalence relation on **R**. Describe the equivalence classes in **R** under R.

Abstract Algebra

This appendix presents definitions and examples of rings, fields, and groups from abstract algebra. Such structures are alluded to in Chapters 2 and 5. The development here is intended to be rather concise; more detailed discussions of these topics may be found, for example, in Fraleigh (1989). For a discussion of the algebraic concepts of rings and fields in the number-theoretic context, see Niven, Zuckerman, and Montgomery (1991).

C.1 Rings and Fields

Definition 1: A *ring* is a nonempty set R together with two binary operations on R, called *addition* (denoted +) and *multiplication* (denoted $\cdot$ or by juxtaposition), such that

(a) if $a, b \in \mathrm{R}$, then $a + b \in \mathrm{R}$ (closure property of addition);

(b) if $a, b, c \in \mathrm{R}$, then $(a + b) + c = a + (b + c)$ (associative property of addition);

(c) there exists $e \in \mathrm{R}$ such that $a + e = e + a = a$ for all $a \in \mathrm{R}$ (identity property of addition);

(d) if $a \in \mathrm{R}$, then there exists $a' \in \mathrm{R}$ such that $a + a' = a' + a = e$ (inverse property of addition);

(e) if $a, b \in \mathrm{R}$, then $a + b = b + a$ (commutative property of addition);

(f) if $a, b \in \mathrm{R}$, than $ab \in \mathrm{R}$ (closure property of multiplication);

(g) if $a, b, c \in \mathrm{R}$, then $(ab)c = a(bc)$ (associative property of multiplication);

(h) if $a, b, c \in \mathrm{R}$, then $a(b + c) = ab + ac$ (left distributivity of multiplication over addition) and $(a + b)c = ac + bc$ (right distributivity of multiplication over addition).

Any ring as above will be denoted $[\mathrm{R}, +, \cdot]$. A ring $[\mathrm{R}, +, \cdot]$ is said to be a *field* if, in addition to (a) through (h), we have

(i) there exists $u \in \mathrm{R}$ such that $au = ua = a$ for all $a \in \mathrm{R}$ (identity property of multiplication);

(j) if $a \in \mathrm{R}$ with $a \neq e$, then there exists $a'' \in \mathrm{R}$ such that $aa'' = a''a = u$ (inverse property of multiplication);

(k) if $a, b \in \mathrm{R}$, then $ab = ba$ (commutative property of multiplication).

Example 1:

(a) Let + and $\cdot$ denote usual addition and multiplication in **Z**. Then $[\mathbf{Z}, +, \cdot]$ is a ring. (Compare the properties of **Z** in the Introduction with the properties of a ring in Definition 1 above!) Here $e = 0$ and $a' = -a$. Furthermore, properties (i) (with $u = 1$) and (k) hold in $\mathbf{Z}$. Note, however, that $[\mathbf{Z}, +, \cdot]$ is *not* a field; indeed, property (j) fails in $\mathbf{Z}$ for two reasons: The reciprocal of zero does not exist, and reciprocals of nonzero integers other than ± 1 are not integers (so such reciprocals do not exist *in* $\mathbf{Z}$).

(b) Let $+$ and $\cdot$ denote usual addition and multiplication in $\mathbf{R}$. Then $[\mathbf{R}, +, \cdot]$ is a field (and thus a ring). Here $e = 0$, $a' = -a$, $u = 1$, and $a'' = \frac{1}{a}$.

(c) Let $\mathbf{Z}_4$ be as in Example 4 of Chapter 2. Then $\mathbf{Z}_4$ is a ring. Here $e = [0]$, $[0]' = [0]$, $[1]' = [3]$, $[2]' = 2$, and $[3]' = 1$. Furthermore, properties (i) (with $u = [1]$) and (k) hold in $\mathbf{Z}_4$. Note, however, that $\mathbf{Z}_4$ is *not* a field; indeed, property (j) fails in $\mathbf{Z}_4$ since $[2]''$ does not exist.

(d) Let $\mathbf{Z}_5$ be as in Example 4 of Chapter 2. Then $\mathbf{Z}_5$ is a field (and thus a ring). Here $e = [0]$, $[0]' = [0]$, $[1]' = [4]$, $[2]' = 3$, $[3]' = 2$, $[4]' = [1]$, $u = [1]$, $[1]'' = [1]$, $[2]'' = 3$, $[3]'' = 2$, and $[4]'' = 4$.

The set of equivalence classes in $\mathbf{Z}$ induced by the equivalence relation of congruence modulo m on $\mathbf{Z}$ under the operations of addition and multiplication modulo m is denoted $\mathbf{Z}_m$. Proposition 2.3 shows that the m classes of $\mathbf{Z}_m$ are given by $[0], [1], [2], \ldots, [m-1]$. Exercise 1(a) establishes that $\mathbf{Z}_m$ is a ring for all m; furthermore, (c) and (d) of Example 1 above as well as Exercise 1(b) show that $\mathbf{Z}_m$ is sometimes, but not always, a field.

C.2 Groups

Definition 2: A *group* is a nonempty set G together with a binary operation $*$ on G such that

(a) if $a, b \in G$, then $a * b \in G$ (closure property of $*$);

(b) if $a, b, c \in G$, then $(a * b) * c = a * (b * c)$ (associative property of $*$);

(c) there exists $e \in G$ such that $a * e = e * a = a$ for all $a \in G$ (identity property of $*$);

(d) if $a \in G$, then there exists $a' \in G$ such that $a * a' = a' * a = e$ (inverse property of $*$).

Any group as above will be denoted $[G, *]$. A group $[G, *]$ is said to be *abelian* if, in addition to (a) through (d), we have

(e) if $a, b \in G$, then $a * b = b * a$ (commutative property of $*$).

Note immediately that (a) through (e) in Definition 2 are identical to (a) through (e) in Definition 1 (with the replacement of $*$ with $+$ or vice versa). It follows that if $[R, +, \cdot]$ is a ring, then $[R, +]$ is an abelian group. By virtue of this remark, Example 1 immediately yields four examples of groups, namely, $[\mathbf{Z}, +]$, $[\mathbf{R}, +]$, $\mathbf{Z}_4$ with the operation of addition only, and $\mathbf{Z}_5$ with the operation of addition only. In fact, all of these groups are abelian.

Example 2:

(a) Let $\cdot$ denote usual multiplication in $\mathbf{Z}$. Then $[\mathbf{Z}, \cdot]$ is *not* a group; although properties (a) through (c) hold in $\mathbf{Z}$ [with $e = 1$ in (c)], property (d) fails in $\mathbf{Z}$ since the reciprocal of zero does not exist and since reciprocals of nonzero integers need not exist in $\mathbf{Z}$.

(b) Let $\cdot$ denote usual multiplication in $\mathbf{R}$. Then $[\mathbf{R}, \cdot]$ is *not* a group; although properties (a) through (c) hold in $\mathbf{R}$ [with $e = 1$ in (c)], property (d) fails in $\mathbf{R}$ since the reciprocal of zero does not exist. However, zero is the only element in $\mathbf{R}$ for which property (d) fails; accordingly, it is left for the reader to verify that $[\mathbf{R} - \{0\}, \cdot]$ *is* a group.

(c) With the operation of multiplication only, $\mathbf{Z}_4$ is *not* a group. (Why?) Let $\mathbf{Z}_4^\times$ denote the equivalence classes [1] and [3] of $\mathbf{Z}_4$ with the operation of multiplication only. Then $\mathbf{Z}_4^\times$ *is* a group. Here $e = [1]$, $[1]' = [1]$, and $[3]' = [3]$.

(d) With the operation of multiplication only, $\mathbf{Z}_5$ is *not* a group. (Why?) Let $\mathbf{Z}_5^\times$ denote the equivalence classes [1], [2], [3], and [4] of $\mathbf{Z}_5$ with the operation of multiplication only. Then $\mathbf{Z}_5^\times$ *is* a group. Here $e = [1]$, $[1]' = [1]$, $[2]' = [3]$, $[3]' = [2]$, and $[4]' = [4]$.

Parts (c) and (d) of Example 2 above motivate the following definition.

Definition 3: Let $m \in \mathbf{Z}$ with $m > 1$. Then the *group of units* of $\mathbf{Z}_m$, denoted $\mathbf{Z}_m^\times$, is the set

$$\{[x] \in \mathbf{Z}_m : x \in \mathbf{Z}, 0 \le x \le m - 1, (x, m) = 1\}$$

under the operation of multiplication (in $\mathbf{Z}_m$) only.

The group of units $\mathbf{Z}_m^\times$ consists precisely of those distinct equivalence classes in $\mathbf{Z}_m$ represented by integers relatively prime to m under the usual multiplication in $\mathbf{Z}_m$. For brevity, we have implicitly assumed the group structure of $\mathbf{Z}_m^\times$ in Definition 3 above. While parts (c) and (d) of Example 2 suggest that $\mathbf{Z}_m^\times$ might be a group for all m, *we have not proven this.* To be complete, Definition 3 should be preceded by a theorem that establishes the group structure of $\mathbf{Z}_m^\times$. We trust that the interested reader will provide such a theorem along with its proof.

Definition 4: Let $[G, *]$ be a finite group. Then $[G, *]$ is said to be *cyclic* if there exists $g \in G$ such that

$$G = \{g, g * g, g * g * g, \ldots\}$$

Such a g is said to be a *generator* of $[G, *]$.

Example 3:

(a) $\mathbf{Z}_4$ with the operation of addition only is a cyclic group with generator [1]. Here $\{g, g * g, g * g * g, \ldots\}$ becomes $\{[1], [1] + [1], [1] + [1] + [1], \ldots\}$, which is equal to $\{[1], [2], [3], [0]\}$ (since repetitions within sets are not counted). Similarly, $\mathbf{Z}_m$ with the operation of addition only is a cyclic group with generator [1] for all m.

(b) $\mathbf{Z}_4^\times$ [see Example 2(c)] is a cyclic group with generator [3]. Here $\{g, g * g, g * g * g, \ldots\}$ becomes $\{[3], [3] \cdot [3], [3] \cdot [3] \cdot [3], \ldots\}$, which is equal to $\{[3], [1]\}$.

(c) $\mathbf{Z}_5^\times$ [see Example 2(d)] is a cyclic group with generator [2]. Here $\{g, g * g, g * g * g, \ldots\}$ becomes $\{[2], [2] \cdot [2], [2] \cdot [2] \cdot [2], \ldots\}$, which is equal to $\{[2], [4], [3], [1]\}$.

(d) $\mathbf{Z}_8^\times$ is *not* a cyclic group since no single class among $\{[1], [3], [5], [7]\}$ will generate all four classes upon repeated multiplication. (Try it!) (Note, however, that $\mathbf{Z}_8^\times$ *is* a group.)

Note that Definition 4 treats the case of *finite* cyclic groups only. The corresponding definition for infinite cyclic groups is a bit different; the interested reader is referred to any introductory abstract algebra book for such a definition.

Exercise Set C

1. *(a)* Prove that $\mathbf{Z}_m$ is a ring for all m.
   (b) Which of the rings $\mathbf{Z}_8$, $\mathbf{Z}_9$, $\mathbf{Z}_{10}$, and $\mathbf{Z}_{11}$ are fields? (See also Exercise 7 of Chapter 2.)
2. Which of the groups $\mathbf{Z}_9^\times$, $\mathbf{Z}_{10}^\times$, $\mathbf{Z}_{11}^\times$, and $\mathbf{Z}_{12}^\times$ are cyclic?

The Binomial Theorem

This appendix presents a statement, proof (using mathematical induction), and example of the Binomial Theorem, which is used in Chapter 5.

D.1 The Binomial Theorem

Definition 1: Let n be a positive integer. The *factorial of n* or *n factorial,* denoted $n!$, is

$$n! = (1)(2)(3) \cdots (n-1)(n)$$

$0! = 1$ by convention.

Definition 2: Let n and k be integers with $0 \le k \le n$. The symbol $\binom{n}{k}$ is given by

$$\binom{n}{k} = \frac{n!}{k!\,(n-k)!}$$

[Those readers who have had experience with a course involving elementary combinatorics (such as probability) will recognize the symbol $\binom{n}{k}$ as being the number of different combinations of n objects taken k at a time.]

Lemma 1: Let n and k be integers with $1 \leq k \leq n$. Then

$$\binom{n}{k} + \binom{n}{k-1} = \binom{n+1}{k}$$

Proof: The straightforward algebraic verification is left to the reader. ■

Theorem 2: (The Binomial Theorem) Let n be a nonnegative integer and let $x, y \in \mathbf{R}$. Then

$$(x+y)^n = \binom{n}{0}x^n + \binom{n}{1}x^{n-1}y + \binom{n}{2}x^{n-2}y^2 + \cdots + \binom{n}{n-1}xy^{n-1} + \binom{n}{n}y^n$$

Proof: We use induction on n. We leave it to you to verify that the desired statement is true if $n = 0$ and $n = 1$. Assume that $k \geq 1$ and that the desired statement is true for $n = k$ so that

$$(x+y)^k = \binom{k}{0}x^k + \binom{k}{1}x^{k-1}y + \binom{k}{2}x^{k-2}y^2 + \cdots + \binom{k}{k-1}xy^{k-1} + \binom{k}{k}y^k$$

We must show that

$$\begin{aligned}(x+y)^{k+1} = \binom{k+1}{0}x^{k+1} + \binom{k+1}{1}x^k y + \binom{k+1}{2}x^{k-1}y^2 \\ + \cdots + \binom{k+1}{k}xy^k + \binom{k+1}{k+1}y^{k+1}\end{aligned}$$

so that the desired result holds for $n = k + 1$. We have

$$\begin{aligned}(x+y)^{k+1} &= (x+y)^k(x+y) \\ &= \left[\binom{k}{0}x^k + \binom{k}{1}x^{k-1}y + \binom{k}{2}x^{k-2}y^2 \right. \\ &\quad \left. + \cdots + \binom{k}{k-1}xy^{k-1} + \binom{k}{k}y^k\right](x+y) \\ &= \binom{k}{0}x^{k+1} + \left[\binom{k}{1} + \binom{k}{0}\right]x^k y + \left[\binom{k}{2} + \binom{k}{1}\right]x^{k-1}y^2 \\ &\quad + \cdots + \left[\binom{k}{k-1} + \binom{k}{k-2}\right]xy^k + \binom{k}{k}y^{k+1} \\ &= \binom{k}{0}x^{k+1} + \binom{k+1}{1}x^k y + \binom{k+1}{2}x^{k-1}y^2 \\ &\quad + \cdots + \binom{k+1}{k}xy^k + \binom{k}{k}y^{k+1} \quad \text{(by Lemma 1)}\end{aligned}$$

The desired result is proven upon noting that $\binom{k}{0} = \binom{k+1}{0}$ and $\binom{k}{k} = \binom{k+1}{k+1}$. ■

Example 1:

Let $x, y \in \mathbf{R}$. Then

$$(x + y)^5 = \binom{5}{0}x^5 + \binom{5}{1}x^4y + \binom{5}{2}x^3y^2 + \binom{5}{3}x^2y^3 + \binom{5}{4}xy^4 + \binom{5}{5}y^5$$

$$= \frac{5!}{0!\,5!}x^5 + \frac{5!}{1!\,4!}x^4y + \frac{5!}{2!\,3!}x^3y^2 + \frac{5!}{3!\,2!}x^2y^3 + \frac{5!}{4!\,1!}xy^4 + \frac{5!}{5!\,0!}y^5$$

$$= x^5 + 5x^4y + 10x^3y^2 + 10x^2y^3 + 5xy^4 + y^5$$

Exercise Set D

1. Prove Lemma 1.
2. Use the Binomial Theorem to expand $(x + y)^{10}$ completely.

Tables

Table 1 Prime Numbers

(See also Table 3.) The grid inside the back cover of this book can be used to determine the primality or compositeness of any given integer between 101 and 9999 inclusive. Obviously, any multiple of 100 is composite. Select any integer between 101 and 9999 inclusive other than a multiple of 100. If the integer is between 101 and 999 inclusive, prefix an initial zero to get a four-digit integer (for example, 347 becomes 0347). Using the vertical axis, locate the row corresponding to the first two digits of the integer (these two digits will be between 01 and 99 inclusive). Similarly, using the horizontal axis, locate the column corresponding to the last two digits of the integer. Now the square at the intersection of this row and column will be black if and only if the integer is prime. Note that the grid may be used to visually address distribution questions concerning the prime numbers (for example, twin primes may be located at a glance!).

Table 2 Arithmetic Functions

This table gives the values of the arithmetic functions $\phi(n)$, $\nu(n)$, $\sigma(n)$, and $\mu(n)$ for integers n with $1 \leq n \leq 100$.

n	$\phi(n)$	$\nu(n)$	$\sigma(n)$	$\mu(n)$	n	$\phi(n)$	$\nu(n)$	$\sigma(n)$	$\mu(n)$
1	1	1	1	1	42	12	8	96	−1
2	1	2	3	−1	43	42	2	44	−1
3	2	2	4	−1	44	20	6	84	0
4	2	3	7	0	45	24	6	78	0
5	4	2	6	−1	46	22	4	72	1
6	2	4	12	1	47	46	2	48	−1
7	6	2	8	−1	48	16	10	124	0
8	4	4	15	0	49	42	3	57	0
9	6	3	13	0	50	20	6	93	0
10	4	4	18	1	51	32	4	72	1
11	10	2	12	−1	52	24	6	98	0
12	4	6	28	0	53	52	2	54	−1
13	12	2	14	−1	54	18	8	120	0
14	6	4	24	1	55	40	4	72	1
15	8	4	24	1	56	24	8	120	0
16	8	5	31	0	57	36	4	80	1
17	16	2	18	−1	58	28	4	90	1
18	6	6	39	0	59	58	2	60	−1
19	18	2	20	−1	60	16	12	168	0
20	8	6	42	0	61	60	2	62	−1
21	12	4	32	1	62	30	4	96	1
22	10	4	36	1	63	36	6	104	0
23	22	2	24	−1	64	32	7	127	0
24	8	8	60	0	65	48	4	84	1
25	20	3	31	0	66	20	8	144	−1
26	12	4	42	1	67	66	2	68	−1
27	18	4	40	0	68	32	6	126	0
28	12	6	56	0	69	44	4	96	1
29	28	2	30	−1	70	24	8	144	−1
30	8	8	72	−1	71	70	2	72	−1
31	30	2	32	−1	72	24	12	195	0
32	16	6	63	0	73	72	2	74	−1
33	20	4	48	1	74	36	4	114	1
34	16	4	54	1	75	40	6	124	0
35	24	4	48	1	76	36	6	140	0
36	12	9	91	0	77	60	4	96	1
37	36	2	38	−1	78	24	8	168	−1
38	18	4	60	1	79	78	2	80	−1
39	24	4	56	1	80	32	10	186	0
40	16	8	90	0	81	54	5	121	0
41	40	2	42	−1	82	40	4	126	1

n	$\phi(n)$	$\nu(n)$	$\sigma(n)$	$\mu(n)$	n	$\phi(n)$	$\nu(n)$	$\sigma(n)$	$\mu(n)$
83	82	2	84	−1	92	44	6	168	0
84	24	12	224	0	93	60	4	128	1
85	64	4	108	1	94	46	4	144	1
86	42	4	132	1	95	72	4	120	1
87	56	4	120	1	96	32	12	252	0
88	40	8	180	0	97	96	2	98	−1
89	88	2	90	−1	98	42	6	171	0
90	24	12	234	0	99	60	6	156	0
91	72	4	112	1	100	40	9	217	0

Table 3 Primitive Roots/Prime Numbers

This table gives the least positive primitive root r modulo n for each integer n with $1 \le n \le 1000$ for which such a primitive root exists. Prime moduli are marked with an asterisk so that this table may function also as a prime number table.

n	r	n	r	n	r	n	r
1	1	*73	5	*179	2	*311	17
*2	1	74	5	*181	2	*313	10
*3	2	*79	3	*191	19	314	5
4	3	81	2	*193	5	*317	2
*5	2	82	7	194	5	326	3
6	5	*83	2	*197	2	*331	3
*7	3	86	3	*199	3	334	5
9	2	*89	3	202	3	*337	10
10	3	94	5	206	5	338	7
*11	2	*97	5	*211	2	343	3
*13	2	98	3	214	5	346	3
14	3	*101	2	218	11	*347	2
*17	3	*103	5	*223	3	*349	2
18	5	106	3	226	3	*353	3
*19	2	*107	2	*227	2	358	7
22	7	*109	6	*229	6	*359	7
*23	5	*113	3	*233	3	361	2
25	2	118	11	*239	7	362	21
26	7	121	2	*241	7	*367	6
27	2	122	7	242	7	*373	2
*29	2	125	2	243	2	*379	2
*31	3	*127	3	250	3	382	19
34	3	*131	2	*251	6	*383	5
*37	2	134	7	254	3	386	5
38	3	*137	3	*257	3	*389	2
*41	6	*139	2	262	17	394	3
*43	3	142	7	*263	5	*397	5
46	5	146	5	*269	2	398	3
*47	5	*149	2	*271	6	*401	3
49	3	*151	6	274	3	*409	21
50	3	*157	5	*277	5	*419	2
*53	2	158	3	278	3	*421	2
54	5	162	5	*281	3	422	3
58	3	*163	2	*283	3	*431	7
*59	2	166	5	289	3	*433	5
*61	2	*167	5	*293	2	*439	15
62	3	169	2	298	3	*443	2
*67	2	*173	2	302	7	446	3
*71	7	178	3	*307	5	*449	3

n	*r*	*n*	*r*	*n*	*r*	*n*	*r*
454	5	578	3	718	7	*857	3
*457	13	586	3	*719	11	*859	2
458	7	*587	2	722	3	862	7
*461	2	*593	3	*727	5	*863	5
*463	3	*599	7	729	2	*877	2
466	3	*601	7	*733	6	878	15
*467	2	*607	3	734	11	*881	3
478	7	*613	2	*739	3	*883	2
*479	13	614	5	*743	5	886	5
482	7	*617	3	746	5	*887	5
486	5	*619	2	*751	3	898	3
*487	3	622	17	*757	2	*907	2
*491	2	625	2	758	3	*911	17
*499	7	626	15	*761	6	914	13
502	11	*631	3	766	5	*919	7
*503	5	634	3	*769	11	922	3
*509	2	*641	3	*773	2	926	3
514	3	*643	11	778	3	*929	3
*521	3	*647	5	*787	2	934	5
*523	2	*653	2	794	5	*937	5
526	5	*659	2	*797	2	*941	2
529	5	*661	2	802	3	*947	2
538	3	662	3	*809	3	*953	3
*541	2	*673	5	*811	3	958	13
542	15	674	15	818	21	961	3
*547	2	*677	2	*821	2	*967	5
554	5	*683	5	*823	3	*971	6
*557	2	686	3	*827	2	974	3
562	3	*691	3	*829	2	*977	3
*563	2	694	5	838	11	982	7
566	3	698	7	*839	11	*983	5
*569	3	*701	2	841	2	*991	6
*571	3	706	3	842	23	*997	7
*577	5	*709	2	*853	2	998	7

Table 4 Continued Fractions

This table gives the eventually periodic infinite simple continued fraction expansions of $\sqrt{d}$ for nonsquare integers d with $2 \le d \le 99$.

d	$\sqrt{d}$	d	$\sqrt{d}$
2	$[1,\overline{2}]$	53	$[7,\overline{3,1,1,3,14}]$
3	$[1,\overline{1,2}]$	54	$[7,\overline{2,1,6,1,2,14}]$
5	$[2,\overline{4}]$	55	$[7,\overline{2,2,2,14}]$
6	$[2,\overline{2,4}]$	56	$[7,\overline{2,14}]$
7	$[2,\overline{1,1,1,4}]$	57	$[7,\overline{1,1,4,1,1,14}]$
8	$[2,\overline{1,4}]$	58	$[7,\overline{1,1,1,1,1,1,14}]$
10	$[3,\overline{6}]$	59	$[7,\overline{1,2,7,2,1,14}]$
11	$[3,\overline{3,6}]$	60	$[7,\overline{1,2,1,14}]$
12	$[3,\overline{2,6}]$	61	$[7,\overline{1,4,3,1,2,2,1,3,4,1,14}]$
13	$[3,\overline{1,1,1,1,6}]$	62	$[7,\overline{1,6,1,14}]$
14	$[3,\overline{1,2,1,6}]$	63	$[7,\overline{1,14}]$
15	$[3,\overline{1,6}]$	65	$[8,\overline{16}]$
17	$[4,\overline{8}]$	66	$[8,\overline{8,16}]$
18	$[4,\overline{4,8}]$	67	$[8,\overline{5,2,1,1,7,1,1,2,5,16}]$
19	$[4,\overline{2,1,3,1,2,8}]$	68	$[8,\overline{4,16}]$
20	$[4,\overline{2,8}]$	69	$[8,\overline{3,3,1,4,1,3,3,16}]$
21	$[4,\overline{1,1,2,1,1,8}]$	70	$[8,\overline{2,1,2,1,2,16}]$
22	$[4,\overline{1,2,4,2,1,8}]$	71	$[8,\overline{2,2,1,7,1,2,2,16}]$
23	$[4,\overline{1,3,1,8}]$	72	$[8,\overline{2,16}]$
24	$[4,\overline{1,8}]$	73	$[8,\overline{1,1,5,5,1,1,16}]$
26	$[5,\overline{10}]$	74	$[8,\overline{1,1,1,1,16}]$
27	$[5,\overline{5,10}]$	75	$[8,\overline{1,1,1,16}]$
28	$[5,\overline{3,2,3,10}]$	76	$[8,\overline{1,2,1,1,5,4,5,1,1,2,1,16}]$
29	$[5,\overline{2,1,1,2,10}]$	77	$[8,\overline{1,3,2,3,1,16}]$
30	$[5,\overline{2,10}]$	78	$[8,\overline{1,4,1,16}]$
31	$[5,\overline{1,1,3,5,3,1,1,10}]$	79	$[8,\overline{1,7,1,16}]$
32	$[5,\overline{1,1,1,10}]$	80	$[8,\overline{1,16}]$
33	$[5,\overline{1,2,1,10}]$	82	$[9,\overline{18}]$
34	$[5,\overline{1,4,1,10}]$	83	$[9,\overline{9,18}]$
35	$[5,\overline{1,10}]$	84	$[9,\overline{6,18}]$
37	$[6,\overline{12}]$	85	$[9,\overline{4,1,1,4,18}]$
38	$[6,\overline{6,12}]$	86	$[9,\overline{3,1,1,1,8,1,1,1,3,18}]$
39	$[6,\overline{4,12}]$	87	$[9,\overline{3,18}]$
40	$[6,\overline{3,12}]$	88	$[9,\overline{2,1,1,1,2,18}]$
41	$[6,\overline{2,2,12}]$	89	$[9,\overline{2,3,3,2,18}]$
42	$[6,\overline{2,12}]$	90	$[9,\overline{2,18}]$
43	$[6,\overline{1,1,3,1,5,1,3,1,1,12}]$	91	$[9,\overline{1,1,5,1,5,1,1,18}]$
44	$[6,\overline{1,1,1,2,1,1,1,12}]$	92	$[9,\overline{1,1,2,4,2,1,1,18}]$
45	$[6,\overline{1,2,2,2,1,12}]$	93	$[9,\overline{1,1,1,4,6,4,1,1,1,18}]$
46	$[6,\overline{1,3,1,1,2,6,2,1,1,3,1,12}]$	94	$[9,\overline{1,2,3,1,1,5,1,8,1,5,1,1,3,2,1,18}]$
47	$[6,\overline{1,5,1,12}]$	95	$[9,\overline{1,2,1,18}]$
48	$[6,\overline{1,12}]$	96	$[9,\overline{1,3,1,18}]$
50	$[7,\overline{14}]$	97	$[9,\overline{1,5,1,1,1,1,1,1,5,1,18}]$
51	$[7,\overline{7,14}]$	98	$[9,\overline{1,8,1,18}]$
52	$[7,\overline{4,1,2,1,4,14}]$	99	$[9,\overline{1,18}]$

Hints and Answers to Selected Exercises

Exercise Set 1.1

1. ***(a)*** Since $42 = 6 \cdot 7$ and $7 \in \mathbf{Z}$, the statement is proven.
(b) Since $50 = 4(\frac{50}{4})$ and $\frac{50}{4} \notin \mathbf{Z}$, the statement is disproven.
3. ***(a)*** $q = 7, r = 5$ ***(c)*** $q = 7, r = 0$ ***(e)*** $q = -16, r = 27$
4. $a = \pm b$
8. Use the binomial theorem to expand $(n + 1)^n$.
10. ***(a)*** Use induction.
12. Use Exercise 11(b).
14. ***(b)*** 166
15. ***(b)*** $q = -7, r = 5$ ***(d)*** $q = -8, r = -1$

Exercise Set 1.2

16. ***(a)*** prime ***(b)*** composite
18. ***(a)*** $14! + 2, 14! + 3, \ldots, 14! + 14$ (for example)
(b) $114, 115, \ldots, 126$
20. ***(a)*** $\pi(10) = 4$, $\pi(100) = 25$, $\pi(200) = 46$
21. ***(a)*** $30 = 7 + 23$ (for example) ***(c)*** $114 = 5 + 109$ (for example)
23. Disprove the conjecture by proving that the only prime number p for which $p + 2$ and $p + 4$ are also prime numbers is $p = 3$.
26. ***(b)*** Use the fact that if a and b are positive integers, then

$$2^{ab} - 1 = (2^a - 1)(2^{a(b-1)} + 2^{a(b-2)} + \cdots + 2^a + 1)$$

30. Disprove the conjecture by using factorization to prove that the only prime number expressible in the desired form is 2.

Exercise Set 1.3

32. *(a)* 7 *(c)* 1 *(e)* 3
33. *(a)* a *(c)* 1 if a is odd, 2 if a is even
34. *(a)* 9 *(c)* 51 *(e)* 6
35. For example, 30, 42, 70, and 105. (A method for constructing appropriate integers in problems like this will be motivated later.)
36. *(a)* No
(b) If n is an integer such that $29 \nmid n$, consider $x = 3n$ and $y = 3(29 - n)$.
38. Let $(a + b, a - b) = d$ and note that $d \mid (a + b) + (a - b)$ and $d \mid (a + b) - (a - b)$.
39. 1, 3
40. $(a + b, 4) = 4$
43. *(a)* By Proposition 1.11, there exist $m, n \in \mathbf{Z}$ such that $ma + nb = 1$. Then $mac + nbc = c$; rewrite the left-hand side of this equation so that ab can be factored out of each term.
46. All integers a, b, c, and d subject to the given hypotheses and such that $a = -c$ and $b = d$.
49. Try to figure out the significance of the fact that there are *eight* composite positive integers not exceeding *360*. (Would the given statement be true if "eight" were replaced by "seven" or if 360 were replaced by 361?)

Exercise Set 1.4

54. *(a)* $1; 1 = 13 \cdot 37 - 8 \cdot 60$ *(c)* $21; 21 = 21 \cdot 441 - 8 \cdot 1155$
(e) $29; 29 = 4 \cdot 2581 - 5 \cdot 2059$
55. $(18209, 19043) = 139$; use Proposition 1.11.
57. *(a)* $1; 1 = 1192 \cdot 247 + 6 \cdot 323 - 1341 \cdot 221$
58. *(b)* $(204, 228) = 12$

Exercise Set 1.5

59. *(a)* $3 \cdot 17$ *(c)* 19^2 *(e)* $2 \cdot 211$ *(g)* $7 \cdot 11 \cdot 13$
60. *(a)* 18 *(c)* 91 *(e)* 2100
61. *(a)* gcd: $2^2 3^2 5 \cdot 7$: lcm: $2^2 3^3 5 \cdot 7^2$
(c) gcd: 1; 1cm: $2^2 3^2 5^7 7^5 11^{13} 13^{11}$
62. For example, 210, 330, 462, 770, and 1155. (If you had difficulty constructing appropriate integers in this problem, consideration of the prime factorizations of these five integers should motivate a procedure for solving problems such as this. See also Exercise 35.)
63. *(a)* 4199 *(c)* 1252
64. 12 and 360, 24 and 180, 36 and 120, 60 and 72
65. *(a)* 60 *(c)* 504
66. Reread Theorem 1.16. Would this theorem still be true if 1 were a prime number?
70. *(b)* If $b^2 = a^2 c$ with $c \in \mathbf{Z}$, prove that c is a perfect square.
72. *(a)* $\min\{a, b\}$ if $a \neq b$; if $a = b$, no general solution is possible.
73. $n!$

74. $a = 2$ and $b = 4$, $a = 4$ and $b = 2$, and all positive integers for which $a = b$

75. *(a)* 83 *(c)* 167

76. *(a)* Referring to Exercise 14, what does $[n/p]$ count? What does $[n/p^2]$ count? What does $[n/p^3]$ count? Continue.

(c) 24

79. *(b)* $a = b$

83. If the term $a + nb = p$ is a prime number, consider the term $a + (n + p)b$.

84. The case for $k = 1$ follows immediately from Dirichlet's Theorem upon taking $a = 1$ and $b = 10$. (Why?) The case for $k = 2$ follows similarly by taking $a = 11$ and $b = 100$. (Why?) Now generalize.

Exercise Set 2.1

1. *(a)* Since $2 \mid 7 - 5$, the statement is proven.

(b) Since $3 \nmid 8 - 12$, the statement is disproven.

2. *(a)* $1, 2, 4, 8$ *(c)* $1, 13$

3. *(a)* Modulo 7, the integers -39, 72, -23, 50, -15, 63, and -52 are congruent, respectively, to the integers 3, 2, 5, 1, 6, 0, and 4, from which the desired result is proven.

4. *(a)* 8 *(c)* 3 *(e)* 4

5. *(a)* 6 *(c)* 12

6. *(a)* 1 *(c)* 1

7. *(a)* For any nonzero class $[a]$ of $\mathbf{Z}_5$, there exists a nonzero class $[b]$ of $\mathbf{Z}_5$ such that $[a][b] = [1]$. This property does *not* hold for the nonzero classes of $\mathbf{Z}_4$. (Consider the nonzero class $[2]$ of $\mathbf{Z}_4$. There is no nonzero class $[b]$ of $\mathbf{Z}_4$ such that $[2][b] = [1]$.) This property, called the *inverse property of multiplication*, is required for a structure to be a field but is not required for a structure to be a ring.

(d) a prime number

13. *(c)* Assume that n can be written as the sum of two squares of integers and derive a contradiction by viewing the equation modulo 4 and using parts (a) and (b).

(d) Consider $n = 6$.

14. It suffices to prove that $n^2 \equiv 1 \bmod 8$ and $n^2 \equiv 1 \bmod 3$. (Why?)

21. Use Exercise 18.

22. *(d)* The five known prime repunits are R_2, R_{19}, R_{23}, R_{317}, and R_{1031}.

23. The formula in Example 1 of Appendix A is useful here.

24. Use the formula in Exercise 1(a) of Appendix A.

Exercise Set 2.2

28. *(a)* $9, 19, 29$ *(c)* no solutions

(e) 3, 100, 197, 294, 391, 488, 585

29. *(a)* 21 *(c)* 79 *(d)* does not exist

30. The least nonnegative solutions here are given as ordered pairs (x, y).
(a) $(2,0)$, $(4,1)$, $(6,2)$, $(1,3)$, $(3,4)$, $(5,5)$, $(0,6)$
(c) no solutions

Exercise Set 2.3

33. *(a)* 7 *(c)* 11 *(e)* 209
34. *(a)* $x \equiv 89 \bmod 105$ *(c)* $x \equiv 2, 86, 170, 254, 338, 422 \bmod 504$
35. 119
37. 79
39. *(a)* $x \equiv 7 \bmod 12$ *(c)* no solutions
41. *(b)* $x \equiv 27 \bmod 30$

Exercise Set 2.4

42. *(a)* 30 *(c)* 1 *(e)* 1
43. *(a)* Prove that $2(p-3)! \equiv (p-1)! \bmod p$ and use Wilson's Theorem (Theorem 2.11).
47. *(a)* Note that $(\frac{p-1}{2}!)^2 = 1 \cdot 2 \cdot 3 \cdot \cdots \cdot (\frac{p-1}{2})(\frac{p-1}{2}) \cdot \cdots \cdot 3 \cdot 2 \cdot 1$. Now rewrite the last $\frac{p-1}{2}$ terms of the right-hand side so that, modulo p, the right-hand side involves $(p-1)!$.
48. Prove that $2 \cdot 4 \cdot 6 \cdot \cdots \cdot (p-1) \equiv (-1)^{(p-1)/2} 1 \cdot 3 \cdot 5 \cdot \cdots \cdot (p-2) \bmod p$ and use Wilson's Theorem (Theorem 2.11).
49. Use Exercise 47(c).

Exercise Set 2.5

51. *(a)* 3 *(c)* 16
52. Prove that $456^{654} \equiv -123^{321} \bmod 11$.
53. *(a)* 10
56. *(a)* all composite integers
57. *(a)* It suffices to prove that $2 \mid n^{21} - n$, $3 \mid n^{21} - n$, and $5 \mid n^{21} - n$. (Why?)
59. It suffices to prove that $p^{q-1} + q^{p-1} \equiv 1 \bmod p$ and $p^{q-1} + q^{p-1} \equiv 1 \bmod q$. (Why?)
61. Where have you recently seen $(p-1)!$?

Exercise Set 2.6

68. *(a)* 1 *(c)* 11
69. *(a)* $\{1, 2, 4, 7, 8, 11, 13, 14\}$ *(c)* $\{1, 2, 3, \ldots, p-1\}$
70. *(a)* 19
71. *(a)* & *(b)* It suffices to prove that $7 \mid n^7 - n$ and $9 \mid n^7 - n$. (Why?)
72. *(b)* 32760
75. The case $m = 1$ is easy. For $m > 2$, prove and use the following lemma: ***Lemma:*** Let m be a positive integer with $m > 2$. If a is a positive integer less than m with $(a, m) = 1$, then $(m - a, m) = 1$.
76. Use the fact that $a^{\phi(m)} - 1 = (a-1)(a^{\phi(m)-1} + a^{\phi(m)-2} + \cdots + a^2 + a + 1)$.

Exercise Set 3.1

1. It should appear that the arithmetic functions ϕ, ν, and σ are multiplicative. Consider, for example, the relatively prime integers $m = 2$ and $n = 9$. We have $\phi(2 \cdot 9) = \phi(18) = 6$ and $\phi(2)\phi(9) = 1 \cdot 6 = 6$, from which

$$\phi(2 \cdot 9) = \phi(2)\phi(9)$$

Similarly,

$$\nu(2 \cdot 9) = \nu(2)\nu(9) = 6$$

and

$$\sigma(2 \cdot 9) = \sigma(2)\sigma(9) = 39$$

2. The arithmetic functions ϕ, ν, and σ are *not* completely multiplicative. Consider, for example, the integers $m = 2$ and $n = 4$. We have $\phi(2 \cdot 4) = \phi(8) = 4$ and $\phi(2)\phi(4) = 1 \cdot 2 = 2$, from which

$$\phi(2 \cdot 4) \neq \phi(2)\phi(4)$$

Similarly,

$$4 = \nu(2 \cdot 4) \neq \nu(2)\nu(4) = 6$$

and

$$15 = \sigma(2 \cdot 4) \neq \sigma(2)\sigma(4) = 21$$

3. If $f(n) \neq 0$, consider $f(n \cdot 1)$.
4. *(b)* $f(n) = (1 + 2a_1)(1 + 2a_2) \cdots (1 + 2a_m)$
5. *(a)* completely multiplicative and thus multiplicative
   (c) not multiplicative [for example, $2 = f(2 \cdot 3) \neq f(2)f(3) = 4$] and thus not completely multiplicative
   (e) completely multiplicative if and only if $k = 0$ or $k = 1$; multiplicative if and only if $k = 0$ or $k = 1$
   (g) not multiplicative [for example, $46656 = f(2 \cdot 3) \neq f(2)f(3) = 108$] and thus not completely multiplicative

Exercise Set 3.2

10. *(a)* $p - 1$ *(c)* 48 *(e)* 2520 *(g)* $2^{17}3^{7}5^{3}7$
12. Rewrite each $1 - \frac{1}{p}$ in Theorem 3.4 as $\frac{p-1}{p}$.
13. *Note*: When characterizing positive integers n for which a certain property of $\phi(n)$ holds (or when simply proving the existence of such a property), it is sometimes helpful to use the result of Exercise 12.
    (a) $n = 1$ or $n = 2$
    (c) At least one of the following must be true:
       (i) $8 \mid n$

(ii) $4 \mid n$ and $p \mid n$ where p is an odd prime number
(iii) $p \mid n$ and $q \mid n$ where p and q are distinct odd prime numbers
(iv) $p \mid n$ where p is a prime number with $p \equiv 1 \bmod 4$
(e) $n = 2^j 3^k$ (j and k positive integers) where $j > 0$ if $k > 0$

14. Use Exercise 12.

16. *(a)* Use Exercise 12 and Theorem 3.4 for the two inequalities (respectively).
(b) Let p be the least prime divisor of n. Then use Proposition 1.7 along with the fact that $\phi(n) \leq n(1 - \frac{1}{p})$.

23. Use the lemma in the hint for Exercise 75 of Chapter 2.

24. Assume that p is a prime number with $p^2 \mid n$ and derive a contradiction.

26. Write n as $2^k m$ with m odd (Exercise 68 of Chapter 1). If $k > 0$, prove that

$$\sum_{d \mid n, d>0} (-1)^{n/d} \phi(d) = \sum_{d \mid 2^{k-1}m, d>0} \phi(d) - \sum_{d \mid m, d>0} \phi(2^k d)$$

Exercise Set 3.3

30. *(a)* 2 *(c)* 8 *(e)* 18 *(g)* 4032

31. *Note*: In what follows, p and q denote distinct prime numbers.
(a) $n = 1$ *(c)* $n = p^2$ *(d)* $n = p^3$ or $n = pq$

32. perfect squares

33. Use Exercise 32 above.

36. $n^{\nu(n)/2}$

37. Partition the divisors of n into two sets, those less than or equal to $\sqrt{n}$ and those greater than $\sqrt{n}$.

Exercise Set 3.4

42. *(a)* $p + 1$ *(c)* 192 *(e)* 8892 *(g)* $2^5 3^5 5 \cdot 7^2 13^2 19 \cdot 1093$

44. All odd prime numbers dividing n must occur in the prime factorization of n with even exponent.

46. The first inequality is easy. For the second inequality, note that $\sigma(n) \leq 1 + 2 + \cdots + n$ (why?) and prove that $1 + 2 + \cdots + n \leq n^2$.

48. Use Exercise 47.

51. *(a)* $\sigma_3(12) = 2044$

Exercise Set 3.5

52. 8589869056 ($p = 17$ in the characterization of Theorem 3.12) and 137438691328 ($p = 19$ in the characterization of Theorem 3.12)

54. *(b)* By Theorem 3.12, any even perfect number takes the form $2^{p-1}(2^p - 1)$ where $2^p - 1$ is a prime number (and so p is a prime number). Consider the cases $p = 2$, $p \equiv 1 \bmod 4$, and $p \equiv 3 \bmod 4$ separately.

58. *(b)* $2^3 3 \cdot 5$ and $2^5 3 \cdot 7$

60. $n = 1$ or $\nu(n) = 4$

Exercise Set 3.6

63. $f(12) - f(6) - f(4) + f(2)$

65. First prove that $\sum_{d|n,d>0} \Lambda(d) = \ln n$. Now apply the Möbius Inversion Formula.

66. The function $F(n) = \sum_{d|n,d>0} \mu(d)f(d)$ is multiplicative by Theorem 3.1. So it suffices to prove the desired result for powers of prime numbers.

67. *Note*: In what follows, we assume that the prime factorization of n is $p_1^{a_1}p_2^{a_2}\cdots p_m^{a_m}$.
(a) $\prod_{i=1}^{m}(2 - p_i)$
(c) $(-1)^m p_1 p_2 \cdots p_m$

68. Use the Möbius Inversion Formula and Exercise 8.

Exercise Set 4.1

1. *(a)* 1, 3, 4, 9, 10, and 12 are quadratic residues: 2, 5, 6, 7, 8, and 11 are quadratic nonresidues.
(c) 1, 4, 5, 6, 7, 9, 11, 16, and 17 are quadratic residues; 2, 3, 8, 10, 12, 13, 14, 15, and 18 are quadratic nonresidues.

2. 2088

4. 1, 3, and 9

5. *(a)* $x \equiv 10, 32, 45, 67 \bmod 77$
(b) no solutions

7. *(a)* Use the formula in Exercise 1(a) of Appendix A.

10. *(a)* $x \equiv 6 \bmod 13$ *(c)* no solutions

11. *(e)* If $(a, m) = 1$, then the congruence $x^2 \equiv a \bmod m$ is solvable if and only if
(i) $\left(\frac{a}{p_i}\right) = 1, i = 1, 2, \ldots, n$
(ii) $a \equiv 1 \bmod 4$ if $k = 2$
(iii) $a \equiv 1 \bmod 8$ if $k \geq 3$
In the case of solvability, the number of incongruent solutions modulo m is 2^n if $k \leq 1$, 2^{n+1} if $k = 2$, and 2^{n+2} if $k \geq 3$.

Exercise Set 4.2

12. *(a)* -1

13. *(a)* 1

14. *(a)* -1 *(c)* -1 *(e)* -1

17. *(b)* Consider $b = 2$, $p = 3$, and $q = 5$.

16. *(a)* Rewrite the given question in congruence notation.
(c) There exists a positive integer n such that $n^2 + 1$ is evenly divisible by the odd prime number p if and only if $p \equiv 1 \bmod 4$.

19. If a and b are both quadratic nonresidues modulo p, then the given congruence is solvable; if one of a and b is a quadratic residue modulo p and the other is a quadratic nonresidue modulo p, then the given congruence is not solvable.

22. Use Exercise 15.

24. *(a)* & *(b)* See Exercise 18.

25. Use Exercise 49 of Chapter 2.

Exercise Set 4.3

28. *(a)* 1 *(c)* −1 *(e)* 1

29. Relate the symbol $(\frac{p}{2})$ to the solvability of the congruence $x^2 \equiv p \bmod 2$.

30. Use the law of quadratic reciprocity (Theorem 4.9).

34. Prove that $(\frac{q}{p}) = (\frac{-q}{p})$.

35. *(a)* Note that $(\frac{-2}{p}) = 1$ if and only if $(\frac{-1}{p})(\frac{2}{p}) = 1$ if and only if $(\frac{-1}{p}) = (\frac{2}{p}) = 1$ *or* $(\frac{-1}{p}) = (\frac{2}{p}) = -1$.

(b) Assume that $p \equiv 1 \bmod 4$. Then $(\frac{3}{p}) = (\frac{p}{3}) = 1$ if and only if $p \equiv 1 \bmod 3$. Now assume that $p \equiv 3 \bmod 4$. Then $(\frac{3}{p}) = -(\frac{p}{3}) = 1$ if and only if $p \equiv 2 \bmod 3$.

36 *(a)* $p \equiv \pm 1 \bmod 5$ *(c)* $p \equiv \pm 1, \pm 3, \pm 9 \bmod 28$

37. *(a)* $(\frac{-79}{105}) = 1$

Exercise Set 5.1

1. *(a)* 10 *(c)* 4

2. *(a)–(d)* Parallel Example 5.

3. *(a)* 3 and 5 are the two incongruent primitive roots

(b) 3, 5, 6, 7, 10, 11, 12, and 14 are the eight incongruent primitive roots

5. *(b)* If $(\text{ord}_m a, \text{ord}_m b) \neq 1$, then $\text{ord}_m(ab) = [\text{ord}_m a, \text{ord}_m b]$.

6. Consider $m = 8$ and $d = 4$.

Exercise Set 5.2

10. *(a)* 16 *(c)* 22

11. *(a)* 6, 26 *(c)* 2, 8, 22, 27, 32, 39

15. In conjunction with Exercise 23 of Chapter 3, prove and use the following lemma:

Lemma: Let p be a prime number. If r is a primitive root modulo p and k is an integer with $(k, p-1) = 1$, then r^k is a primitive root modulo p.

16. *(a)* Use Euler's Criterion (Theorem 4.4) and Exercise 12(a).

18. *(a)* If a is a quadratic nonresidue modulo q, then $-1 = (\frac{a}{q}) \equiv a^{2p} \bmod q$. What are the possibilities for $\text{ord}_q a$?

(c) For example, 13, 29, 53, 173, and 269.

22. *(a)* Use Lagrange's Theorem (Theorem 5.7).

Exercise Set 5.3

23. *(a)* no primitive roots exist *(c)* 8 *(e)* no primitive roots exist

24. *(a)* 3 (for example) *(c)* 2 (for example)

25. *(a)* 3 (for example) *(c)* 15 (for example)

26. *(b)* Let r be a primitive root modulo p^m and let $n = \text{ord}_p r$. Now use $r^n \equiv 1 \bmod p$ to show that $r^{p^{m-1}n} \equiv 1 \bmod p^m$.

28. *(a)* Let r be a primitive root modulo m. Then $(m-1)! \equiv r^{1+2+3+\cdots+\phi(m)} \bmod m$.

Exercise Set 5.4

29. *(a)* $\text{ind}_2 1 = 12$, $\text{ind}_2 2 = 1$, $\text{ind}_2 3 = 4$, $\text{ind}_2 4 = 2$, $\text{ind}_2 5 = 9$, $\text{ind}_2 6 = 5$, $\text{ind}_2 7 = 11$, $\text{ind}_2 8 = 3$, $\text{ind}_2 9 = 8$, $\text{ind}_2 10 = 10$, $\text{ind}_2 11 = 7$, $\text{ind}_2 12 = 6$

30. *(a)* $x \equiv 12 \bmod 13$ *(c)* no solutions

(d) $x \equiv 10, 16, 57, 59, 90, 99, 115, 134, 144, 145, 149, 152 \bmod 156$

(f) $x \equiv 2, 8, 9, 15 \bmod 17$

(h) $x \equiv 0, 1, 9, 13, 17, 18, 21, 25, 33, 34, 35, 49, 51, 52, 65, 67, 68, 69, 81, 85, 86, 89, 97, 101, 102, 103, 113, 119, 120, 121, 129, 135, 136, 137, 145, 149, 153, 154, 157, 161, 169, 170, 171, 177, 185, 187, 188, 193, 203, 204, 205, 209, 217, 221, 222, 225, 237, 238, 239, 241, 255, 256, 257, 271 \bmod 272$

31. *(a)* $a \equiv 4, 6, 7, 9 \bmod 13$

32. *(a)* $b \equiv 4, 6, 7, 9 \bmod 13$ *(d)* $b \equiv 0 \bmod 17$

33. *(a)* not solvable *(d)* solvable; six incongruent solutions

36. *(a)* Use Theorem 5.21.

39. Use Theorem 5.21.

40. Convert the given congruence into an equivalent linear congruence using indices.

Exercise Set 6.1

1. 84

2. *(a)* $x = -1 - 14n$, $y = 1 + 9n$ $(n \in \mathbf{Z})$ *(b)* no solutions

3. 11 apples and 8 oranges

5. *(a)* Fourteen 12¢ stamps and one 15¢ stamp, nine 12¢ stamps and five 15¢ stamps, or four 12¢ stamps and nine 15¢ stamps.

(c) The package cannot be mailed with exact postage using only 12¢ and 15¢ stamps.

6. *(a)* $x_1 = -3 - 4n_1$, $x_2 = 2 + 3n_1 - 4n_2$, $x_3 = 1 + 3n_2$ $(n_1, n_2 \in \mathbf{Z})$

7. 9 apples, 10 oranges, and 12 plums (this solution is not unique)

Exercise Set 6.2

11. *(a)* View the equation modulo 7.

(b) Consider $x = 2$ and $y = 1$.

12. *(a)* View the equation modulo 4.

Exercise Set 6.3

13. *(a)* The primitive Pythagorean triples x, y, z with y even and $z \leq 50$ are:

$$3,4,5;\ 5,12,13;\ 15,8,17;\ 7,24,25;\ 21,20,29;\ 35,12,37;\ 9,40,41$$

Clearly, the roles of x and y can be interchanged.

14. *(a)* Assume, by way of contradiction, that $3 \nmid x$ and $3 \nmid y$. Now view the equation $x^2 + y^2 = z^2$ modulo 3.

16. *(a)* $x = (m^2 - 2n)^2/2$, $y = mn$, $z = (m^2 + 2n^2)/2$, where m and n are positive integers with m even and $m > \sqrt{2}\,n$; $x = (2m^2 - n^2)/2$,

$y = mn$, $z = (2m^2 + n^2)/2$, where m and n are positive integers with n even and $m > n/\sqrt{2}$.

(c) $x = m^2 - n^2$, $y = mn$, $z = m^2 + n^2$, where m and n are positive integers with $m > n$.

19. $x = 2mn(m^2 + n^2)$, $y = m^4 - n^4$, $z = 2mn(m^2 - n^2)$, where m and n are positive integers with $(m, n) = 1$, one of m and n even, and $m > n$.

Exercise Set 6.4

22. Assume that there exists a Pythagorean triple containing at least two perfect squares and derive a contradiction using Theorem 6.4 and Exercise 21.

24. Rewrite the given equation in a form to which Theorem 6.3 would apply.

Exercise Set 6.5

28. *(a)* no *(c)* yes *(e)* no

29. *(a)* $7^2 + 7^2$ *(c)* $9^2 + 21^2$ *(e)* $12^2 + 31^2$

30. *(a)* yes *(c)* no *(e)* no

31. 103, 111, 112, 119, 124, 127, 135, 143, 151, 156, 159, 167, 175, 183, 188, 191, 199

32. *(a)* $3^2 + 6^2 + 9^2 + 9^2$ *(c)* $9^2 + 12^2 + 18^2 + 33^2$

33. *(a)* 9 *(c)* 1

34. For the "if" direction, explicitly express $4n$, $4n + 1$, and $4n + 3$ as the difference of two squares of integers [as in $4n = (n + 1)^2 - (n - 1)^2$, for example]. For the "only if" direction, prove that the difference of two squares of integers cannot be congruent to 2 modulo 4.

38. Note that if $n = a^2 + b^2 + c^2 + d^2$, then

$$8(n + 1) = (2a + 1)^2 + (2a - 1)^2 + (2b + 1)^2 + (2b - 1)^2 + (2c + 1)^2 + (2c - 1)^2 + (2d + 1)^2 + (2d - 1)^2.$$

Exercise Set 7.1

1. *(a)* $0.\overline{857142}$ *(c)* $0.\overline{45}$ *(e)* $0.\overline{5294117647058823}$

2. *(a)* $\frac{6}{5}$ *(c)* $\frac{37}{30}$ *(e)* $\frac{1111}{900}$

3. *(a)* $x^5 - 3$ (for example) *(c)* $x^2 - 2x - 5$ (for example)

7. Reread the first half of the proof of Proposition 7.4 under the additional assumptions that $(a, b) = 1$ and $(b, 10) = 1$. Show that the sequence of remainders satisfies

$$r_{i+j} \equiv 10^j r_i \bmod b$$

and that each remainder is relatively prime to b. Conclude that if the decimal representation of $\frac{a}{b}$ has period length p, then $\text{ord}_b 10 = p$. Now apply these results to $\frac{1}{m}$ as given.

Exercise Set 7.2

9. *(a)* $[1, 2, 3, 4]$ *(c)* $[1, 1, 1, 1, 1, 1, 1, 2]$ *(e)* $[3, 5, 2, 4]$

10. *(a)* $[-2, 1, 1, 3, 4]$ *(c)* $[-2, 2, 1, 1, 1, 1, 2]$ *(e)* $[-4, 1, 4, 2, 4]$

11. *(a)* $\frac{43}{10}$ *(c)* $\frac{13}{8}$ *(e)* $\frac{204}{457}$

12. *(a)* $\frac{-37}{10}$ *(c)* $\frac{-3}{8}$ *(e)* $\frac{204}{457}$

15. $f_{n+1}/f_n = [\underbrace{1, 1, \ldots, 1}_{n}]$. (If $n > 2$, then f_{n+1}/f_n may be written $[\underbrace{1, 1, \ldots, 1}_{n-2}, 2]$.)

Exercise Set 7.3

17. *(a)* $1, \frac{3}{2}, \frac{10}{7}, \frac{43}{30}$;
$1 < \frac{10}{7} < \frac{43}{30} < \frac{3}{2}$

(c) $1, 2, \frac{3}{2}, \frac{5}{3}, \frac{8}{5}, \frac{13}{8}, \frac{21}{13}, \frac{34}{21}, \frac{55}{34}$;
$1 < \frac{3}{2} < \frac{8}{5} < \frac{21}{13} < \frac{55}{34} < \frac{34}{21} < \frac{13}{8} < \frac{5}{3} < 2$

(e) $3, \frac{16}{5}, \frac{35}{11}, \frac{156}{49}$;
$3 < \frac{35}{11} < \frac{156}{49} < \frac{16}{5}$

18. For the first equality, use $p_i = a_i p_{i-1} + p_{i-2}$ from Proposition 7.6 to prove that

$$\frac{p_i}{p_{i-1}} = a_i + \frac{1}{\left(\dfrac{p_{i-1}}{p_{i-2}}\right)}$$

Exercise Set 7.4

20. *(a)* $[2, \overline{1, 1, 1, 4}]$ *(c)* $[3, \overline{1, 1, 1, 1, 6}]$ *(e)* $[4, \overline{2, 1, 3, 1, 2, 8}]$

21. *(a)* $1, \frac{4}{3}, \frac{5}{4}, \frac{29}{23}, \frac{34}{27}$ *(c)* $7, \frac{15}{2}, \frac{22}{3}, \frac{37}{5}, \frac{133}{18}$

22. *(a)* $\frac{22}{7}$ *(b)* $\frac{333}{106}$

25. For example, the 0th convergent of the infinite simple continued fraction expansion of $\sqrt{3}$ is $p_0/q_0 = 1/1$, and this convergent does not satisfy the desired inequality.

Exercise Set 7.5

33. *(a)* $\frac{1+\sqrt{5}}{2}$ *(c)* $\frac{2+\sqrt{6}}{2}$ *(d)* $\frac{\sqrt{15}}{3}$

36. Use Proposition 7:18.

38. *(a)* Use the procedure of Example 18. *(c)* $\sqrt{730} = [27, \overline{54}]$

39. *(c)* $\sqrt{531} = [23, \overline{23, 46}]$

Exercise Set 7.6

41. *(a)* no *(c)* yes *(e)* no

42. $[4, \overline{1, 3, 1, 8}]$

Exercise Set 8.2

6. *(a)* 1413 284 1003 1738 2655 855 2292
(b) THE EAGLE HAS LANDED

9. *(a)* 1113 1647 966 0 1691 966 1691 966 1979 202 1785 966 1113 1647 183
(b) ZERO HOUR
Note: In both parts of this exercise, two is chosen as the block length in the formatting stage of the encryption scheme, since it is the maximal even

number with the property that *all* blocks of this length, when viewed as single positive integers, are less that 2419. Upon decryption in part (b), this is easily recognized. (The message deciphered with blocks of length four would be AZ AE AR AO AH AO AU AR; the occurrence of A as the first component of each block is easily recognized as incorrect.) If blocks of length four are desired in all messages, one need only choose appropriate p and q so that m is larger than 2525, the largest possible block of length four that could appear in a message (since 2525 corresponds to plaintext ZZ). Similar comments apply to general m.

11. ***(b)*** The component primes of m are 419 and 421.

Exercise Set 8.3

13. ***(a)*** prime (take $a = 11$ in Theorem 8.1 or Corollary 8.2, for example)
(c) composite (divisible by 3, for example)
(e) composite (divisible by 11, for example)

14. ***(c)*** $F_4 = 65537$ is prime, while $F_5 = 4294967297$ and $F_6 = 18446744073709551617$ are composite.

Exercise Set 8.4

15. ***(a)*** $x = \pm 1$ and $y = \pm 2$ ***(c)*** no solutions ***(e)*** no solutions
(g) $x = \pm 3$ and $y = 0$; $x = \pm 2$ and $y = \pm 1$

16. ***(a)*** $x = p_{6n-1}$, $y = q_{6n-1}$ ***(b)*** $x = p_{10n-1}$, $y = q_{10n-1}$

17. ***(a)*** Fundamental solution: $x = 55$, $y = 12$
Next positive solution: $x = 6049$, $y = 1320$
(b) Fundamental solution: $x = 9801$, $y = 1820$
Next positive solution: $x = 192119201$, $y = 35675640$

18. If ρ is even, the Pell Equation $x^2 - dy^2 = -1$ has no solutions; if ρ is odd, the positive solutions of the Pell Equation $x^2 - dy^2 = -1$ are given precisely by $x = p_{(2n-1)\rho-1}$ and $y = q_{(2n-1)\rho-1}$ where n is a positive integer.

19. ***(a)*** no solutions
(b) $x = p_{10n-6}$, $y = q_{10n-6}$ (in the notation of the answer to Exercise 18 above)

Bibliography

T. M. Apostol, *Introduction to Analytic Number Theory.* New York: Springer-Verlag, 1976.

W. W. R. Ball, *Mathematical Recreations and Essays.* New York: Macmillan, 1944.

N. G. W. H. Beeger, "On Even Numbers m Dividing $2^m - 2$," *American Mathematical Monthly, 58* (1951), 553–555.

S. K. Berberian, "Number-Theoretic Functions via Convolution Rings," *Mathematics Magazine, 65* (1992), 75–90.

D. M. Bressoud, *Factorization and Primality Testing.* New York: Springer-Verlag, 1990.

D. M. Burton, *Elementary Number Theory* (2nd ed.). Dubuque, Iowa: Wm. C. Brown, 1989.

B. A. Cipra, "Archimedes Andrews and the Euclidean Time Bomb," *Mathematical Intelligencer, 9* (1987), 44–47.

J. Dyer-Bennet, "A Theorem on Partitions of the Set of Positive Integers," *American Mathematical Monthly, 47* (1940), 152–154.

H. M. Edgar, *A First Course in Number Theory.* Belmont, California: Wadsworth, 1988.

H. M. Edwards, *Fermat's Last Theorem: A Genetic Introduction to Algebraic Number Theory.* New York: Springer-Verlag, 1977.

H. Eves, *An Introduction to the History of Mathematics* (6th ed.). Philadelphia: Saunders, 1990.

D. Fendel, "Prime-Producing Polynomials and Principal Ideal Domains," *Mathematics Magazine, 58* (1985), 204–210.

P. Fletcher and C. W. Patty, *Foundations of Higher Mathematics* (2nd ed.). Boston: PWS-KENT, 1992.

J. B. Fraleigh, *A First Course in Abstract Algebra* (4th ed.). Reading, Massachusetts: Addison-Wesley, 1989.

D. Garbanati, "Class Field Theory Summarized," *Rocky Mountain Journal of Mathematics, 11* (1981), 195–225.

M. Gardner, *Mathematical Carnival.* New York: Random House, 1975.

M. Gerstenhaber, "The 152nd Proof of the Law of Quadratic Reciprocity," *American Mathematical Monthly, 70* (1963), 397–398.

A. Granville, "Primality Testing and Carmichael Numbers," *Notices of the American Mathematical Society, 39* (1992), 696–700.

R. K. Guy, "The Strong Law of Small Numbers," *American Mathematical Monthly, 95* (1988), 697–712.

R. K. Guy, *Unsolved Problems in Number Theory.* New York: Springer-Verlag, 1981.

G. H. Hardy and E. M. Wright, *An Introduction to the Theory of Numbers* (5th ed.). Oxford, England: Oxford University Press, 1979.

T. W. Hungerford, *Algebra.* New York: Springer-Verlag, 1974.

K. Ireland and M. I. Rosen, *Elements of Number Theory: Including an Introduction to Equations over Finite Fields.* New York: Bogden & Quigley, 1972.

B. W. Jones, *The Theory of Numbers.* New York: Rinehart & Co., 1955.

H. L. Keng, *Introduction to Number Theory.* New York: Springer-Verlag, 1982.

J. O. Kiltinen and P. B. Young, "Goldbach, Lemoine, and a Know/Don't Know Problem," *Mathematics Magazine, 58* (1985), 195–203.

D. E. Knuth, *Art of Computer Programming: Semi-Numerical Algorithms* (2nd ed.). Reading, Massachusetts: Addison-Wesley, 1981.

A. G. Konheim, *Cryptography: A Primer.* New York: John Wiley & Sons, 1981.

R. Laatsch, "Measuring the Abundancy of Integers," *Mathematics Magazine, 59* (1986), 84–92.

E. Landau, *Handbuch der Lehre von der Verteilung der Primzahlen.* Leipzig, Germany: Teubner Gesellschaft, 1909.

M. R. Murty, "Artin's Conjecture for Primitive Roots," *Mathematical Intelligencer, 10* (1988), 59–67.

I. Niven, H. S. Zuckerman, and H. L. Montgomery. *An Introduction to the Theory of Numbers* (5th ed.). New York: John Wiley & Sons, 1991.

Penn & Teller, "Penn & Teller's Impossible Number Prediction," *GAMES, 16* (1992), 9–11.

C. A. Pickover, *Computers, Pattern, Chaos and Beauty.* New York: St. Martin's Press, 1990.

C. A. Pickover, *Mazes for the Mind: Computers and the Unexpected.* New York: St. Martin's Press, 1992.

P. Ribenboim, *The Little Book of Big Primes.* New York: Springer-Verlag, 1991.

P. Ribenboim, *13 Lectures on Fermat's Last Theorem.* New York: Springer-Verlag, 1979.

I. Richards, "Continued Fractions Without Tears," *Mathematics Magazine, 54* (1981), 163–171.

R. Rivest, A Shamir, and L. Adleman, "A Method for Obtaining Digital Signatures and Public-Key Cryptosystems," *Communications of the ACM, 21* (1978), 120–126.

K. H. Rosen, *Elementary Number Theory and Its Applications* (2nd ed.). Reading, Massachusetts: Addison-Wesley, 1988.

R. D. Silverman, "A Perspective on Computational Number Theory," *Notices of the American Mathematical Society, 38* (1991), 562–568.

B. K. Spearman & K. S. Williams, "Representing Primes by Binary Quadratic Forms," *American Mathematical Monthly, 99* (1992), 423–426.

S. Wagon, *Mathematica® in Action.* New York: W. H. Freeman & Co., 1991.

L. Wenlin and Y. Xiangdong, "The Chinese Remainder Theorem," *Ancient China's Technology and Science* (1983), 99–110.

Index